Post-harvest Management of Fruits and Plantation Crops
(Volume 1: Basic Concepts)

NIPA® GENX ELECTRONIC RESOURCES & SOLUTIONS P. LTD.
New Delhi-110 034

About the Book

Substantial amount of horticultural produce is lost from production to consumption chain globally. Post-harvest losses in fruit crops is the burning issue in India. Large proportion of produce is spoiled/ wasted due to many factors ranging from harvesting to handling, packaging, transportations, loading and unloading, storage, marketing and even during consumption. Approximately 45 to 50 million tons of horticulture produce is wasted annually. This leads to significant loss in the country's economy when food is wasted after its production. This book (Volume 2: Applied Concepts) will play an important role to understand the reasons of losses and their management. The volume 2 contains eleven chapters on applied knowledge to have in depth know how about the factors affecting shelf-life and quality of fruits. I hope this book will be highly useful to the students, researchers and extension scientists of the country.

Post-harvest Management of Fruits and Plantation Crops
(Volume 1: Basic Concepts)

Edited by
WS Dhillon
Former Assistant Director General, Horticulture, ICAR, New Delhi
Former Director, Punjab Horticultural Post-harvest Technology Centre
Former Director PAU-Fruit Research Station, Gangian
Former Professor of Horticulture, PAU, Ludhiana

NIPA® GENX ELECTRONIC RESOURCES & SOLUTIONS P. LTD.
New Delhi-110 034

NIPA® GENX ELECTRONIC
RESOURCES & SOLUTIONS P. LTD.

101,103, Vikas Surya Plaza, CU Block
L.S.C.Market, Pitam Pura, New Delhi-110 034
Ph : +91 11 4386 0225, 9717133558, 9540816132
E-mail: newindiapublishingagency@gmail.com
Website: www.niparesources.com

Print ISBN: 978-93-58876-82-6

ebook ISBN: 978-93-58876-02-4

Composed and Designed by NIPA®.

This book is dedicated to my

Father-in-law: **Late Sardar Aminder Singh Randhawa**

Ex-Chief Wild Life Warden, Punjab

Mother-in-law: **Late Sardarni Jagjit Kaur Randhawa**

Who always inspired and motivate me to achieve success in life

प्रो. संजय कुमार सिंह
उप महानिदेशक (बागवानी विज्ञान)

Prof. S.K. Singh
FNAAS, FIAHS, FDBT
Deputy Director General (Horticultural Science)

भारतीय कृषि अनुसंधान परिषद
कृषि अनुसंधान भवन – II,
पूसा, नई दिल्ली– 110012
INDIAN COUNCIL OF AGRICULTURAL RESEARCH
KRISHI ANUSANDHAN BHAVAN-II,
PUSA, NEW DELHI-110 012 (INDIA)

Foreword

The post-harvest management of fruits and vegetables in India presents both challenges and opportunities. India is one of the world's second largest producer of fruits, yet the post-harvest losses are substantial, estimated to range from 20 to 40 per cent of total production. This is primarily due to inadequate infrastructure, including cold storage facilities, transportation systems, and packaging technologies. Cold storage facilities in India are insufficient and unevenly distributed across the country. While certain states have relatively well-developed cold chains, others lack the necessary infrastructure, leading to increased spoilage and waste during storage and transportation. Additionally, the lack of proper transportation infrastructure, including refrigerated trucks and poor road conditions, further exacerbates post-harvest losses.

Traditional post-harvest practices also contribute to losses. Many farmers still rely on age-old methods of harvesting, handling, and storage, which are not conducive to preserving the quality and shelf life of fruits. Moreover, there is a lack of awareness and access to modern post-harvest technologies and practices among smallholder farmers. However, amidst these challenges, there are also promising developments and initiatives aimed at improving post-harvest management in India. The government has launched various schemes and programs such as the Pradhan Mantri Kisan SAMPADA Yojana and the Mission for Integrated Development of Horticulture (MIDH), which focus on creating modern infrastructure, including cold storage facilities, pack houses, and processing units.

Private sector investment in cold chain infrastructure is also increasing, with companies investing in modern cold storage facilities, refrigerated

E-mail: ddghort.icar@gov.in Phone: (Off): 91-11-25842068

transportation, and packaging solutions. Additionally, the rise of e-commerce platforms provides new marketing channels for farmers, reducing the reliance on traditional supply chains and intermediaries. Furthermore, there is a growing emphasis on farmer education and capacity building. Various government and non-governmental organizations conduct training programs to educate farmers on best practices in post-harvest handling, packaging, and storage. Agricultural extension services play a crucial role in disseminating knowledge and technologies to farmers.

Research and development efforts are also underway to develop innovative solutions to reduce post-harvest losses. Collaborations with international organizations bring in global expertise and technologies to improve post-harvest management practices in India. While challenges persist, concerted efforts by the researchers, government, private sector and other stakeholders offer hope for improving the post-harvest management of fruits. By addressing infrastructure gaps, promoting technology adoption and empowering farmers with knowledge and resources, India can minimize post-harvest losses, enhance food security and unlock the full potential of its agricultural sector.

The book on "Post-harvest management of fruits and plantation Crops (Volume I: Basic Concept)" is being published and edited by Prof. (Dr) WS Dhillon, Former Assistant Director General Horticulture, ICAR, New Delhi. It consists of eleven chapters on different aspects of basic in nature, will be useful to the students, researchers, teachers, extension workers and policy planners. The chapters are contributed by highly experienced researchers is highly praiseworthy.

(Sanjay Kumar Singh)

E-mail: ddghort.icar@gov.in Phone: (Off): 91-11-25842068

Preface

India is one of the leading fruits producing country in the world. In spite of that there is a significant gap between per capita demand and supply due to substantial waste during post-harvest storage and handling. This waste is caused by improper bagging without crating, lack of temperature controlled vehicles, and unavailability of cold chain facilities in various parts of the country. Therefore, it is essential to improve the supply chain management for fruits in India to meet the country's demand. This can be achieved by adopting best global practices in storage, packaging, handling, transportation and value-added services.

India's post-harvest management of fruits faces significant challenges that impact food security and economic stability. A recent study by NABCONS estimates a monetary loss of 1.5 trillion rupees (USD 18.5 Billion) to post-harvest losses in 54 crops in India annually. The study highlights the significant losses in fruits and vegetables, with fruits contributing 19.34 per cent and vegetables 17.97 per cent to the total agricultural losses. Among individual commodities, guava faces the highest loss at 15.05 per cent followed by apple at 9.51 per cent in fruit crops. These losses are driven by inadequate infrastructure, particularly in cold storage and transportation. The country's cold storage capacity stands at around 37 million metric tons, far short of the 70 million metric tons needed. Moreover, these facilities unevenly distributed, with over 75 per cent in states like UP, WB and Punjab, primarily for potato storage.

The logistics sector also struggles, with only about 10,000 refrigerated trucks available against a requirement of 60,000 (NCCD). Poor road infrastructure exacerbates the problem, with freight vehicles averaging speed of just 25-30 km/h, leading to further delays and spoilage (FICCI). Despite the availability of modern technologies such as controlled atmosphere storage, modified atmosphere packaging and nano technologies, adoption remains slow and is mainly limited to larger enterprises. Traditional post-harvest practices are still prevalent among a significant portion of farmers, resulting in inefficiencies and higher losses.

To combat these problems, an effort has been made to come out with a publication in the form of a book to understand all the factors of the post-harvest handling and management of produce to minimize the losses. All the chapters in this book are designed in such a way that it covers basic knowledge of post-harvest handling of fruits and other crops. It is highly painful that we in developing country like India loosing substantial amount of commodities after producing the crops. I am sure that the book will act as catalyst for all the stakeholders including students and professionals in horticulture to make concerted efforts to reduce the post-harvest losses in fruits.

I greatly acknowledge the contribution of contributors, who are experts on their own specialized fields that helped me getting this book (Volume 1) in the present shape. No academic journey is complete without constructive criticisms and comments from the readers of this book. I urge upon the readers to share their candid views on wasakhasingh@yahoo.com regarding the coverage of this book so that it would value for the improvement in future editions.

(WS Dhillon)

Contents

1

Ethylene Evolution and Management in Post-harvest Storage of Fruits

Anirudh Thakur and Uday Raj Patial

Department of Fruit Science, Punjab Agricultural University Ludhiana-141 004, Punjab

Plant hormone ethylene is involved in a number of physiological activities within the plant system, ranging from seed germination to fruit development. Ethylene acts as a key player in accelerating fruit ripening by regulating the activities of different enzymes, and is responsible for shortening the shelf life and associated post-harvest losses during the storage of fruit crops. However, all fruit crops have different ethylene production capacities and vary in their response to ethylene. Hence, effective management of post-harvest ethylene is imperative for addressing the global challenge of reducing fruit losses after harvesting. In recent periods, numerous technologies and compounds have been developed for the suppression of ethylene and for enhancing in-situ oxidation of ethylene or inhibiting its production at a receptor level. Among these emerging techniques, the use of sachets, fumigating agents, catalysis and composite films embedded with nanoparticles has been found to be trending, but the use of some of compounds has been limited due to health risks or consumer unacceptance. However, most of these technologies are very costly and are not affordable for marginal farmers and developing nations. In order to address these issues, long term breeding approaches for shelf-life improvement, targeting the tfs, ERFs have been involved. This manuscript fills the knowledge gap by encompassing all the recent progress made in enhancing the shelf life of fruits by managing ethylene during storage.

Introduction

Ethylene is a naturally occurring simple gaseous molecule, comprising of two carbon atoms linked together with a double bond (C_2H_4). The initial evidence of ethylene affecting the plant tissues dates back to 19th century when senescence and defoliation symptoms were observed in the trees growing close to illuminating streetlights (Martínez-Romero *et al.*, 2007). In 1901, Neljubov recognized ethylene as the compound responsible behind this phenomenon, earning the recognition of pioneer to discoverer this hormone. Later, after the establishment of procedures of gas chromatography, Gane (1934) proved the hypothesis of ethylene production naturally by apple (Osborne, 1989).

Ethylene regulates various physiological activities in plant system from germination, leaf abscission, fruit ripening to senescence by regulating the production of certain respiratory enzymes (Lee *et al.*, 2015). Additionally. The hyperactive nature of ethylene makes it trigger certain reactions like breakdown, modification, absorption and adsorption of certain chemical components. It promotes the respiration rate and then senescence in the fruit crops. The continual respiration and ethylene emission hampers the nutritional, sensory qualities and reduces the shelf life of the fruit crop (Kostekli *et al.*, 2016). Since, the ripening and senescence of certain fruits is totally dependent on the ethylene evolution, hence, it can be used as a marker for the fruit ripening (Lang and Hubert, 2012).

Depending on the ripening procedure or capacity of individual fruit to produce ethylene, they are classified as climacteric or non-climacteric (Saltveit, 1999). A burst in the ethylene production followed by rise in respiration rate has been observed in the climacteric fruits whereas, no rise in the respiration rate is observed with the ethylene production in case of non-climacteric fruits. The climacteric fruits (peach, banana, mango, apple) ripe even after they are detached from the plant followed by some indications like change in colour, firmness, fruit shape like, rising of shoulders in mango (Shamim *et al.*, 2015), disappearance of angles in case of banana (Bolfarini *et al.*, 2020), TSS etc. However, the non-climacteric (grapes, citrus, persimmon etc.) fruits do not ripe significantly after their detachment from the plant (Raghava Rao and Chundawat, 1986; Chen *et al.*, 2018).

The maturity stage of fruit highly influences the rate of ethylene production in the fruit. Bhutia *et al.*, (2011) observed a high rate of CO_2 and ethylene production in fruit attaining ripening stage as compared to those in mature of half ripe stage. Besides this endogenous ethylene production inside fruit tissues, certain micro-organism (fungi and bacteria) which are present freely in the environment ferments biomass and releases ethylene into environment as result of pyrolysis of hydrocarbons (Cape, 2003). This free ethylene present in the air, exerts a similar effect like that of endogenously produced ethylene on the shelf life of fruit. It is crucial to control the ethylene concentration in storage conditions in order to slow down the ripening process and increase the shelf-life of the fruits (Kerry and Butler, 2008). Hence, to accomplish this goal, it is necessary to understand the pathway involved in the synthesis of ethylene, its action so as to develop effective techniques for managing ethylene from the vicinity of fruits during their post-harvest storage conditions.

Biosynthesis pathway

Since the identification of plant hormone 'ethylene', comprehensive efforts have been made by a number of scientists around the world to reveal its biosynthesis pathway. A number of studies focusing on biosynthesis, signal transduction, biochemical and molecular regulatory mechanisms have been published (Bleecker and Kende, 2000; Ecker, 2002; Liu *et al.*, 2020c). Throughout the development of fruits, many compounds have capacity to undergo their conversion into ethylene. The biosynthesis of ethylene is catalysed by ACO and ACS activities. During the initiation of ripening, 'system 1' regulates ethylene pathway which is characterized as an autocatalytic process featuring basal phytohormone levels and low ethylene response. Subsequently, the catalytic phase *i.e.* 'system 2' begins with an increased ethylene production attributed to higher phytohormonal sensitivity (Kumar *et al.,* 2014).

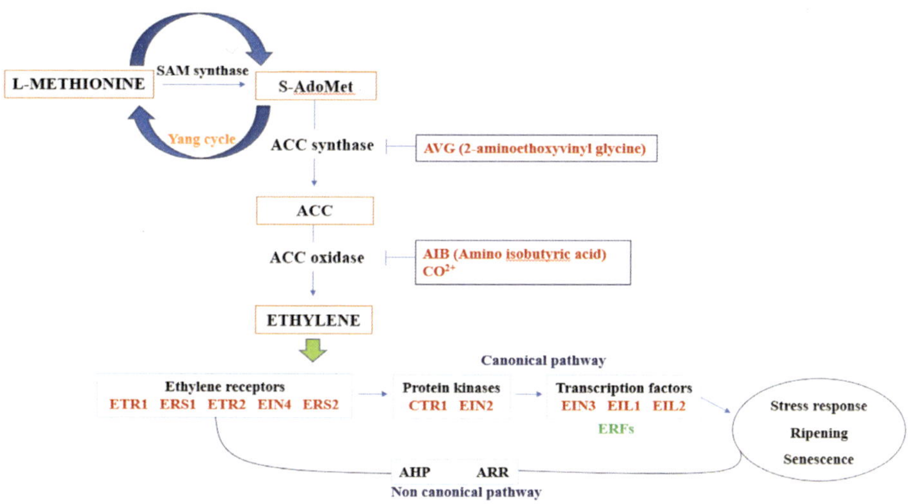

Fig. 1: Ethylene biosynthesis pathway involved in fruit ripening.

Ethylene biosynthesis involves a two-step reaction (Fig.1). In the first step SAM (S-adenosylmethionine) converts into ACC (1- aminocyclopropane-1-carboxylic acid) by the action of ACS (ACC-synthase) followed by its conversion into ethylene in the second step by the action of ACC oxidase (Houben and Van De Poel, 2019). The understanding of components affecting ethylene signalling is highly important to avoid ethylene induced losses. Binder (2020) reported two (canonical and non-canonical) ethylene signalling pathways (Fig. 1.). The canonical pathway encompasses ethylene receptors (*ETR1, ETR2, ERS1, ERS2, EIN4*), protein kinases (*CTR1* & *EIN2*) and some transcription factors which activates TFs of ethylene responsive factors (*ERF*) family. The alternative 'non canonical' pathway involves AHP and ARR as regulation proteins.

Ethylene responsiveness of fruit crops

Different fruit species vary in their susceptibility to ethylene (Table 1), thus making it a challenging task to decide a universal threshold limit for the fruit sensitivity to ethylene (Wills *et al.*, 2001). The climacteric fruits (kiwifruit, apple, banana, pear) exhibit a higher ethylene production rate and are highly ethylene sensitive (0.03-0.1 µl L^{-1}). Conversely, the non-climacteric fruits (persimmon, strawberry) produce minimal amount of ethylene but exhibit a higher sensitivity to ethylene levels, sometimes even at lower concentrations (0.01-0.02 µl L^{-1}) than climacteric fruits (Wills and Warton, 2000). Many fruits viz., grapes, cherry, pineapple, berries etc. exhibit both lower ethylene production and low sensitivity to hormone (0.2 µl L^{-1}). Wills and Warton (2000) studied the corelation between the atmospheric ethylene concentration and storage life of fruits and reported concentrations above 0.10 µl L^{-1} to be detrimental for the shelf-life of fruits. These detrimental effects (Table 1) are triggered by certain factors like tissue effected, exposure time, temperature, atmospheric concentration etc. (Saltveit, 1999). The prevalence of higher temperature (0-20°C) exacerbates the sensitivity of fruits to ethylene (Nanos *et al.*, 2002). A simulation model designed by Genard and Gouble (2005) revealed 30 per cent increase in ethylene production with a 5 per cent rise in temperature. However, endogenous ethylene production and sensitivity to exogenous ethylene decreases if temperature exceeds 35°C because of the inactivation of ethylene receptors (Martínez-Romero *et al.*, 2014).

Table 1: Ethylene dynamics in different fruit crops (production, sensitivity and effects).

Fruit crop	Ethylene production	Sensitivity	Adverse effects	References
Apple	*****	H	Scald, loose crunch.	Nissen *et al.*, 2018
Kiwifruit	***	H	Softness, white inclusions.	Kader, 1985
Pear	***	H	Scald, Internal browning.	He *et al.*, 2022
Peach	***	H	Fruit softening, internal browning.	Yang *et al.*, 2016
Passion fruit	*****	M	Shrivelling.	Paull and Chen, 2014
Avocado	****	M	Mesocarp discoloration.	Hershkovitz *et al.*, 2009

Apricot	***	H	Flesh softening.	Fan *et al.*, 2018a
Banana	**	H	Off flavour, crown rot.	Lassois & de Bellaire, 2014
Mango	**	H	Fruit softening.	Li *et al.*, 2022
Papaya	**	M	Fruit softening, decay, Phomopsis rot.	Abeywickrama *et al.*, 2012; Gao *et al.*, 2020
Plum	**	M	Flesh translucency.	Wang *et al.*, 2016
Persimmon	*	H	Softening, decay.	Zeng *et al.*, 2021
Strawberry	*	H	Fruit softening, mould.	Kader, 1985; Tosetti *et al.*, 2020
Citrus	*	M	Rot, mould, Black centre rot.	Kader, 1985; Mojerlou and Safaie, 2012
Cherry	*	L	Browning, blue mould, Alternaria rot.	Qin *et al.*, 2003; Ansari & Tuteja, 2015
Grape	*	L	Grey mould.	Ozkana *et al.*, 2011
Pineapple	*	L	Internal browning.	Li *et al.*, 2018

Ethylene production (μL $kg^{-1}h^{-1}$): * = Very low (< 0.1); ** = low (0.1-1.0); *** = Moderate (1.0- 10);
**** = High (10-100); ***** = very high (>100)
Ethylene sensitivity (ppm): H = high (0.01– 0.5); M = moderate (0.5–3); L = low (3–5)
Scale by Keller *et al.*, (2013)

Ethylene management for post-harvest quality of fruit crops

Certain cases of ethylene release are observed in the supply chain, like during packaging, storage, transportation or refrigeration. Therefore, implementation of additional strategies is necessary to minimize the ethylene levels (Martínez-Romero *et al.*, 2007). Wills and Warton, 2004 has recognized 0.1 μl L^{-1} as a detrimental concentration of ethylene present in the air for inducing physiological changes. Hence, in order to sustain the quality of climacteric fruits during extended transport and subsequent storage, it becomes essential to maintain low ethylene levels (typically below 1.0–0.01 μL L−1) in the environment (Terry *et al.*, 2007). Different techniques employed to inhibit ethylene effects and increase shelf-life of fruits includes supressing ethylene synthesis, oxidising ethylene into CO_2 and blocking ethylene receptors (Zhang *et al.*, 2017).

Suppressing ethylene synthesis: Ethylene synthesis can be supressed by lowering down the respiration rates and oxygen levels in the storage conditions, which will control the ATP breakdown and then ACC oxidation, ultimately. It has been observed to be achieved using CTHH (Controlled temperature and high humidity), CA (controlled atmosphere) storage, DCA (dynamic controlled atmosphere), SCA (static controlled atmospheres), ULO (ultra-low oxygen), MAP (Modified atmosphere packing) and silver treatment (Zhang *et al.*, 2017;

Saquet and Streif, 2017; Falagan and Terry 2018; Mditshwa *et al*., 2018; Park *et al*., 2018). Certain chemical compounds have been identified, which possess the capacity to hinder ethylene activity and block its biosynthesis, block receptors sites thereby altering the activities of ACC synthase and ACC oxidase (Bregoli *et al*., 2002; Valero, 2003). AOA (Aminooxyacetic acid) and AVG (Aminoethoxyvinylglycine) are well known ACS inhibitors whereas Co2+, AIB (α-aminoisobutyric acid), acetaldehyde and ethanol vapours inhibit ACO activity (Martínez-Romero *et al*., 2007).

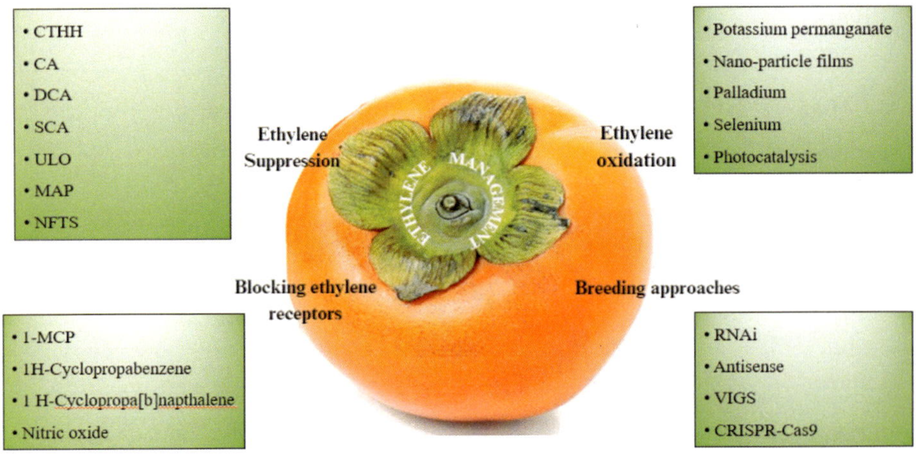

Fig. 2: Ethylene management strategies for improving post-harvest shelf life of fruits.

***CTHH*:** Controlled temperature conditions are maintained with high humidity for storing fruits in order to minimize the weight loss and extend the shelf-life. The low temperature decreases the respiratory activity of the fruit, thereby reducing ATP breakdown and supresses ethylene synthesis (Fan *et al*., 2018a). Meanwhile, maintaining high humidity levels reduces the moisture evaporation, thus effectively alleviating the fruit weight loss. Fruits are preserved in ULO, CA, DCA, SCA, MAP by maintaining carbon dioxide and oxygen levels below 5-15 and 3 per cent, respectively (Falagan and Terry, 2018). Since, the ACC oxidase activates in the presence of oxygen and low CO_2 levels, the basic idea behind these techniques is to lower down O_2 levels and reduce ethylene levels (Zhang *et al*., 2017).

***DCA*:** An alternative non-chemical approach, DCA is capable of reducing the ethylene production levels far below than conventional CA approach (Brizzolara *et al*., 2017). Oxygen concentrations can be customized in DCA at the beginning or any time in the storage. Initially, oxygen level is reduced to a point which could be tolerable by the fruit and then adjusted as per the metabolic requirements of the fruit during the storage (Mditshwa *et al.*, 2018). These hypoxic conditions maintain minimal ethylene levels and, in turn supresses ethylene mediated disorders like superficial scald along with maintaining the organoleptic and physiochemical properties of the fruits. However, high CO_2 and low O_2 levels may trigger anaerobic respiration and lead to production of certain unexpected metabolites (Thewes *et al.*, 2017). The effectiveness of DCA lacked in some cultivars viz., Topaz (Gasser & Von Arx, 2013), Gala and Fuji (Zanella and Rossi, 2015) as compared to ULO. This could be due to presence of variable hypoxia related regulatory mechanisms of different cultivars. Hence, further investigations are required to refine and enhance these technologies for the long-term storage of cultivar specific fruit crops.

***NFTS*:** Harvested fruits using NFTS approach are stored at a non-freezing temperature maintained at a range amid the super-cooling and freezing point of the fruits (Liu *et al.*, 2020a). This technique has demonstrated efficacy in supressing respiration rate, enzymatic activity and ethylene release (Fan *et al.*, 2018b). Moreover, it hinders multiplication of pathogenic agents, prevents water and nutrient loses during storage, thereby preserving fruit quality (Ma *et al.*, 2022). Successful implementation of NFTS has been reported in improving post-harvest storage of different fruit crops viz., apricot (Li *et al.*, 2022), apple (Shu *et al.*, 2020), cherry (Zhao *et al.*, 2019) etc.

Salicylic acid: Salicylic acid (SA) and its derivatives (MeJA & ASA), are extensively used in preservation of fruits, in the form of pre-harvest sprays, post-harvest dipping, fumigation and coating (Chen *et al.*, 2023). ASA, though not a natural product, spontaneously converts to SA, providing similar effects with enhanced accessibility and cost-effectiveness (Fan *et al.*, 2021) while, MeJA, derived from SA, acts as an odorous volatile compound inducing plant defence (Poitou *et al.*, 2021). SA delays fruit ripening by supressing the expression of ERFs (*ACO, ACS*), maintains cell wall integrity and fruits WHC by inhibiting PME, PME, CEL activity (Kumar *et al.*, 2021). The application of SA has been observed to lower the extent of oxidative damage (ROS, O_2^-) by raising activity of antioxidants and phenolic compounds, thereby improving the shelf-life of lemon (Serna-Escolano *et al.*, 2021), cherry (Pan *et al.*, 2022), plum (Davarynejad *et al.*, 2015), pomegranate (García-Pastor *et al.*, 2020b). However, there is a limited acceptance of SA for post-harvest treatment of fruits, because of its poor water solubility and off-smell of chitosan coating, impacting consumer satisfaction (Chen *et al.*, 2023). Since, pre-harvest application is also practiced in many cases, the SA solution is mixed with Tween-20 for improving its solubility and absorption by fruit plants (Valverde *et al.*, 2015).

Table 2: Strategies for supressing post-harvest ethylene production in fruit crops.

Technique	Crop	Conditions	Effects	Reference
CTTH	Apricot	Stored at near freezing temperature (-2.1 ~ -1.4°C) for 90 days and with humidity levels > 95 per cent.	Supressed respiration rate and ethylene production, maintained antioxidant activity and delayed the production of phenolic compounds.	Fan *et al.*, 2018a
	Durian	33 °C with 77 per cent humidity.	Low ethylene and CO_2 production.	Ketsa and Pangkool, 1995
CA	Apple (cv. Jonagold), Pear (cv. Conference)	O2 @ 0.5 kPa, CO2 @ 6.0 kPa at 0°C with 95 per cent RH for 6 months and shelf life checked at 20 °C.	Supressed respiration and ethylene production along with fruit metabolism.	Saquet and Streif, 2017
	Kiwifruit	O_2 @ 2 per cent, CO_2 @ 5 per cent at 0 °C with 95 per cent RH (13 weeks)	Reduced the fruit sensitivity to ethylene and enhanced the shelf life.	Han, 2021
	Plum	O_2 @ 1 kPa + CO_2 @1 kPa at 0.5 ± 0.1°C and RH 96 ± 2 per cent (55 days).	Reduced flesh browning, maintained flesh consistency and delayed ripening.	Steffens *et al.*, 2018
DCA	Apple cv. Holsteiner Cox', 'Boskoop, Elstar	O_2 concentration ranged 1.4-0.5 per cent, CO_2 2.5 per cent at 2-4 °C with 95 per cent humidity.	Retain firmness and acidity, reduced skin spots.	Hennecke *et al.*, 2008
	Apple cv. Ariane	Initial O_2 concertation 1.0 per cent and then reduced further, CO_2 @ 1.5 per cent at 1 °C.	Better fruit firmness and acidity levels.	Gasser and Von Arx, 2013
	Apple cv. Braeburn	O_2 levels maintained <0.7 kPa, CO_2 @ 1.0 kPa at 1.3 °C with 98 per cent humidity.	Found more effective and improved fruit firmness.	Zanella and Rossi, 2015
	Apple cv. Granny Smith	O_2 @ 0.4 kPa, CO_2 levels between 0.7-1.3 kPa at 0.8 °C with 95 per cent humidity.	Reduces superficial scald with no off flavour in the fruits.	Zanella *et al.*, 2004

Technique	Crop	Conditions	Effects	Reference
NFTS	Plum	-1 ~ -1.5 °C	Reduced weight loss, respiration rate with increased firmness and ascorbic acid.	Ma *et al*., 2022
	Sweet cherry	-3.0 ± 0.1 °C	Improved firmness, sugar, colour and reduced H_2O_2, O_2^- and prevented flesh browning.	Zhao *et al*., 2019
	Apricot	−2.1 ~ −1.7 °C	Delayed decline of antioxidants, phenolic compounds and senescence.	Fan *et al*., 2018b
		−1 ~ −1.5 °C	Mitigated CI, supressed cell wall degradation and improved quality.	Li *et al*., 2022
	Apple	-1.7 ± 0.2 °C	delayed respiration peak, maintained fruit colour, TSS, TA, antioxidants and retarded the increasing membrane permeability and MDA content.	Shu *et al*., 2020

Table 3. Application of ethylene supressing compounds for post-harvest management of fruit crops.

Compound	Crop	Dose	Storage condition	Effects	Reference
SA	Pistachio	2 mM L^{-1}	4°C for 28 days	Exhibited superior sensory attributes (color, texture) accompanied by diminished peroxide levels, free fatty acids and weight loss.	Molamohammadi *et al.*, 2020
	Papaya	1.5 mM L^{-1}	13°C for 28 days	Enhanced antioxidant enzymatic activity, resulting in reduced fruit decay (5.15 %), weight loss (5.19 %) and sustained fruit quality (firmness, TSS, TA, ascorbic acid).	Hanif *et al.*, 2020
	Peach	2 mM	2°C (8 days) + 20°C (2 days)	Exhibited better quality characteristics (weight, fruit phenol and antioxidants) with reduced firmness loss.	Erogul *et al.*, 2020
	Winter jujube	3 mM L^{-1}	4 °C with an RH of 80–90 per cent	Maintained fruit quality (color, TSS, TA, firmness) with low respiration rate up to 12 days.	Yang *et al.*, 2022
	Longan	0.3 g L^{-1}	28 ± 1 °C (5 days)	Reduced pericarp browning, fruit decay with enhanced antioxidant activity.	Chen *et al.*, 2020
	Pomegranate	2mM	6 °C, 90 ± 5 per cent RH (180 days)	Enhanced phenolic compounds and antioxidant activity in fruits.	Koyuncu *et al.*, 2019
	Pineapple	2.5 and 5.0 mM	13 ± 1 °C (10days)	Reduced chilling injury and browning incidence.	Sangprayoon *et al.*, 2019
	Mandarin	4mM	5±1 °C, 90±5 per cent RH (90 days)	Maintained prolonged enzymatic activity with very low decay and almost no disease incidence.	Haider *et al.*, 2021
	Grape	1~2 mmol L^{-1}	23.8 ± 5 °C, RH 80 ± 2 per cent (12 days)	Reduced berry drop and fruit decay at the time of storage.	Gomes *et al.*, 2021

Compound	Crop	Dose	Storage condition	Effects	Reference
MeJA	Blood orange	3mM pre and 1.5mM post-harvest	6 ± 1 °C, RH 80 ± 5 per cent (45 days)	Improved rind colour with increased anthocyanin and flavonoid content of fruits.	Vithana *et al.*, 2024
	Blueberry	0.4mM	(0°C, 90 ± 5 per cent RH (14 days)	Maintained fruit quality with reduced weight loss.	Yılmaz (2024)
	Chestnut	0.2 mM	0 ± 0.5 °C, RH 90 ± 5 per cent (80 days)	Reduced moisture, weight and firmness loses with proper level of bioactive compounds	Uzun (2023)
	Strawberry	250 µmol L^{-1}	20°C, RH 40 per cent (3 days)	Improved CAT, anthocyanin content, maintained firmness and fruit weight with reduced ROS production.	Zuñiga *et al.*, 2020
	Pear	100 µmol L^{-1}	20±1°C, RH 60–70 per cent (12 days)	Suppressed weight loss, respiration rate and maintained flesh firmness.	LI *et al.*, 2021
	Cherry	0.05 mmol L^{-1}	25°C (6 days)	Enhanced activity of antioxidant and disease resistance enzymes, reduced degree of decay.	Pan *et al.*, 2022
	Persimmon	(PH) 16, and 24 µL L^{-1}	0–1 °C (120 days)	Preserved fruit physio-chemical characteristics and reduced chilling injury.	Bagheri *et al.*, 2022
ASA	Peach	1 mM	1 °C, RH 85 ± 5 per cent (9 weeks)	Preserved fruit quality (firmness, weight) and delayed decay in fruit quality.	Saif (2022)
	Loquat	2 mM	25°C (6 days)	Enhanced fruit quality with low incidence of physiological disorders.	Hadjipieri *et al.*, 2021
	Kiwifruit	1.0 mM	20 °C (10 days)	Expedited the ripening by inhibiting ethylene receptors modulation.	Yin *et al.*, 2013
	Grape	1 mM	25 ± 1 °C, RH 80-85 per cent (5 days)	Maintained fruit quality with reduced grey mould infection.	García-Pastor *et al.*, 2020a
	Pomegranate	10 mM	10°C, RH 85-90 per cent (90 days)	Enhanced aril anthocyanin content and maintained quality characters.	García-Pastor *et al.*, 2020b
	Cherry	1 mmol L^{-1}	2°C (28 days) + 20°C (1 day)	Enhanced colour development with improved firmness and quality characteristics.	Giménez *et al.*, 2017

Compound	Crop	Dose	Storage condition	Effects	Reference
Melatonin	Blueberry	100 μM for 30 min	4 °C, RH 45 per cent (21 days)	Delayed wax decline (15.17 per cent), reduced water loses, induced triterpenoid and alkanes accumulation.	Li *et al.*, 2024
	Mango	200 mM	5 ± 1 °C for 28 d.	Reduced chilling injury, weight loss, MAD levels and improved membrane stability.	Khedr *et al.*, 2023
	Peach	100 μM (2 hours)	2 ± 1°C, RH 90–95 per cent (21 days)	Reduced chilling injury, delayed respiration rate, lower MAD levels and higher proline levels (1.5 times).	Sati *et al.*, 2023
	Litchi	0.5 mM (10 min)	25±3 °C, RH 75 ± 5 per cent	Reduced POD, PPO activity, delayed pericarp browning and upregulated anthocyanin levels (LcUFGT, LcDFR).	Marak *et al.*, 2023
	Kiwifruit	0.1 mM (10 min)	2°C, RH 90 per cent (84 days) + 20°C (5 days)	Reduced chilling injury, electrolyte leakage, delayed chlorophyll degradation, and ripening.	Ruiz-Aracil *et al.*, 2023
Ethanol	Banana	Dipping in 4.5 per cent Ethanolic extract of propolis	20 ± 2 °C, 60–70 per cent RH (13 days)	Improved fruit TSS, membrane stability with higher free radical scavenging capacity.	Awad and Al-Qurashi, 2019
	Orange	Dipping in 5 per cent Ethanolic extract of propolis	5-7°C, 90±5 per cent RH (10 weeks)	Prolonged storage period with reduced fruit deterioration.	Badawy, 2016
	Papaya	Dipping in 25 ml/L for 2 hours	25°C (14 days)	Delayed ripening, decay, improved firmness and antioxidant enzymes activity in fruits.	Liu *et al.*, 2022
	Blueberry	Vapor exposure (500 μL L^{-1}) for 18 hours	0 ± 0.5 °C, RH 90 per cent (40 days)	Inhibited CWDEs activity, delayed cell wall degradation and improved shelf life.	Ji *et al.*, 2021
	Mulberry	Vapor sachet	10 °C, RH 60–70 per cent	Reduced fruit decay, improved aroma and shelf life by 3 days.	Choosung *et al.*, 2019
	Kiwifruit	Vapor exposure (250–500 μL L^{-1}) for	1 ± 1 °C (14 days)	Delayed adverse physiological changes (softening, scald, rot), enhanced antioxidant activity and consumption life up to 14 days.	Xiong *et al.*, 2024

***Melatonin*:** N-acetyl-5-methoxytryptamine commonly known as Melatonin, is a signalling molecule commonly present in a variety of fruit plants and is involved in a variety of physiological activities from seed and root development to fruit ripening (Wang *et al.*, 2018). As a derivative of tryptophan, it is synthesised in different parts of fruit encompassing seeds, pulp and peel (Zhang *et al.*, 2018). Several precursors involved in melatonin synthesis like tryptophan, serotonin, tryptamine, 5-methoxytryptamine, N-acetylserotonin and genes encoding them have been identified across different fruit crops (Liu *et al.*, 2018; Guo *et al.*, 2020). The ROS produced during the post-harvest storage, cause an oxidative loss and hampers the fruit quality. Melatonin raises the activity of the enzymatic (APX, CAT, GPX, POD, PPO, SOD) and non-enzymatic (AsA, DHA, GSH) system to deal with these free oxygen atoms (Ze *et al.*, 2021). The melatonin application promotes the endogenous melatonin accumulation, induces nitric oxide (NO) production, this NO decreases ROS levels and suppresses *ACO* and *ACS* expressions and inhibits ethylene production (Zhang *et al.*, 2020). The exogenous application of melatonin has been observed to effectively decelerate the post-harvest ripening (cherry (Tijero *et al.*, 2019); mango (Liu *et al.*, 2020b)), delay senescence, inhibit deterioration (peach (Gao *et al.*, 2016); pear (Liu *et al.*, 2019)) and chilling injury in fruits stored at low temperature (pomegranate (Jannatizadeh, 2019); sapota (Mirshekari *et al.*, 2020)). Furthermore, melatonin application has enhanced disease resistance and reduced fruit decay symptoms during storage in kiwi (Hu and Rao, 2018), plum (Bal, 2019), strawberry (Aghdam and Fard, 2017).

Ethanol: Being a GRAS (Generally Recognized as Safe substance) certified product, ethanol, has long been used for the post-harvest preservation of many fruits (Ji *et al.*, 2021). Owing to the high permeability and volatility of ethanol, it is used via immersion or fumigation techniques. Recently, the efficacy of the ethanol dipping has been reported in delaying softening, improving flavour and avoiding fruit deterioration during post-harvest storage of papaya (Liu *et al.*, 2022). The ethanolic extracts of ethanol are also trending and are observed to improve physio-biochemical fruit properties along with the enhancement of shelf life in orange (Badawy, 2016) and banana (Awad and Al-Qurashi, 2019). However, application of ethanol vapor revealed marked potential in enhancing fruit quality, mitigating cold injury and decay symptoms during storage in blueberry (Ji *et al.*, 2021) and kiwifruit (Xiong *et al.*, 2024).

Oxidation of ethylene: In particular conditions, when we fail to supress the synthesis of ethylene, the techniques focusing on the immediate oxidation of the released ethylene have been followed. Several technologies and chemicals ($KMnO_4$, TiO_2, O_3) have been developed which result in the oxidation of the released ethylene from the storage conditions.

$KMnO_4$ a major scavenging agent: The high oxidizing capacity of the $kMnO_4$ has led to its emergence as a major ethylene scavenger. Forsyth *et al.* (1967) reported $KMnO_4$ for the first time, to be efficient agent for the removal of ethylene from the environment containing fresh apple fruits. Kim *et al*., (2019) outlined the reactions involved in the decomposition of ethylene by $KMnO_4$ as follow:

Step 1 $3CH_2CH_2 + 2KMnO_4 + H_2O \rightarrow 2MnO_2 + 3CH_3CHO + 2KOH$

Step 2 $CH_3CHO + 2KMnO_4 + H_2O \rightarrow 3CH_3COOH + 2MnO_2 + 2KOH$

Step 3 $3CH_3COOH + 8KMnO_4 \rightarrow 6CO_2 + 8KOH + 2H_2O$

The moisture accumulated from the respiration and transpiration of the harvested produce, activates this reaction and the purple hue of $KMnO_4$ changes to brown (Janjarasskul and Suppakul, 2018). Since, H_2O and CO_2 are released in the end of this chain reaction, the CO_2 concentration rises and the endogenous ethylene production is also blocked, consequently slowing down the ripening process (Kostekli *et al.* 2016). Since, the $KMnO_4$ and ethylene reacts by the means of convection and diffusion, the oxidant requires exposure of a large surface area for the efficient reduction of ethylene. Immobilization of $KMnO_4$ into inert materials (vermiculite, celite, clay, alumina) is done for this (Ebrahimi *et al*., 2021).

However, toxic effect of $KMnO_4$ have been reported on its direct contact with fruit commodities. Hence, $KMnO_4$ is encapsulated inside highly permeable sachets, blankets, or tubes and these encapsulated units are then placed inside the packages (Janjarasskul and Suppakul, 2018). Various such products are available in the market with different commercial names such as Air Repair, Befresh, Bi-On, BIOPAC, BRYSORB 508, GK3, Retarder, KEEPFRESH, Ethysorb, Purafl, Bloomfresh, Circul- Aire (E brahimi *et al*., 2021) with 3.5–12 per cent content of $KMnO_4$ (Álvarez-Hernández *et al*., 2019). Meanwhile, several challenges arise with the use of KMn04 as a scavenger. One of the primary issues is its reduced efficiency over time, necessitating its replacement for long term storage. This reduction in the efficiency is because of the ineffectiveness of $KMnO_4$ inside sachet to bind with ethylene and hence large number of sachets are required (Pathak, 2018). Furthermore, $KMnO_4$ occasionally interacts with the chemical compounds within the sachet and shifts its characteristic purple colour onto the commodities (Yildirim *et al*., 2019).

Table 4: Application of ethylene oxidizing compounds for post-harvest management of fruits.

Compounds	Fruit	$KMnO_4$ concentration	Result	Reference
$KMnO_4$	Sugar apple *cv.* Pinha	$KMnO_4$ sachet (3g $KMnO_4$) and stored for 20 days at 13°C.	Delayed ripening, preserved physio-chemical traits without affecting the visual appearance.	Sanches *et al.*, 2019
	Nectarine	10g $KMnO_4$ sachets inside polypropylene baskets in MAP at 0 – 1 Â°C and 90 per cent for 40 days.	Better biochemical properties and inhibited CI symptoms on fruits.	Bal, 2018
	Banana	6g KMnO4 kept in tissue paper and stored at room temperature of 30.35 °C and 82 per cent RH.	Improved shelf life up to 13.75 days and delayed post-harvest losses.	Dhakal *et al.*, 2021
	Guava	Sachets placed inside 76.2 µm LDPE film and kept at 8±2°C.	Maintained textural, physico-chemical properties of the fruits with less deacy and increased shelf life upto 4 weeks.	Ranjeet and Giri, 2014
	Pear *cv.* Bartlett	15 per cent $KMnO_4$ sachet in 50.8 µm LDPE film cold storage (2 ± 1°C with 95 ± 2 %, and shelf-life holding periods under ambient conditions (24 ± 3°C; 68 ± 5 %).	Better fruit quality with reduced spoilage and respiration rate.	Kumar and Thakur, 2020
	Pear *cv.* Nijisseiki	20g $KMnO_4$ with Aluminium oxide placed inside 50 µm LDPE bag for 252 days at 0°C.	Reduced brown heart incidence with low ethylene levels.	Giraldo *et al.*, 2005
	Mango	Fruits treated with wax containing 2g KMnO4 and stored at 18 ± 1°C with 85-90 per cent RH.	Delayed ripening upto 4 days and reduced weight loss upto 46 per cent.	Elzubeir *et al.*, 2018
	Avocado	0.03 g KMnO4 added into an alumina base nano fibre membrane for 6 days.	No ethylene concentration was recorded for 5 days.	Tirgar *et al.*, 2018
	Blueberry *cv.* Duke	0.7g $KMnO_4$ sachets in polypropylene trays kept at 2 for 46 days.	Reduced decay, weight and firmness losses with optimum quality.	Álvarez-Hernández *et al.*, 2019
	Blueberry cv. Lanfeng	10 g sachet inside 20 µm thick PE bag kept at 0 °C (85 per cent RH) for 60 days and room temperature (20 °C, 85 per cent RH) for 8 days.	Delayed firmness and weight loss. Reduced decay and maintained sugars and phenolic content.	Srivastava *et al.*, 2012

Compounds	Fruit	$KMnO_4$ concentration	Result	Reference
	Datepalm *cv.* Barhee	10g $KMnO_4$ sachet in PVC-PE bags for 27 days at 5°C and 75 per cent RH.	Reduced incidence of brown spots and firmness loss.	Mortazavi *et al.*, 2015
	Papaya cv. 'Sunrise Golden'	1.5g $KMnO_4$ sachets in 28 µm hermetic LDPE bags kept at 10.4°C for 25 days.	Rise in SSC with delayed pulp consistency and weight loss.	Silva *et al.*, 2009
	Persimmon *cv.* Fuyu	Two alumina-$KMnO_4$ sachets (8.5 g) in 40 µm PE box placed at 22±1ºC for 9 days.	Delayed flesh softness.	Argenta *et al.*, 2009
	Quince *cv.* Isfahan	One 5g KMnO4 sachet inside 60 µm PE bag kept at 0ºC and 85-95 per cent for 75 days.	Less astringent fruits, reduced accumulation of β-carotene with delayed firmness losses.	Akbari and Ebrahimpour, 2014
	Apple *cv.* Golden Delicious	10g zeolite-$KMnO_4$ sachets placed in carton at 0 °C and 90 per cent RH for 5 months.	Delayed firmness loss and greening with reduced pH, SSC and TA losses.	Sardabi *et al.*, 2014
	Sapodilla *cv.* "Itapirema-31'	0.3 g $KMnO_4$ sachet in PVC coated Styrofoam trays kept at 25 °C and 54 per cent RH for 10 days.	Delayed ascorbic acid degradation and firmness loss.	de Souza Freitas *et al.*, 2017
	Sapota *cv.* 'Kallipati'	Four $KMnO_4$ sachets inserted in CFB box kept at 27 – 32 °C and 70 per cent RH for 13 days.	Reduced decay and weight loss incidence with delayed ethylene production.	Bhutia *et al.*, 2011
	Loquat *cv.* Precoce de Itaquera	One $KMnO_4$ sachet inserted in 10 µm PVC coated PET tray kept at 6°C and 90 per cent RH for 18 days.	Improved shelf life of fruits.	Campos *et al.*, 2007
	Pineapple *cv.* Gold Extra Sweet	$KMnO_4$ absorbent on waxed/unwaxed fruits kept at 7.5°C and 85-90 per cent RH for 25 days.	Reduced peduncle mould incidence with simulation of shelf life up to 3 days.	Sabater-Vilar *et al.*, 2016
	Strawberry *cv.* Torrey	One 10 g alumina-$KMnO_4$ sachet held in 9 µm PE wrapped punnet kept at 20°C for 4 days.	Reduced deterioration and ethylene production.	Kim *et al.*, 1998
	Fig cv. Brown Turkey	KMnO4 incorporated PE film wrapped on Plastic tray kept at 2°C for 21 days.	Delayed fruit firmness and weight loss.	Bouzo *et al.*, 2012

Compounds	Fruit	$KMnO_4$ concentration	Result	Reference
Palladium	Banana cv. Gros Michel	Palladium (1 per cent) incorporated activated carbon at 28 ± 1 °C, 70–75 per cent RH for 10 days.	Delayed fruit ripening, colour development with reduced weight loss and starch-sugar digestion.	Tepamatr *et al.*, (2023)
	Sapota cv. Kalipatti	1-2 per cent PdCl2 encapsulated in nanomats, 26 ± 2 °C for 8 days.	Reduced respiration rate, physiological weight loss with improved quality indices.	Gundewadi *et al.*, 2022
Selenium	Pear *cv.* Nanhong	0.1 per cent nano-Se + 40 per cent ethephon + 1 per cent sodium selenite.	Decreased stone cell content, ethylene production and inhibited ERFs expression.	Yuan *et al.*, 2023
	Apple cv. Starking Delicious	1.5 mg Se L^{-1} g, kept at 0 ± 1□ for 6 months.	Enhanced TSS, TA, firmness, antioxidants with reduction in ROS production.	Babalar *et al.*, 2019
	Guava *cv.* Hisar Safeda	Selenium (0.02 ppm) + chitosan 1.5 per cent.	Maintained physio-biochemical properties of fruits and prolonged shelf life up to 12 days.	Choudhary *et al.*, 2020

Palladium: The scavenging action of palladium-based compounds have been observed even after the initiation of climacteric peak (Terry *et al.* 2007). Pd has six times high ethylene adsorption capacity relative to $KMnO_4$ in high humid conditions, whereas under low humidity, it escalates to sixty times that of $KMnO_4$ (Ebrahimi *et al.*, 2021). Palladium is impregnated into activated carbon and zeolite to avoid the toxicity issues. Tepamatr *et al.*, (2023) observed that use of palladium (1%) impregnated activated carbon delayed the colour development and ripening in 'Gros Michel' banana with increased ethylene removal efficacy. The use of $PdCl_2$ incorporated nanopore mats prolonged the shelf of sapota fruits by reducing the respiration rate, physiological weight loss, PLW with improved ethylene scavenging capacity (47-68%) up to 8 days (Gundewadi *et al.*, 2022).

Selenium: As a major trace element, Se has a major role to play in the plant system for promoting growth and ameliorate stress conditions. Recently, selenium has emerged as a topic of interest in altering fruit physiology by reducing certain enzymatic activity and ethylene production (Pezzarossa *et al.*, 2014). The activity of key controlling ethylene genes *PuERF2, PuACSs* and lignin biosynthesis gene *PuC4H* was inhibited on pre-harvest application of Se in Nanhong pear (Yuan *et al.*, 2023). Similarly, Babalar *et al.*, (2019) observed an improvement in the shelf-life of apple fruits with better flesh firmness, increased activity of antioxidant enzymes and membrane integrity by reducing ROS and MDA production. However, combined application of selenium with chitosan has reported to improve the shelf life of guava fruits by 12 days under room conditions along with delayed weight loss and low disease incidence throughout the storage (Choudhary *et al.*, 2020).

Ethylene scavenging/ nano micro particle films: The incorporation of scavenging agents into packaging films holds promise for addressing challenges associated with sachet packaging. Ethylene scavengers can be embedded onto a solid matrix, dispersed within plastic, or integrated into distinct layers of the packaging material (Firouz *et al.*, 2021). ZMS (zeolite molecular sieve) is most commonly incorporated into packing films owing to its three-dimensional crystalline porous structure (Yildirim *et al.*, 2018) which promotes the gas permeability and helps in ethylene absorption (Zi *et al.*, 2011). The requirement of thin films for packaging has necessitated the incorporation of nanotechnology in film products (Wuttke *et al.*, 2018). Li *et al.* (2009) recorded better effect on sensory and physio-chemical parameters of Chinese jujube fruits on blending nano-powder (kaolin, nano-Ag, TiO_2) with PE as compared to normal packing materials. High surface-volume ratio of nano-chitosan (Nasr *et al.*, 2021) makes it an effective agent for developing coatings which will protect the cell wall structures and delay the post-harvest aging. Lustriane *et al.* (2018) coated banana fruits with nano-chitosan film and recorded an improvement in the shelf-life and fruit quality. Similarly, Resende *et al.*, (2018) observed an increased antioxidant activity with a reduction in weight loss of nano-chitosan coated strawberries as compared to normal chitosan treated ones. A variety of alginate-based films are also used in the market which are observed to retard senescence and protect fruits from certain pathogens (Senturk Parreidt *et al.*, 2018). Alginate-$CaCl_2$ based coatings are observed to efficiently reduce the mould yeast growth and delay browning in 'Red Fuji' sliced apples (Sarengaowa *et al.*, 2018). The efficacy of a single layer of coatings/film is poor many a times because of the high hydrophobicity. To overcome this issue, a layer-by-layer assembly technique has been developed which creates bilayer films with better mechanical properties (Nian *et al.*, 2022). A reduced endogenous ethylene production, suppressed respiration with delayed degreening has been observed in 'Tommy Atkins' mango on using Nano silica/chitosan-based coating (Kassem *et al.*, 2022).

Table 5: Application of ethylene scavenging composite films in post-harvest fruit preservation.

Film	Fruit	Effect	Reference
CS/HNT composite film	Banana	Maintained freshness and extended the shelf life.	Wang *et al.*, 2023a
Pumice, KMnO4 based LDPE film	Avocado	Reduced firmness loss and preserved fruits upto 20 days at 25 °C.	Wang & Ajji, 2022
ASKG nanoparticle coated fabricated film	Banana	Reduced fruit browning, improved hardiness and shelf life of fruits.	Fan *et al.*, 2022
Nano-powder (Kaolin, nano Ag, TiO_2) blended PE film	Chinese jujube	Increased sugar, malondialdehyde content followed by a decrease in ascorbic acid and TA.	Li *et al.*, 2009
Nano silica/chitosan film	Mango	Reduced weight loss, decay, delayed degreening and ethylene production.	Kassem *et al.*, 2022
Nano-SiO_2/TiO_2 chitosan coating	Blueberry	Maintained nutrient status with reduced microbial growth and delayed shelf life.	Li *et al.*, 2021
Chitosan/ Cellulose nanofibril nanocomposite	Strawberry	Reduced firmness loses with better fruit colour and biochemical properties.	Resende *et al.*, 2018
Alginate-based edible coating	Sliced apple	Inhibited browning, microbial growth and firmness losses.	Sarengaowa *et al.*, 2018

Photocatalysis: The photocatalytic oxidation (PCO) involves utilization of a catalyst (ZnO, TiO2, ZnS, Fe2O3, CdS etc.) and UV light within a range of 200 to 380 nm, depending on the catalyst (Kartheuser and Boonaert, 2007).

The catalyst surface is irradiated by UV light, generating electron–hole pairs that produce reactive oxygen species and oxidize ethylene into CO_2 and water (Hussain *et al.*, 2011a). Though it's a cost-effective technique and applicable at room pressure/temperature, but it faces certain challenges like catalyst deactivation and reduced efficiency due to humidity effects (Huang *et al.*, 2011; Jeong *et al.*, 2013). To overcome these issues, vacuum ultraviolet (VUV) photolysis can be used. VUV is devoid of a catalyst, produces ROS in the gas phase and displays a higher efficiency than PCO in ethylene removal studies (Hussain *et al.*, 2011b). VUV photolysis remains effective even at high humidity, demonstrating its higher potential for ethylene removal in fruit storage. Pathak *et al.*, (2017a) reported VUV to be a better technique for the removal of ethylene and observed 76 per cent ethylene removal from the storage chamber of kiwifruit and apple. The combined effect of photocatalysis along with $KMnO_4$ application was evaluated by Alonso-Salinas *et al.*, (2022) on peaches stored at two different temperatures (1°C and 25°C) and extension of shelf life up to 7 days was recorded in fruits stored at 25°C. the use of photocatalytic QDs@ZIF-8 nanoparticles has been demonstrated (Wang *et al.*, 2023b) by for developing nano-composite film incorporating them onto CS/SA matrix by LBL technique in kiwifruit. This film extended the shelf-life under illumination exhibiting antibacterial properties and the ability to scavenge ethylene.

Blocking ethylene receptors: Use of chemical compounds is common to block the interaction of ethylene with its receptors on the cells and prolong the shelf-life of fruits. Certain chemicals like 2,5-NBD, CP and DACP are observed to block these ethylene-receptor interactions in apple (Blankenship and Sisler, 1989), strawberry (tian *et al.*, 1997) and other fruits. While 2,5-NBD is constrained by an unpleasant odour and carcinogenic potential, DACP faces limitations due to its explosive behaviour in intense light, thereby necessitating exploration for more novel chemicals to effectively block ethylene-receptor reaction (Zhang *et al.*, 2017). In past three decades, researchers have focused on employing CP and CP based derivatives (3,3-DMCP, 1-MCP) for blocking ethylene receptors. Among these, CP and 1-MCP have exhibited better efficiency than 3,3-DMCP in many studies (Sisler and Serek, 1997). Furthermore, the better stability of 1-MCP makes it more preferable choice over CP (Sisler *et al.*, 2003).

1-MCP: 1-MCP has extensively been used in the post-harvest handling of fruits owing to its ease of handling, better efficiency and non-toxic nature. It inhibits the fruit ripening by binding to the ethylene receptors (*AdERS1a, Ad-ETR3*) and inhibiting the function of key TFs (*Ad-ERFs*) (Liu *et al.*, 2019). The working mechanism of 1-MCP can be better understood by the lock and key model (Fig. 3) suggested by Zhang *et al.*, (2017), where each receptor functions as a lock and the ethylene molecule serves as a corresponding key to it. The ethylene molecule interacts with the metals present on the receptor, opens the lock (by electron withdrawal) and initiates ethylene induced ripening. However, 1-MCP outcompetes ethylene by releasing electrons from the metal, binds to the receptor and thereby unable the ethylene key to interact with metal (receptor) and open the lock. Once the ethylene-receptor interaction is inhibited, the ethylene activity weakens and this inhibitory effect remains persistent even after the removal of 1-MCP from the environment (Pathak *et al.*, 2017b). Some other structural analogues (1-OCP, 1-PentCP) of 1-MCP have also been reported to supress the ethylene-receptor interaction in the same way (Wang *et al.*, 2015). Many studies are available in the literature suggesting 1-MCP to be a potential agent for preserving fruit crops. Zhao *et al.* (2024) observed alleviation of chilling injury by lowering the expression of PG, Exp, PLY genes during the post-harvest storage of peach fruits using 1 $\mu l L^{-1}$ 1-MCP. Zhou *et al.* (2017) observed an inhibition of scald incidence in 1-MCP (1 $\mu l L^{-1}$) treated 'Wujiuxiang' pears after 120d of storage along with a variable expression of glutathione peroxidase, glutathione S transferase and ethylene related genes. Similarly, Giné-Bordonaba *et al.* (2020) reported decreased scald incidence by repression of PPO gene and lower ethylene production.

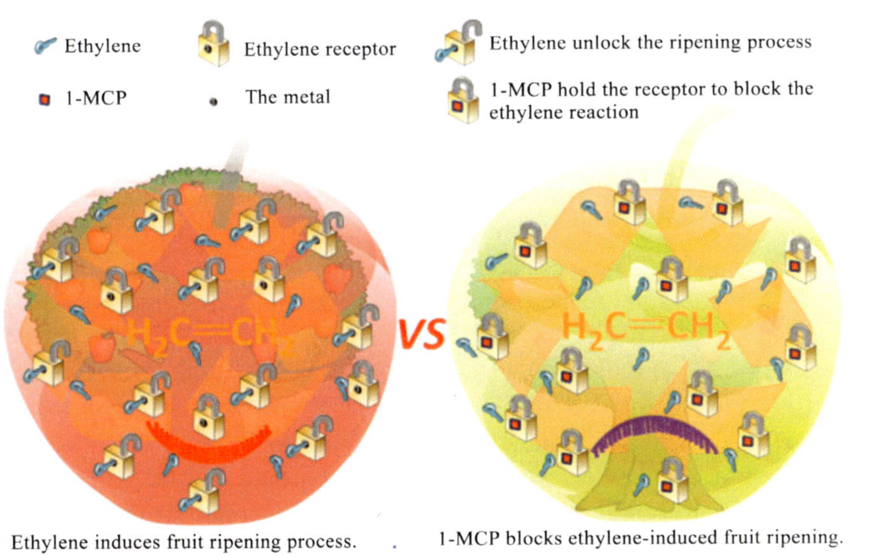

Fig. 3: Lock and Key model given by Zhang *et al.* (2017), suggesting altered fruit ripening behaviour in the absence and presence of 1-MCP.

However, an abnormal ripening behaviour was observed in the banana fruits on using higher 1-MCP concentrations, attributing to inconsistent sucrose metabolism and ethylene signalling expression (Song *et al.*, 2020). Hence, an ethylene absorbent (EA) like $KMnO_4$ is used along with the reduced concentrations of 1-MCP to mitigate these adverse effects (Wang *et al.*, 2024). Similarly, Xu *et al.*, (2019) observed that combined application of isoenzyme coatings with 1-MCP exerted a pronounced effect on the post-harvest storage of kiwifruit rather than 1-MCP alone.

Some other new ethylene antagonistic compounds like 1H-Cyclopropabenzene (BC) and its derivative 1H-cyclopropa[*b*]naphthalene (NC) are growing popular over 1-MCP due to their better water solubility and efficient ethylene inhibition capacity as reported by Kyaw *et al.*, 2021; 2023 in Japanese plum.

***Nitric oxide*:** NO is produced naturally in the plant system and acts as a signalling molecule during different physiological activities. There are various reports suggesting the role of NO in blocking the ethylene production among different fruits peaches (Cai *et al.*, 2020), kiwi (Zheng *et al.*, 2017), mango (Barman *et al.*, 2014) etc. Nitric oxide blocks the ethylene production by regulating transcriptional and post translational changes in ACO and MAT genes along with reduction of ACC to MACC followed by the formation of ACC–ACC oxidase–NO complex (da Veiga *et al.*, 2023). The NO application can be done by immersion, fumigation or spraying using different sources of NO (SNP, GSNO, DETANO), characterized by varying chemical structure and capacity to release NO. Zhang *et al.*, (2022) proposed, dipping of rambutan fruits in 200 µmol/L SNP solution prolonged the firmness and reduce the browning incidence till 8 days under room temperature. A significant reduction in respiration rate, ethylene production and mass loss has been observed on treating sweet cherry fruit with 60 mmol L^{-1} GSNO for one minute (Ma *et al.*, 2019). NO (10 µL L^{-1}) alleviated the loss of VOCs in peach after a cold storage (4°C) of 21 days followed by shelf-life (20°C) for 3 days with recovery of ethylene production and respiration rate. However, there are reports revealing the reduction in the efficiency of these NO donors in response to environmental factors (light, temperature, pH), causing adverse effects on the fruits (Silveira *et al.*, 2019). Hence, to avoid these losses, these NO donors are combined with nanoparticles to enhance their stability and bioactivity during storage. The efficient potential of NO releasing chitosan nanoparticles has been observed in the post-harvest preservation of sweet cherries by reducing the respiration rate, ethylene production and mass loss during the cold storage (Ma *et al.*, 2019).

Table 6: Application of ethylene receptor blocking compounds for post-harvest preservation of fruit crops.

Compound	Crop	Concentration	Storage condition	Effects	Reference
1-MCP	Peach (Lv Hua 9)	1 $\mu l L^{-1}$	4±0.5°C	Reduced incidence of internal browning, chilling injury and delayed fruit softening.	Zhao *et al.*, 2024
	Banana (Fenjiao)	7 g L^{-1} + EA (2 g kg^{-1}	20 ± 1 °C	Enhanced shelf life with the uniform ripening of fruits.	Wang *et al.*, 2024
	Apple (Empire)	1 μL L^{-1}	9 months CA (3°C + 2.5 kPa O_2 + 2kPa CO_2) + 7d room temp.	Improved firmness retention upto 7d at room temp., low core browning incidence.	DeEll *et al.*, 2016
	Pear (Bartlett)	0.15 μL L^{-1}	150d RA storage (− 1.1 ± 0.5 °C) + 5d room temp. (20 ± 0.5 °C)	Prolonged storage with delayed and reduced sunlight induced core browning.	Dong and Zhi, 2024
	Guava (Arka Mridula)	500 ppb	8 & 12 °C ±1 °C	Improved fruit physical and biochemical properties with extension of shelf life up to 2 weeks.	Sachin *et al.*, 2023
	Pomegranate	1.0 μL L^{-1}	4 ± 0.5 °C for 90 days	Supressed ethylene production in cold storage and halted peel browning	Wan *et al.*, 2023
	Kiwifruit (Xuxiang)	0.5 μL L^{-1}	23 ± 1 □, 80–90 per cent RH	Inhibited hard core incidence with enhancement of shelf life up to 15 days.	Chai *et al.*, 2023
	Apple (Santana)	625 μL L^{-1}	DCA (1-3 □) 6 months + 7d (20□)	Reduced fungal decay and core browning incidence.	Büchele *et al.*, 2023
	Avocado	0.93 & 9.3 mmol m^{-3}	2□ for 18d	Delayed ethylene production, softening, accumulation of total soluble, flavonoids, phenolics and antioxidants.	Zhang *et al.*, 2013
	Pear (Blanquilla)	300 nL L^{-1}	0.5□, 95 per cent RH for 4 months	Repressed superficial scald incidence by down regulating the expression of polyphenol oxidase.	Giné-Bordonaba *et al.*, 2020

Compound	Crop	Concentration	Storage condition	Effects	Reference
1H-Cyclopropabenzene (BC)	Japanese plum cv. Tegan Blue	1 µM BC fumigation	0 ± 1 °C, RH 90 ± 5 per cent (40 days) + 20 ± 1 °C (10 days)	Supressed ethylene production, weight loss, anthocyanin levels, retained firmness and ascorbic acid levels.	Kyaw *et al.*, 2021
1 *H*-cyclopropa[*b*] naphthalene (NC)	Japanese plum cv. Angeleno	2 µM spray + 1 µM fumigation at 20±1 °C (18 hrs)	0±1 °C, RH 90±5 per cent (40 days)	Retained firmness and fruit weight with delayed ripening and reduced SSC, SSC: TA content.	Kyaw *et al.*, 2023
Nitric Oxide	Rambutan	SNP, 200 µmol/L for 2 hrs	25□ (8 days)	Reduced weight loss, pericarp browning and maintained fruit firmness for long period.	Zhang *et al.*, 2022
	Pear	SNP, 0.002 mol L^{-1} for 2 hrs	0–1 °C, RH 90–95 per cent (60 days)	Reduced respiration rate, weight loss, browning incidence, retained colour, TSS, TA and other quality parameters for long duration.	Adhikary *et al.*, 2021
	Blueberry	SNP 0.10 mmol L^{-1} (10 min)	20 ± 0.5 °C, RH 80 per cent (8 d)	Exhibited high fruit firmness, lower decay incidence and increased activity of fatty acid synthesis genes.	Dai *et al.*, 2021
	Peach	NO gas fumigation (10 µL L^{-1}) for 3 h	4°C (21 days), shelf life at 20°C (3 days)	Promoted expression of aroma volatile related genes and emission of VOCs.	Cai *et al.*, 2020
	Cherry	GSNO-CS NPs , 60 mmol L^{-1}	4 ◦C (25 days)	Reduced ROS production, increased antioxidant activity and glutathione levels.	Ma *et al.*, 2019

Evolving breeding techniques for post-harvest management: As discussed above, a number of technologies have been devised to enhance post-harvest handling and prolong shelf life of fruits, yet their adoption is predominantly limited to developed nations only, due to the substantial costs involved (Jarman *et al.*, 2023). Therefore, the genetic modification or breeding technologies can be applied to extend the shelf-life of the fruits. Over a long time, there was pre-dominance of physiological and biochemical aspects in fruit ripening and shelf-life related studies. However, the emergence of the genome sequences has geared forward in understanding the ripening mechanisms and developing effective technologies for fruit crop improvement. Several candidate genes viz., *MdPAE10, MdACS1-2, MdACO1-1, PpPG, PpYUC11, FaPG* (Wu *et al.*, 2021) have been identified on studying ripening mechanisms in different fruits. These candidate genes can be used for TILLING or manipulated using genome editing. Biotechnological approaches, such as RNAi, VIGS, co-suppression, antisense, CRES-T, are being explored to target such genes with the aim of altering their expression and further extending the shelf-life of fruits (Mithapara and Butani, 2023).

The varied expression of *PpPG* has been observed to govern textural and cell wall degradation changes in the melting and non-melting peaches (Cheng *et al.*, 2022). Qian *et al.* (2021) identified a high expression of *PpPG21, 22* in melting peaches. Meanwhile, suppression of this highly expressing factor (*PpPG21, 22*) in MF peaches by VIGS, resulted in increased firmness during late ripening stages and suppressed the melting stage by reducing polygalacturonase activity. The role of B-box proteins was illustrated for the first time, in governing textural changes taking place during the storage of apple fruits and a BBX TF (*MdBBX25*) was identified as a negative regulator of ethylene biosynthesis (Zhang *et al.*, 2024); silencing of *MdBBX25* has accelerated the fruit softening. The depolymerization of HGA during fruit ripening in strawberries is mediated by PG (*FaPG1*) and PL (*FaPLC*) gene. However, a greater reduction in the softening was observed on suppression of *FaPG1* as compared to *FaPLC* (Pose *et al.*, 2015). Nevertheless, the application of RNA interference (RNAi) or antisense techniques in the post-harvest domain has been discouraged in certain parts of world for fruit and vegetable crops particularly (Friedman, 2019). Hence, these methodologies are now employed for the functional analysis of candidate genes only. A recent, target genome-oriented editing technology (CRISPR cas9) has successfully been applied in many fruit crops. Hu *et al.* (2021) used CRISPR Cas9 approach for editing of *MaACO1* gene and recorded an increase in shelf life with reduced ethylene production levels in Brazilian banana.

Table 7: Application of breeding approaches to supress ethylene production in fruit crops.

Crop	**Technique**	**Gene**	**Effect**	**Reference**
Strawberry	Antisense	FaPLC, FaPG1	Reduced fruit softening.	Pose *et al*., 2015
Peach	VIGS	*PpPG21, PpPG22*	Hindered texture change and delayed fruit ripening.	Qian *et al*., 2021
Apple	VIGS	*MdBBX25*	Negative suppressor of ethylene biosynthesis.	Zhang *et al*., 2024
Strawberry	RNAi	FaRGLyase1	Increased RGL pectins in cell well, reducing degradation and delayed senescence.	Ric-Varas *et al*., 2024
Banana	RNAi	*MaMADS, MaMADS36*	Delayed colour development and softening with enhanced shelf life.	Elitzur *et al*., 2016
Papaya	RNAi	*ACO1, ACO2*	Delayed color development and extended shelf life upto 20 days.	Sekeli *et al*., 2014
Banana	CRISPR/Cas9	*MaACO1*	Extended shelf-life.	Hu *et al*., 2021
Strawberry	CRISPR/Cas9	*FaPG1*	Reduced post-harvest softening (70 per cent) and water loses.	López-Casado *et al*., 2023

This article reviewed the ethylene evolution mechanism in both climacteric and non-climacteric fruits. It discussed the characteristics and effectiveness of different ethylene management techniques. Active packaging, utilizing ethylene absorbents and $KMnO_4$ sachets, as well as the application of 1-MCP, have been demonstrated to prolong storage life and delay the ripening period in fruit crops. The incorporation of nanocomposite films and nanomaterials into polymer matrices has shown promise in enhancing ethylene absorption. Additionally, breeding approaches such as CRISPR-Cas9 hold significant potential for reducing ethylene production by targeting ethylene-specific candidate genes in the near future.

References

Abeywickrama K P, Wijerathna C, Rajapaksha N, Sarananda K, and Kannangara S, 2012. Disease control strategies for extending storage life of papaya (*Carica papaya*), cultivars 'Red Lady'and 'Rathna'.

Adhikary T, Gill, P P S, Jawandha S K, Bhardwaj R D and Anurag R K, 2021. Efficacy of postharvest sodium nitroprusside application to extend storability by regulating physico-chemical quality of pear fruit. *Food chemistry*, *346*, 128934.

Aghdam M S and Fard, J R, 2017. Melatonin treatment attenuates postharvest decay and maintains nutritional quality of strawberry fruits (*Fragaria* × *anannasa* cv. Selva) by enhancing GABA shunt activity. *Food Chemistry*, 221, 1650-1657.

Akbari H and Ebrahimpour H, 2014. Potassium permanganate and packing types impacts on postharvest quality and storage period of quince fruit (*Cydonia oblonga* Mill). *Int J Adv Life Sci* 7:267–27.

Alonso-Salinas R, Acosta-Motos J R, Núñez-Delicado E, Gabaldón, J A and López-Miranda S, 2022. Combined effect of potassium permanganate and ultraviolet light as ethylene scavengers on post-harvest quality of peach at optimal and stressful temperatures. *Agronomy*, 12(3): 616.

Álvarez-Hernández M H, Martínez-Hernández G B, Avalos-Belmontes F, Castillo-Campohermoso M A, Contreras-Esquivel J C and Artés-Hernández F, 2019. Potassium permanganate-based ethylene scavengers for fresh horticultural produce as an active packaging. *Food Engineering Reviews*, 11, 159-183.

Álvarez-Hernández M H, Martínez-Hernández G B, Avalos-Belmontes F, Rodríguez-Hernández A M, Castillo-Campohermoso M A and Artés-Hernández F, 2019. An innovative ethylene scrubber made of potassium permanganate loaded on a protonated montmorillonite: A case study on blueberries. *Food and Bioprocess Technology*, 12:524-538.

Ansari M W and Tuteja N, 2015. Post-harvest quality risks by stress/ethylene: management to mitigate. *Protoplasma*, 252:21-32.

Argenta L, Vieira M and Scolaro A, 2009. Conservação da qualidade de caqui 'Fuyu' em ambiente refrigerado pela combinação de 1- MCP e atmosfera modificada. Rev Bras Frutic 31:323–333. https://doi.org/10.1590/S0100-29452009000200006

Awad M A and Al-Qurashi, A D, 2019. Quality and biochemical changes of 'Sukkari'bananas during shelf life as affected by postharvest dipping in ethanolic extract of propolis. *Philippine Agricultural Scientist*, 102(2), 132-140.

Babalar M, Mohebbi S, Zamani Z and Askari M A, 2019. Effect of foliar application with sodium selenate on selenium biofortification and fruit quality maintenance of 'Starking Delicious' apple during storage. *Journal of the Science of Food and Agriculture*, 99(11), 5149-5156.

Badawy I F, 2016. Effect of ethanol-extracted propolis on fruit quality and storability of balady oranges during cold storage. *Assiut Journal of Agricultural Sciences*, 47(4): 156-166.

Bagheri M and Esna-Ashari M, 2022. Effects of postharvest methyl jasmonate treatment on persimmon quality during cold storage. *Scientia Horticulturae*, 294: 110756.

Bal E, 2018. Extension of the postharvest life of nectarine using modified atmosphere packaging and potassium permanganate treatment. *Turkish Journal of Agriculture-Food Science and Technology*, 6(10): 1362-1369.

Bal E, 2019. Physicochemical changes in 'Santa Rosa'plum fruit treated with melatonin during cold storage. *Journal of Food Measurement ànd Characterization*, 13: 1713-1720.

Barman K, Asrey R, Pal R K, Jha S K and Bhatia K, 2014. Post-harvest nitric oxide treatment reduces chilling injury and enhances the shelf-life of mango (*Mangifera indica* L.) fruit during low-temperature storage. *The Journal of Horticultural Science and Biotechnology*, 89(3): 253-260.

Bhutia W, Pal R, Sen S and Jha S, 2011. Response of diferent maturity stages of sapota (*Manilkara achras* Mill.) cv. Kallipatti to in-package ethylene absorbent. *J Food Sci Technol* 48:763–768.

Bhutia W, Pal R, Sen S and Jha S, 2011. Response of different maturity stages of sapota (*Manilkara achras* Mill.) cv. Kallipatti to inpackage ethylene absorbent. *J. Food Sci. Technol* 48,763–768. https://doi.org/10.1007/s13197-011-0360-x

Blankenship S M and Sisler E C, 1989. 2, 5-norbornadiene retards apple softening. *HortScience*, 24(2), 313-314.

Bleecker A B and Kende H, 2000. Ethylene: a gaseous signal molecule in plants. *Annu. Rev. Cell Dev. Biol.*, 16:1–18.

Bolfarin, A C B, Souza J M A, de Souza Silva M, Ferreira R B, Leonel M, Tecchio M A and Leonel S, 2020. Ideal harvest stage and quality descriptors of 5 banana cultivars based on 5 fruit diameters. *Emirates Journal of Food and Agriculture*, 220-228.

Bouzo C A, Travadelo M and Gariglio N F, 2012. Effect of Different Packaging Materials on Postharvest Quality of Fresh Fig Fruit. *International Journal of Agriculture & Biology*, 14(5).

Bregoli A M, Scaramagli S, Costa G, Sabatini E, Ziosi V, Biondi S and Torrigiani P, 2002. Peach (*Prunus persica*) fruit ripening: aminoethoxyvinylglycine (AVG) and exogenous polyamines afect ethylene emission and fesh frmness. *Physiol Plant* 114:472–481

Brizzolara S, Santucci C, Tenori L, Hertog M, Nicolai B, Stürz S, and Tonutti P, 2017. A metabolomics approach to elucidate apple fruit responses to static and dynamic controlled atmosphere storage. *Postharvest Biology and Technology*, 127: 76-87.

Büchele F, Khera K, Wagner R, Thewes F R and Neuwald D A, 2023. Interaction between dynamic controlled atmosphere (DCA-CD), 1-methylcyclopropene and elevated temperatures in the long-term storage of organic 'Santana'apples. *Postharvest Biology and Technology*, 204: 112471.

Cai H, Han S, Yu M, Ma R and Yu Z, 2020. Exogenous nitric oxide fumigation promoted the emission of volatile organic compounds in peach fruit during shelf life after long-term cold storage. *Food Research International*, 133: 109135.

Campos J, Hasegawa P, Purgatto E, Lajolo F and Cordenunsi B, 2007. Qualidade pós-colheita de nêsperas submetidas ao armazenamento sob baixa temperatura e atmosfera modificada. *Food Sci. Technol.* 27:401–407. https://doi.org/10.1590/S0101- 20612007000200033

Cape J N, 2003. Effects of airborne volatile organic compounds on plants. *Environ. Pollut.*, 122: 145-157.

Chai J, Li J, Li L, Liao B and Liu Z, 2023. 1-MCP application in 'Xuxiang' kiwifruit at appropriate firmness avoids hard-core and prolongs the shelf life. *Postharvest Biology and Technology*, 206: 112549.

Chen C, Sun C, Wang Y, Gong H, Zhang A, Yang Y and Li X, 2023. The preharvest and postharvest application of salicylic acid and its derivatives on storage of fruit and vegetables: A review. *Scientia Horticulturae*, 312: 111858.

Chen Y, Grimplet J, David K, Castellarin S D, Terol J, Wong D C and Chervin C, 2018. Ethylene receptors and related proteins in climacteric and non-climacteric fruits. *Plant Science*, 276: 63-72.

Chen Y, Sun J, Lin H, Lin M, Lin Y, Wang H and Hung Y C, 2020. Salicylic acid reduces the incidence of Phomopsis longanae Chi infection in harvested longan fruit by affecting the energy status and respiratory metabolism. *Postharvest Biology and Technology*, 160: 111035.

Cheng C, Liu J, Wang X, Wang Y, Yuan Y and Yang S, 2022. PpERF/ABR1 functions as an activator to regulate PpPG expression resulting in fruit softening during storage in peach (*Prunus persica*). *Postharvest Biology and Technology*, 189: 111919.

Choosung P, Utto W, Boonyaritthongchai P, Wasusri T and Wongs-Aree C, 2019. Ethanol vapor releasing sachet reduces decay and improves aroma attributes in mulberry fruit. *Food Packaging and Shelf Life*, 22: 100398.

Choudhary, P, Kumari N and Jain V, 2020. Influence of selenium and chitosan on physico-chemical properties of guava (*Psidium guajava*) under storage. *The Indian Journal of Agricultural Sciences*, 90(2): 351-355.

da Veiga J C, Silveira N M, Seabra A B and Bron I U, 2023. Exploring the power of nitric oxide and nanotechnology for prolonging postharvest shelf-life and enhancing fruit quality. *Nitric Oxide*.

Dai H, Ji S, Zhou X, Wei B, Cheng S, Zhang F and Zhou Q, 2021. Postharvest effects of sodium nitroprusside treatment on membrane fatty acids of blueberry (*Vaccinium corymbosum*, cv. Bluecrop) fruit. *Scientia Horticulturae*, 288, 110307.

Davarynejad G H, Zarei M, Nasrabadi M E and Ardakani E, 2015. Effects of salicylic acid and putrescine on storability, quality attributes and antioxidant activity of plum cv.'Santa Rosa'. *Journal of Food Science and Technology*, 52: 2053-2062.

de Souza Freitas W E, Bezerra Almeida M L, Dantas de Morais P L, da Curnha Moura A K and Sales Júnior R, 2017. Potassium permanganate effects on the quality and post-harvest conservation of sapodilla (*Manilkara zapota* (L.) P. Royen) fruits under modified atmosphere. *Acta Agronómica*, 66(3): 331-337.

DeEll J R, Lum G B and Ehsani-Moghaddam B, 2016. Effects of multiple 1-methylcyclopropene treatments on apple fruit quality and disorders in controlled atmosphere storage. *Postharvest Biology and Technology*, 111: 93-98.

Dhakal S, Aryal S, Khanal P, Basnet B and Srivastava A, 2021. Effect of KMnO4 on shelf life and quality of banana (*Musa paradisiaca* L.). *Fundamental and Applied Agriculture*, 6(3): 225-231.

Dong Y and Zhi H, 2024. Applications of 1-methylcyclopropene concentrations and timing in relation to sunlight-related core browning in 'Bartlett'pears after extending regular-air storage. *Postharvest Biology and Technology*, 207: 112624.

Ebrahimi A, Zabihzadeh Khajavi M, Ahmadi S, Mortazavian A M, Abdolshahi A, Rafiee S and Farhoodi M, 2022. Novel strategies to control ethylene in fruit and vegetables for extending their shelf life: A review. *International Journal of Environmental Science and Technology*, 19(5): 4599-4610.

Ecker W L, 2002. The ethylene response pathway in Arabidopsis. *Annu. Rev. Plant Physiol. Plant Mol. Biol.*, 48:277–296.

Elitzur T, Yakir E, Quansah L, Zhangjun F, Vrebalov J, Khayat E and Friedman H, 2016. Banana MaMADS transcription factors are necessary for fruit ripening and molecular tools to promote shelf-life and food security. *Plant Physiology*, 171(1): 380-391.

Elzubeir M M, Abu-bakr A, Osman O A and Safi A I A, 2018. Effect of waxing and potassium permanganate on quality and shelf-life of mango fruits. *American Journal of Biology and Life Sciences*, 6(1): 1-7.

Erogul D and Özsoydan İ, 2020. Effect of pre-harvest salicylic acid treatments on the quality and shelf life of the 'Cresthaven'peach cultivar. *Folia Horticulturae*, 32(2): 221-227.

Falagan N and Terry L A 2018. Recent advances in controlled and modified atmosphere of fresh produce. *Johnson Matthey Technology Review*, 62(1): 107-117.

Fan X, Du Z, Cui X, Ji W, Ma J, Li X, and Gong H, 2021. Preharvest methyl salicylate treatment enhance the chilling tolerance and improve the postharvest quality of apricot during low temperature storage. *Postharvest Biology and Technology*, 177:111535.

Fan X, Rong L, Li Y, Cao Y, Kong L, Zhu Z and Huang J, 2022. Fabrication of bio-based hierarchically structured ethylene scavenger films via electrospraying for fruit preservation. *Food Hydrocolloids*, 133: 107837.

Fan X, Shu C, Zhao K, Wang X, Cao J and Jiang W, 2018a. Regulation of apricot ripening and softening process during shelf life by post-storage treatments of exogenous ethylene and 1-methylcyclopropene. *Scientia Horticulturae*, 232:63-70.

Fan X, Xi Y, Zhao H, Liu B, Cao J and Jiang W, 2018b. Improving fresh apricot (*Prunus armeniaca* L.) quality and antioxidant capacity by storage at near freezing temperature. *Scientia Horticulturae*, 231: 1–10.

Firouz M S, Mohi-Alden K and Omid M, 2021. A critical review on intelligent and active packaging in the food industry: Research and development. *Food Research International*, 141: 110-113.

Friedman H, 2019. Candidate genes to extend fleshy fruit shelf life. *Plant breeding reviews 43* 61-94.

Gane R, 1934. Production of ethylene by some ripening fruits. *Nature* 134 1008.

Gao H, Zhang Z K, Chai H K, Cheng N, Yang Y, Wang D N and Cao W, 2016. Melatonin treatment delays postharvest senescence and regulates reactive oxygen species metabolism in peach fruit. *Postharvest Biology and Technology*, 118: 103-110.

Gao Q, Tan Q, Song Z, Chen W, Li X and Zhu X, 2020. Calcium chloride postharvest treatment delays the ripening and softening of papaya fruit. *Journal of Food Processing and Preservation*, 44(8): e14604.

García-Pastor M E Giménez M J Zapata P J Guillén F Valverde J M Serrano M and Valero D, 2020a. Preharvest application of methyl salicylate acetyl salicylic acid and salicylic acid alleviated disease caused by Botrytis cinerea through stimulation of antioxidant system in table grapes *International Journal of Food Microbiology*, 334: 108807

García-Pastor M E, Zapata P J, Castillo S, Martínez-Romero D, Guillén F, Valero D and Serrano M, 2020b. The effects of salicylic acid and its derivatives on increasing pomegranate fruit quality and bioactive compounds at harvest and during storage *Frontiers in Plant Science*, 11: 668.

Gasser F and Von Arx K, 2013. Dynamic CA storage of organic apple cultivars In *XI International Controlled and Modified Atmosphere Research Conference*. 1071: 527-532.

Genard M and Gouble B, 2005. ETHY A theory of fruit climacteric ethylene emission *Plant physiology* 139(1) 531-545.

Giménez M J, Serrano M, Valverde J M, Martínez-Romero D, Castillo S, Valero D and Guillén F, 2017. Preharvest salicylic acid and acetylsalicylic acid treatments preserve quality and enhance antioxidant systems during postharvest storage of sweet cherry cultivars *Journal of the Science of Food and Agriculture*, 97(4): 1220-1228.

Giné-Bordonaba J, Busatto N, Larrigaudière C, Lindo-García V, Echeverria G, Vrhovsek U and Costa F, 2020. Investigation of the transcriptomic and metabolic changes associated with superficial scald physiology impaired by lovastatin and 1-methylcyclopropene in pear fruit (cv"Blanquilla") *Horticulture Research,* 7.

Giraldo E, Szczerbanik M, Scottpez K, Paton J and Best D, 2005. Effects of polyethylene bags ethylene absorbent and 1- methylcyclopropene on the storage of Japanese pears *J Hortic Sci Biotechnol* 80:162–166.

Gomes E P, Borges C V, Monteiro G C, Belin M A F, Minatel I O, Junior A P and Lima G P P, 2021. Preharvest salicylic acid treatments improve phenolic compounds and biogenic amines in 'Niagara Rosada'table grape *Postharvest Biology and Technology,* 176 111505.

Gundewadi G, Rudra S G, Prasanna R, Banerjee T, Singh S K, Dhakate S R and Anand A, 2022. Palladium encapsulated nanofibres for scavenging ethylene from sapota fruits *Frontiers in Nutrition, 9:* 994813.

Guo S H, Xu T F, Shi T C, Jin X Q, Feng M X, Zhao X H and Meng J F, 2020. Cluster bagging promotes melatonin biosynthesis in the berry skins of *Vitis vinifera* cv Cabernet Sauvignon and Carignan during development and ripening *Food Chemistry,* 305: 125502.

Hadjipieri M, Georgiadou E C, Drogoudi P, Fotopoulos V and Mangan,aris G A, 2021. The efficacy of acetylsalicylic acid spermidine and calcium preharvest foliar spray applications on yield efficiency incidence of physiological disorders and shelf-life performance of loquat fruit *Scientia Horticulturae,* 289: 110439.

Haider S T A , Ahmad S, Anjum M A, Naz S, Liaqat M and Saddiq B, 2021. Effects of different postharvest techniques on quality management and shelf life of 'Kinnow'mandarin fruit *Journal of Food Measurement and Characterization,* 15: 2549-2561.

Han Y, 2021. 'Hayward'kiwifruit responses to ethylene in controlled atmosphere and storage performance in modified atmosphere packaging: a thesis presented in partial fulfilment of the requirements for the degree of Master of Science in Horticultural Science at Massey University Palmerston North New Zealand (Doctoral dissertation Massey University)

Hanif A, Ahmad S, Shahzad S, Liaquat M and Anwar R, 2020. Postharvest application of salicylic acid reduced decay and enhanced storage life of papaya fruit during cold storage *Journal of Food Measurement and Characterization,* 14: 3078-3088.

He J, Feng Y, Cheng Y, Wang M and Guan J, 2022. A comprehensive insight on the main physiological biochemical and related genes expression changes during the development of superficial scald in "Yali" pear *Frontiers in Plant Science,* 13: 987240.

Hennecke C, Köpcke D and Dierend W, 2008. Storage of apples in dynamic controlled atmosphere *Erwerbs-Obstbau,* 50: 19-29.

Hershkovitz V, Friedman H, Goldschmidt E E and Pesis E, 2009. The role of the embryo and ethylene in avocado fruit mesocarp discoloration *Journal of experimental botany,* 60(3):791-799.

Hu M, Li J and Rao J, 2018. Effect of melatonin on ripening and senescence of postharvest kiwifruits *Shipin Kexue/Food Science,* 39(19): 226-232.

Huang H, Leung D Y, Li G, Leung M K and Fu X, 2011. Photocatalytic destruction of air pollutants with vacuum ultraviolet (VUV) irradiation *Catalysis Today,* 175(1) 310-315.

Hussain M, Bensaid S, Geobaldo F, Saracco G and Russo N, 2011a. Photocatalytic degradation of ethylene emitted by fruits with TiO2 nanoparticles *Industrial & Engineering Chemistry Research* 50(5): 2536-2543.

Hussain M, Russo N and Saracco G, 2011b. Photocatalytic abatement of VOCs by novel optimized TiO2 nanoparticles *Chemical Engineering Journal,* 166(1) 138-149.

Jannatizadeh A, 2019. Exogenous melatonin applying confers chilling tolerance in pomegranate fruit during cold storage *Scientia Horticulturae,* 246: 544-549.

Jarman A, Thompson J, McGuire E, Reid M, Rubsam S, Becker K and Mitcham E, 2023. Postharvest technologies for small-scale farmers in low-and middle-income countries: A call to action *Postharvest Biology and Technology,* 206: 112491.

Jeong M G, Park E J, Seo H O, Kim K D, Kim Y D and Lim D C, 2013. Humidity effect on photocatalytic activity of TiO2 and regeneration of deactivated photocatalysts *Applied Surface Science,* 271 164-170.

Ji Y, Hu W, Liao J, Xiu Z, Jiang A, Guan Y and Feng K, 2021. Ethanol vapor delays softening of postharvest blueberry by retarding cell wall degradation during cold storage and shelf life *Postharvest Biology and Technology,* 177: 111538.

Kader A A, 1985. Ethylene-induced senescence and physiological disorders in harvested horticultural crops *HortScience,* 20(1): 54-57

Kartheuser B and Boonaert C, 2007. Photocatalysis: a powerful technology for cold storage applications *Journal of Advanced Oxidation Technologies,* 10(1) 107-110.

Kassem H S, Tarabih M E, Ismail H and Eleryan E E, 2022. Influence of nano-silica/chitosan film coating on the quality of 'Tommy Atkins' Mango *Processes* 10(2): 279.

Keller N, Ducamp M N, Robert D and Keller V, 2013. Ethylene removal and fresh product storage: a challenge at the frontiers of chemistry Toward an approach by photocatalytic oxidation *Chemical Reviews,* 113(7) 5029-5070.

Kerry J and Butler P, 2008. Smart packaging technologies for fast moving consumer goods John Wiley & Sons Ltd West Sussex.

Ketsa S and Pangkool S, 1995. The effect of temperature and humidity on the ripening of durian fruits *Journal of Horticultural Science,* 70(5): 827-831.

Khedr E H, Khedr N and Abdel-Haleem M, 2023. Harnessing the metabolic modulatory and antioxidant power of 1-(3-Phenyl-Propyl) cyclopropane and melatonin in maintaining mango fruit quality and prolongation storage life *BMC Plant Biology,* 23(1): 464.

Kim G-H and Wills R, 1998. Interaction of enhanced carbon dioxide and reduced ethylene on the storage life of strawberries *J Hortic Sci Biotechnol* 73:181–184 https://doiorg/101080/14620316 199811510962

Köstekli M, Özdzikicierlev O, Cortés C, Zulueta A, Esteve Más MJ and Frígola Cánoves A, 2016. Role of Potassium Permanganate Ethylene on Physicochemical Properties during Storage of Five Different Tomato Cultivars MOJFPT 3:1–9.

Koyuncu M A, Erbas D, Onursal C E, Secmen T, Guneyli A and Sevinc Uzumcu S, 2019. Postharvest treatments of salicylic acid oxalic acid and putrescine influences bioactive compounds and quality of pomegranate during controlled atmosphere storage *Journal of Food Science and Technology,* 56(1): 350-359.

Kumar N, Tokas J, Raghavendra M and Singal H R, 2021. Impact of exogenous salicylic acid treatment on the cell wall metabolism and ripening process in postharvest tomato fruit stored at ambient temperature *International Journal of Food Science & Technology* 56(6): 2961-2972.

Kumar R, Khurana A and Sharma A K, 2014. Role of plant hormones and their interplay in development and ripening of fleshy fruits *J. Exp. Bot.* 65: 4561– 4575.

Kumar S and Thakur K S, 2020. Active packaging technology to retain storage quality of pear cv"Bartlett" during shelf□life periods under ambient holding after periodic cold storage *Packaging Technology and Science,* 33(7): 239-254.

Kyaw P N, Singh Z and Tokala V Y, 2021. Aqueous formulations of 1H-cyclopropabenzene modulate ethylene production and fruit quality in Japanese plums *Postharvest Biology and Technology,* 180 111625.

Kyaw P N, Singh Z and Tokala V Y, 2023. 1 H-cyclopropa [b] naphthalene: a novel ethylene antagonist for extending storage life and maintaining quality of climacteric and suppressed climacteric plums *Plant Growth Regulation* 1-15.

Lang C and Hübert T, 2012. A color ripeness indicator for apples *Food Bioprocess Tech* 5:3244–3249.

Lassois L and de Bellaire L D L, 2014. Crown rot disease of bananas In *Postharvest decay,* 103-130, Academic Press.

Lee S Y, Lee S J Choi D S and Hur S J, 2015. Current topics in active and intelligent food packaging for preservation of fresh foods *J. Sci. Food Agric.*, 95:2799–2810.

LI C Y, Cheng Y, HOU J B, Jie Z H U, Lei S U N and GE Y H, 2021. Application of methyl jasmonate postharvest maintains the quality of Nanguo pears by regulating mitochondrial energy metabolism *Journal of Integrative Agriculture,* 20(11): 3075-3083.

Li J, Cao Y, Bian S, Hong S B, Xu K, Zang Y and Zheng W, 2024. Melatonin improves the storage quality of rabbiteye blueberry (*Vaccinium ashei*) by affecting cuticular wax profile *Food Chemistry: X 21* 101106.

Li R, Ma J, Gu H, Jia W, Shao Y and Li W, 2022. 1-Methylcyclopropene counteracts ethylene promotion of fruit softening and roles of MiERF2/8 and MiPG in postharvest mangoes *Frontiers in Plant Science* 13 971050.

Li X, Zhu X, Wang H, Lin X, Lin H and Chen W, 2018. Post-harvest application of wax controls pineapple fruit ripening and improves fruit quality *Postharvest Biology and Technology,* 136: 99-110.

Li Y, Rokayya S, Jia F, Nie X, Xu J, Elhakem A and Helal M, 2021. Shelf-life quality safety evaluations of blueberry fruits coated with chitosan nano-material films *Scientific Reports,* 11(1) 55.

Li Y, Zhao Y, Zhang Z, He H, Shi L, Zhu X and Cui K, 2022. Near-freezing temperature storage improves shelf-life and suppresses chilling injury in postharvest apricot fruit (*Prunus armeniaca* L) by regulating cell wall metabolism *Food Chemistry,* 387: 132921.

Liu C, Zheng H, Sheng K, Liu W and Zheng L, 2018. Effects of melatonin treatment on the postharvest quality of strawberry fruit *Postharvest Biology and Technology,* 139: 47-55.

Liu D K, Xu C C, Guo C X and Zhang X X, 2020a. Sub-zero temperature preservation of fruits and vegetables: A review *Journal of Food Engineering,* 275: 109881.

Liu J, Yang J, Zhang H, Cong L, Zhai R, Yang C and Xu L, 2019. Melatonin inhibits ethylene synthesis via nitric oxide regulation to delay postharvest senescence in pears *Journal of Agricultural and Food Chemistry,* 67(8): 2279-2288.

Liu S, Huang H, Huber D J, Pan Y, Shi X and Zhang Z, 2020b. Delay of ripening and softening in 'Guifei'mango fruit by postharvest application of melatonin *Postharvest Biology and Technology,* 163: 111136.

Liu Y, Tang M, Liu M, Su D, Chen J, Gao Y and Li Z, 2020c. The molecular regulation of ethylene in fruit ripening *Small Methods,* 4(8): 1900485.

Liu Z, Jiang F, Mo Y, Liao H, Chen P and Zhang H, 2022. Effects of Ethanol Treatment on Storage Quality and Antioxidant System of Postharvest Papaya *Frontiers in Plant Science,* 13: 856499

López-Casado G, Sánchez-Raya C, Ric-Varas P D, Paniagua C, Blanco-Portales R, Muñoz-Blanco J and Mercado J A, 2023. CRISPR/Cas9 editing of the polygalacturonase FaPG1 gene improves strawberry fruit firmness *Horticulture Research,* 10(3); uhad011

Lustriane C, Dwivany F M, Suendo V and Reza M, 2018. Effect of chitosan and chitosan-nanoparticles on post harvest quality of banana fruits *Journal of Plant Biotechnology,* 45(1): 36-44

Ma Y, Fu L, Hussain Z, Huang D and Zhu S, 2019. Enhancement of storability and antioxidant systems of sweet cherry fruit by nitric oxide-releasing chitosan nanoparticles (GSNO-CS NPs) *Food Chemistry,* 285: 10-21.

Ma Y, Zhang W, Cheng S, Liu Y, Yang W, Wang Y and Chen G, 2022. Postharvest storage at near-freezing temperature maintained the quality and antioxidant properties of *Prunus domestica* L cv Ximei fruit *Scientia Horticulturae,* 293 110720.

Marak K A, Mir H, Singh P, Siddiqui M W, Ranjan T, Singh D R and Irfan M, 2023. Exogenous melatonin delays oxidative browning and improves postharvest quality of litchi fruits *Scientia Horticulturae,* 322: 112408.

Martínez-Romero D, Bailén G, Serrano M, Guillén F, Valverde J M, Zapata P and Valero D, 2007. Tools to maintain postharvest fruit and vegetable quality through the inhibition of ethylene action: a review *Critical Reviews in Food Science and Nutrition,* 47(6) 543-560.

Mditshwa A, Fawole O A and Opara U L, 2018. Recent developments on dynamic controlled atmosphere storage of apples—A review *Food Packaging and Shelf life,* 16: 59-68.

Mirshekari A, Madani B, Yahia E M, Golding J B and Vand S H, 2020. Postharvest melatonin treatment reduces chilling injury in sapota fruit *Journal of the Science of Food and Agriculture,* 100(5): 1897-1903.

Mithapara K D and Butani A M, 2023. Biotechnological approaches for extended shelf life of horticultural commodities *Journal of Pharmacognosy and Phytochemistry,* 12(6): 107-119.

Mojerlou S and Safaie N, 2012. Phylogenetic analysis of Alternaria species associated with citrus black rot in Iran *Journal of Plant Pathology & Microbiology.*

Molamohammadi H, Pakkish Z, Akhavan H R and Saffari V R, 2020. Effect of salicylic acid incorporated chitosan coating on shelf life extension of fresh in-hull pistachio fruit *Food and Bioprocess Technology,* 13: 121-131.

Mortazavi S, Karami Z and Mostaan A, 2015. Use of ethylene scavenger sachet in modified atmosphere packaging to maintain storage stability of khalal date fruit *Int J Postharvest Technol Innov* 5: 52-63.

Nanos G D, Agtsidou E and Sfakiotakis E M, 2002. Temperature and propylene effects on ripening of green and black'Conservolea'olives *HortScience,* 37(7):1079-1081.

Nasr F, Pateiro M, Rabiei V, Razavi F, Formaneck S, Gohari G and Lorenzo J M, 2021. Chitosan-phenylalanine nanoparticles (Cs-Phe Nps) extend the postharvest life of persimmon (*Diospyros kaki*) fruits under chilling stress *Coatings,* 11(7) 819.

Nian L, Wang M, Sun X, Zeng Y, Xie Y, Cheng S and Cao C, 2022 Biodegradable active packaging: Components preparation and applications in the preservation of postharvest perishable fruits and vegetables *Critical Reviews in Food Science and Nutrition,* 1-36.

Nissen R, Bound S, Adhikari R and Cover I, 2018. Factors affecting postharvest management of apples: a guide to optimising quality.

Ozkan R, Smilanick J L and Karabulut O A, 2011. Toxicity of ozone gas to conidia of Penicillium digitatum Penicillium italicum and Botrytis cinerea and control of gray mold on table grapes *Postharvest Biology and Technology* 60(1) 47-51.

Pan L, Chen X, Xu W, Fan S, Wan T, Zhang J and Cai Y, 2022. Methyl jasmonate induces postharvest disease resistance to decay caused by *Alternaria alternata* in sweet cherry fruit *Scientia Horticulturae,* 292: 110624.

Park M H, Sangwanangkul P and Choi J W, 2018. Reduced chilling injury and delayed fruit ripening in tomatoes with modified atmosphere and humidity packaging *Scientia Horticulturae,* 231 66–72.

Pathak N, Caleb O J, Geyer M, Herppich W B, Rauh C and Mahajan P V, 2017b. Photocatalytic and photochemical oxidation of ethylene: Potential for storage of fresh produce—A review *Food and Bioprocess Technology,* 10(6): 982–1001.

Pathak N, Caleb O J, Rauh C and Mahajan P V, 2017a. Effect of process variables on ethylene removal by vacuum ultraviolet radiation: application in fresh produce storage *Biosystems Engineering,* 159: 33-45.

Paull R E and Chen C C, 2014. Passion Fruit: Postharvest Quality-Maintenance Guidelines *Fruits Nuts and Beverage Crops*F_N-44

Pezzarossa B, Rosellini I, Borghesi E, Tonutti P and Malorgio F, 2014. Effects of Se-enrichment on yield fruit composition and ripening of tomato (*Solanum lycopersicum*) plants grown in hydroponics *Scientia Horticulturae,* 165:106-110.

Poitou X, Redon P, Pons A, Bruez E, Delière L, Marchal A and Darriet P, 2021. Methyl salicylate a grape and wine chemical marker and sensory contributor in wines elaborated from grapes affected or not by cryptogamic diseases *Food Chemistry,* 360: 130120.

Pose S, Kirby A R, Paniagua C, Waldron K W, Morris V J, Quesada M A and Mercado J A, 2015. The nanostructural characterization of strawberry pectins in pectate lyase or polygalacturonase silenced fruits elucidates their role in softening *Carbohydrate Polymers,* 132: 134-145.

Qian M, Xu Z, Zhang Z, Li Q, Yan X, Liu H and Zhao C, 2021. The downregulation of PpPG21 and PpPG22 influences peach fruit texture and softening *Planta,* 254 1-12.

Qin G Z, Tian S P, Xu Y and Wan Y K, 2003. Enhancement of biocontrol efficacy of antagonistic yeasts by salicylic acid in sweet cherry fruit *Physiological and Molecular Plant Pathology,* 62(3): 147-154.

Raghava Rao D and Chundawat B, 1986. Extension of shelf life of lacatan bananas stored in cartons *Guj Agric Univ Res J,* 11(2):26–31.

Ranjeet S and Giri S K, 2014. Shelf-life study of guava (*Psidium guajava* L) under active packaging: an experiment with potassium permanganate salt as ethylene absorbent *Archiv für Lebensmittelhygiene* 65(2) 32-39.

Resende N S, Gonçalves G A S, Reis K C, Tonoli G H D and Boas E V B V, 2018. Chitosan/ cellulose nanofibril nanocomposite and its effect on quality of coated strawberries *Journal of Food Quality.*

Ric-Varas P, Paniagua C, López-Casado G, Molina-Hidalgo F J, Schückel J, Knox J P and Mercado J A, 2024. Suppressing the rhamnogalacturonan lyase gene FaRGLyase1 preserves RGI pectin degradation and enhances strawberry fruit firmness *Plant Physiology and Biochemistry,* 206: 108294.

Ruiz-Aracil M C, Guillén F, Ilea M I M, Martínez-Romero D, Lorente-Mento J M and Valverde J M, 2023. Comparative Effect of Melatonin and 1-Methylcyclopropene Postharvest Applications for Extending 'Hayward'Kiwifruit Storage Life *Agriculture,* 13(4): 806.

Sabater-Vilar M, SuÃ±Ã©-Colell E, Castro-Chinchilla J and SÃ¡ enz-Murillo M V, 2016. Reduction of postharvest rotting with an ethylene absorbent: a case study with pineapple In *VIII International Postharvest Symposium: Enhancing Supply Chain and Consumer Benefits-Ethical and Technological Issues,* 1194:721-728.

Sachin A J, Sudhakar Rao D V, Ranjitha K, Vasugi C, Narayana C K, Reddy S V R and Preethi P, 2023. Differential Efficacy of Postharvest Application of Ethylene Inhibitors on Storage Life and Nutritional Quality of Guava (cv'Arka Mridula') *Erwerbs-Obstbau,* 1-13.

Saif M I, 2022. Reducing fruit quality losses in peach fruits and increasing profitability during postharvest storage using salicylic acid and aspirin *Egyptian Journal of Horticulture,* 49(1) 35-57.

Saltveit M E, 1999. Efect of ethylene on quality of fresh fruits and vegetables *Postharvest Biol Technol* 15:279–292.

Sanches A G, da Silva M B, Moreira E G S, dos Santos E X, Menezes K R P and Cordeiro C A M, 2019. Ethylene absorber (KMnO4) in postharvest quality of pinha (*Anona squamosa* L) *Emirates Journal of Food and Agriculture,* 605-612.

Sangprayoon P, Supapvanich S, Youryon P, Wongs-Aree C and Boonyaritthongchai P, 2019. Efficiency of salicylic acid or methyl jasmonate immersions on internal browning alleviation and physicochemical quality of Queen pineapple cv"Sawi" fruit during cold storage *Journal of Food Biochemistry,* 43(12): e13059.

Saquet A A and Streif J, 2017. Respiration rate and ethylene metabolism of 'Jonagold'apple and 'Conference'pear under regular air and controlled atmosphere *Bragantia,* 76: 335-344.

Sardabi F, Mohtadinia J, Shavakhi F and Jafari AA, 2014. The effects of 1-methylcyclopropen (1-MCP) and potassium permanganate coated zeolite nanoparticles on shelf life extension and quality loss of golden delicious apples *J Food Process Preserv*, 38:2176–2182.

Sarengaowa Hu W, Jiang A, Xiu Z and Feng K, 2018. Effect of thyme oil–alginate-based coating on quality and microbial safety of fresh-cut apples *Journal of the Science of Food and Agriculture,* 98(6): 2302-2311.

Sati H, Bhardwaj R, Fawole O A and Pareek S, 2023. Postharvest Melatonin Application Preserves Quality and Imparts Chilling Tolerance in Peaches *Journal of Food Biochemistry,* 1:8126640.

Sekeli R, Abdullah J O, Namasivayam P, Muda P, Bakar U K A, Yeong W C and Pillai V, 2014. RNA Interference of 1-Aminocyclopropane-1-carboxylic Acid Oxidase (ACO 1 and ACO 2) Genes Expression Prolongs the Shelf Life of Eksotika (*Carica papaya* L) Papaya Fruit *Molecules,* 19(6): 8350-8362.

Senturk Parreidt T, Müller K and Schmid M, 2018. Alginate-based edible films and coatings for food packaging applications *Foods,* 7(10): 170.

Serna-Escolano V, Martínez-Romero D, Giménez M J, Serrano M, García-Martínez S, Valero D and Zapata P J, 2021. Enhancing antioxidant systems by preharvest treatments with methyl jasmonate and salicylic acid leads to maintain lemon quality during cold storage *Food Chemistry,* 338: 128044

Shamim F, Khan A R and Ali M A, 2015. Classification of maturity indices of mango fruit for safe harvesting and storage *Ukrainian Food Journal,* 4(3).

Shu C, Liu B, Zhang W Cao J and Jiang W, 2020. Improving postharvest quality of apple (*Malus × domestica* Borkh cv Golden Delicious) fruit by storage at near freezing temperature *Shipin Kexue/Food Science,* 41(1) 244-251,

Silva D, Salomão L, Siqueira D, Cecon P and Rocha A, 2009. Potassium permanganate effects in postharvest conservation of the papaya cultivar Sunrise Golden Pesq Agrop Brasileira 44: 669–675.

Silveira N M, Seabra A B, Machado E C, Hancock J T and Ribeiro R V, 2019. Nitric Oxide-Induced Tolerance in Plants under Adverse Environmental Conditions In *Approaches for Enhancing Abiotic Stress Tolerance in Plants,* (pp 371-386) CRC Press.

Sisler E C, Alwan T, Goren R, Serek M and Apelbaum A, 2003. 1-Substituted cyclopropenes: effective blocking agents for ethylene action in plants *Plant Growth Regulation,* 40: 223-228.

Sisler E, Serek M, 1997. Inhibitors of ethylene responses in plants at the receptor level: recent developments *Physiol Plant*, 100 577–582.

Song Z, Qin J, Yao Y, Lai X, Zheng W, Chen W and Li X, 2020. A transcriptomic analysis unravels key factors in the regulation of stay-green disorder in peel of banana fruit (Fenjiao) caused by treatment with 1-MCP *Postharvest Biology and Technology,* 168: 111290.

Srivastava A, Rao L J M and Shivanandappa T, 2012. A novel cytoprotective antioxidant: 4-Hydroxyisophthalic Acid *Food Chemistry,* 132(4) 1959-1965.

Steffens C A, AMARANTE C V T, Espindola B P, Heinzen A S, Brackmann A and Both V, 2018. 'Laetitia'plums Stored in Controlled Atmospheres Combined with Induction of Mass Loss and Ethylene Manegement *Revista Caatinga,* 31: 225-234.

Tepamatr P, 2023. Efficacy of a palladium-modified activated carbon in improving ethylene removal to delay the ripening of Gros Michel banana *Journal of Agriculture and Food Research,* 12: 100561.

Terry L A, Ilkenhans T, Poulston S, Rowsell L and Smith A W, 2007. Development of new palladium-promoted ethylene scavenger *Postharvest Biology and Technology,* 45(2): 214-220.

Thewes F R, Brackmann A, de Oliveira Anese R, Bronzatto E S, Schultz E E and Wagner R, 2017. Dynamic controlled atmosphere storage suppresses metabolism and enhances volatile concentrations of 'Galaxy'apple harvested at three maturity stages *Postharvest Biology and Technology,* 127: 1-13.

Tian M S, Gong Y and Bauchot A D, 1997. Ethylene biosynthesis and respiration in strawberry fruit treated with diazocyclopentadiene and IAA *Plant Growth Regulation,* 23: 195-200.

Tijero V, Munoz P and Munne-Bosch S, 2019. Melatonin as an inhibitor of sweet cherries ripening in orchard trees *Plant Physiology and Biochemistry,* 140: 88-95.

Tirgar A, Han D and Steckl A J, 2018. Absorption of ethylene on membranes containing potassium permanganate loaded into alumina-nanoparticle-incorporated alumina/carbon nanofibers *Journal of Agricultural and Food Chemistry,* 66(22) 5635-5643.

Tosetti R, Elmi F, Pradas I, Cools K and Terry L A, 2020. Continuous exposure to ethylene differentially affects senescence in receptacle and achene tissues in strawberry fruit *Frontiers in Plant science,* 11 174.

Uzun S, 2023. Postharvest Quality Traits of Chestnut (*Castanea sativa* Mill) Fruit as Affected by Methyl Jasmonate During Cold Storage *Erwerbs-Obstbau,* 1-10

Valero D, Martınez-Romero D, Valverde J M, Guillen F and Serrano M, 2003. Quality improvement and extension of shelf life by 1-methylcyclopropene in plum as affected by ripening stage at harvest *Innovative Food Science & Emerging Technologies,* 4(3) 339-348.

Valverde J M, Giménez M J, Guillen F, Valero D, Martinez-Romero D and Serrano M, 2015. Methyl salicylate treatments of sweet cherry trees increase antioxidant systems in fruit at harvest and during storage *Postharvest Biology and Technology,* 109: 106-113.

Vithana M D K, Singh Z and Ul Hasan M, 2024. Pre-and Post-harvest Elicitation with Methyl Jasmonate and Salicylic Acid Followed by Cold Storage Synergistically Improves Red Colour Development and Health-Promoting Compounds in Blood Oranges *Journal of Plant Growth Regulation,* 1-15

Wan R, Song J, Lv Z, Qi X, Feng Z, Yang Z and Chen Y, 2023. Effects of 1–MCP Treatment on Postharvest Fruit of Five Pomegranate Varieties during Low-Temperature Storage *Horticulturae,* 9(9) 1031.

Wang C and Ajji A, 2022. Ethylene scavenging film based on low-density polyethylene incorporating pumice and potassium permanganate and its application to preserve avocados *LWT,* 172: 114200.

Wang J, Pan H, Wang R, Hong K and Cao J, 2016. Patterns of flesh reddening translucency ethylene production and storability of 'Friar'plum fruit harvested at three maturity stages as affected by the storage temperature *Postharvest Biology and Technology 121* 9-18.

Wang K, Wu L, Li Y and Li H, 2023a. Preparation and characterization of chitosan/halloysite nanotubes composite film with ethylene scavenging and gas resistance for active food packaging *Journal of Food Safety,* 43(2): e13027.

Wang M, Nian L, Wu J, Cheng S, Yang Z and Cao C, 2023b. Visible light-responsive chitosan/ sodium alginate/QDs@ ZIF-8 nanocomposite films with antibacterial and ethylene scavenging performance for kiwifruit preservation *Food Hydrocolloids,* 145: 109073.

Wang R, Zhang L, Rahman F U, Luo J, Liu T, Chen W and Zhu X, 2024. 1-methylcyclopropene combined with ethylene absorbent delays the ripening of 'Fenjiao'banana (Musa ABB Pisang Awak) *Scientia Horticulturae,* 326: 112772.

Wang Y, Reiter R J and Chan Z, 2018. Phytomelatonin: a universal abiotic stress regulator *Journal of Experimental Botany,* 69(5): 963-974.

Wang Y, Xu F, Feng X and MacArthur R L, 2015. Modulation of *Actinidia arguta* fruit ripening by three ethylene biosynthesis inhibitors *Food Chemistry*, 173: 405–413.

Wills R B H and Warton M A, 2000. A new rating scale for ethylene action on postharvest fruit and vegetables *Improving Postharvest Technologies of Fruits Vegetables and Ornamentals Institute International of Refrigeration Murcia Spain,* 43-47.

Wills R B H, Warton M A, Mussa D M D N and Chew L P, 2001. Ripening of climacteric fruits initiated at low ethylene levels *Australian Journal of Experimental Agriculture,* 41(1): 89-92.

Wu B, Shen F, Chen C J, Liu L, Wang X, Zheng W Y and Zhang X Z, 2021. Natural variations in a pectin acetylesterase gene MdPAE10 contribute to prolonged apple fruit shelf life *The Plant Genome,* 14(1); e20084.

Wuttke S, Medina D D, Rotter J M, Begum S, Stassin T Ameloot R *et al.,* 2018. Bringing porous organic and carbon-based materials toward thin-film applications *Advanced Functional Materials*, 28(44): 1801545.

Xiong S, Zhou F, Jiang A, Yang L and Hu W, 2024. Ethanol vapor ameliorates chilling injury and maintains postharvest quality by increasing antioxidant capacity of hardy kiwifruit (*Actinidia arguta*) *Scientia Horticulturae,* 327: 112796.

Xu F, Liu S, Liu Y, Xu J, Liu T and Dong S, 2019. Effectiveness of lysozyme coatings and 1-MCP treatments on storage and preservation of kiwifruit *Food Chemistry,* 288: 201-207.

Yang W, Kang J, Liu Y, Guo M and Chen G, 2022. Effect of salicylic acid treatment on antioxidant capacity and endogenous hormones in winter jujube during shelf life *Food Chemistry,* 397: 133788.

Yang Y, Ma R, Zhang B, Song Z, Zhang C, Guo S and Yu M, 2016. Different expression analysis in fruit softening and ethylene biosynthetic pathways in peaches of different flesh textures *Horticultural Plant Journal,* 2(2): 75-81.

Yildirim S, Röcker B, Pettersen M K, Nilsen-Nygaard J, Ayhan Z, Rutkaite R, *et al.,* 2018. Active packaging applications for food *Comprehensive Reviews in Food Science & Food Safety* 17(1): 165–199.

Yılmaz M, 2024. Enhancing Cold Storage Quality of Blueberry (*Vaccinium corymbosum* 'Bluecrop') with Methyl Jasmonate Treatments *Applied Fruit Science,* 1-8.

Yin X R, Zhang Y, Zhang B, Yang S L, Shi Y N, Ferguson I B and Chen K S, 2013. Effects of acetylsalicylic acid on kiwifruit ethylene biosynthesis and signaling components *Postharvest biology and technology,* 83 27-33.

Yuan C, Bu H, Zhao J, Liu J, Wang G, Yuan H and Wang A, 2023. Effect of Se application on selenium accumulation and fruit quality in pear (*Pyrus ussuriensis*) *Acta Physiologiae Plantarum,* 45(3): 45.

Zanella A and Rossi O, 2015. Post-harvest retention of apple fruit firmness by 1-methylcyclopropene (1-MCP) treatment or dynamic CA storage with chlorophyll fluorescence (DCA-CF) *European Journal of Horticultural Science,* 80(1): 11-17.

Zanella A, Cazzanelli P, Panarese A, Coser M, Chistè C and Zeni F, 2004. Fruit fluorescence response to low oxygen stress: modern storage technologies compared to 1-MCP treatment of apple In *V International Postharvest Symposium,* 682:1535-1542.

Ze Y, Gao H, Li T, Yang B and Jiang Y, 2021. Insights into the roles of melatonin in maintaining quality and extending shelf life of postharvest fruits *Trends in Food Science & Technology,* 109: 569-578.

Zeng L, Shi L, Lin H, Lin Y, Lin Y and Wang H, 2021. containing 1-methylcyclopropene treatment suppresses fruit decay of fresh Anxi persimmons by enhancing disease resistance *Food Quality and Safety 5* fyab007.

Zhang H, Liu X, Chen T, Ji Y, Shi K, Wang L and Kong J, 2018. Melatonin in apples and juice: Inhibition of browning and microorganism growth in apple juice *Molecules,* 23(3): 521.

Zhang J, Cheng D, Wang B, Khan I and Ni Y, 2017. Ethylene control technologies in extending postharvest shelf life of climacteric fruit *Journal of Agricultural and Food Chemistry,* 65(34): 7308-7319.

Zhang K, Zhang D, Wang X, Xu X, Yu W, Wang C and Cheng C, 2024. Integrated analysis of postharvest storage characteristics of seven apple cultivars and transcriptome data identifies MdBBX25 as a negative regulator of fruit softening during storage in apples *Postharvest Biology and Technology,* 207: 112646.

Zhang R, Yuan Z, Jiang Y, Jiang F and Chen P, 2022. Sodium nitroprusside functions in browning control and quality maintaining of postharvest rambutan fruit *Frontiers in Plant Science,* 12: 795671.

Zhang W, Cao J, Fan X and Jiang W, 2020. Applications of nitric oxide and melatonin in improving postharvest fruit quality and the separate and crosstalk biochemical mechanisms *Trends in Food Science & Technology* 99: 531–541.

Zhang Z, Huber D J and Rao J, 2013. Antioxidant systems of ripening avocado (*Persea americana* Mill) fruit following treatment at the preclimacteric stage with aqueous 1-methylcyclopropene *Postharvest Biology and Technology*, 76: 58–64.

Zhao H, Liu B, Zhang W, Cao J and Jiang W, 2019. Enhancement of quality and antioxidant metabolism of sweet cherry fruit by near-freezing temperature storage *Postharvest Biology and Technology,* 147: 113-122.

Zhao Y, Song C, Lin Q and Duan Y, 2024. Exploring the Effects of Different 1-MCP Concentration Treatment on Chilling Injury of Postharvest Peach Fruit *Journal of Food Biochemistry.*

Zheng X, Hu B, Song L, Pan J and Liu M, 2017. Changes in quality and defense resistance of kiwifruit in response to nitric oxide treatment during storage at room temperature *Scientia Horticulturae,* 222: 187-192.

Zhou S, Cheng Y and Guan J, 2017. The molecular basis of superficial scald development related to ethylene perception and α-farnesene metabolism in 'Wujiuxiang' pear *Scientia Horticulturae* 216(216); 76–82.

Zi N Z, Xin X N and Rui W, 2011. Study on the Properties of LDPE/POE/Zeolite Molecular Sieves Composite Film *Advanced Materials Research* 4:194–196.

Zuñiga P E, Castañeda Y, Arrey-Salas O, Fuentes L, Aburto F and Figueroa C R, 2020. Methyl jasmonate applications from flowering to ripe fruit stages of strawberry (*Fragaria × ananassa* 'Camarosa') reinforce the fruit antioxidant response at post-harvest *Frontiers in Plant Science,* 11: 538.

2

Physiology and Biochemistry of Fruit Ripening

Manju Bala, Shivani and Seerat Bhinder

ICAR-Central Institute of Post-Harvest Engineering and Technology Ludhiana-141 004, Punjab, India

Fruit ripening is a phenomena during which horticulture produce undergoes a series of complex physical and biochemical changes initiated through a series of hormones (Ethylene, auxin, abscisic acid (ABA), gibberellins, jasmonic acid, etc.) to attain maximum physiological maturity and favourable organoleptic properties. These changes lead to alterations in colour, texture, aroma, flavour and nutrient composition. In a nut shell, ripening involves increase in pulp to peel ratio, solubilisation of pectic substances, degradation of complex carbohydrates, organic acids and chlorophyll, accumulation sugars, flavour volatiles, polyphenolic compounds and pigments (carotenoids, xanthophylls and anthocynanins). The extent of these changes guide harvesting, readiness for processing and consumption decisions, with physic-chemical alterations serving as indicators of optimum ripeness. Respiration rate, ethylene and ABA are key regulators of fruit ripening in climacteric and non-climacteric fruits, respectively. In climacteric fruits, ABA interacts with ethylene, influencing the synthesis of ethylene itself. Although there is ample documentation on the involvement of hormones in fruit development and ripening, our understanding of the functions of various hormones throughout growth, maturation, and specific ripening processes remains incomplete.

Introduction

Fruits are a vital component of our diet, offering both nutritional value and economic significance. They contribute essential growth-regulating factors necessary for maintaining good health when consumed as part of a balanced diet. Fruit development is a distinctive phenomenon observed in plants, necessitating a sophisticated interplay among genes and signaling pathways. In the case of succulent fruits, it encompasses three discernible phases: fruit set, fruit development, and fruit ripening (Bouzayen *et al*., 2011). Among these, fruit ripening represents a highly intricate developmental process during which a plant organ undergoes significant physiological and bio-chemical changes. In the initial stages of growth and development, fleshy fruits are typically green, accumulate water and nutrients, and possess thick epidermal layers that serve as protective barriers for the seeds. As the fruit reaches maturity, ripening ensues, involving a succession of transformations marked by alterations in color, texture, aroma, and nutrient composition. These changes render the fruit appealing to predators, thereby facilitating seed dispersal, while also enhancing its nutritional value for human consumption. While the structural, physiological, and biochemical mechanisms may differ among various types of fruits, ripening typically involves a sequence of sensory and nutritional changes that have been evolutionarily conserved and adapted across different species for consumption in the human diet (Alos *et al*., 2012).

Fruit ripening, a complex and irreversible process, involves a series of coordinated physiological, biochemical, and sensory changes that culminate in the development of a ripe fruit with desirable quality attributes. However, excessive softening during ripening can lead to spoilage during storage. Numerous intricate biochemical processes occur during fruit ripening, encompassing seed maturation, color alteration, softening of texture, generation of volatile flavours, development of wax on the skin, adjustments in tissue permeability, and fluctuations in carbohydrate content, organic acids, and proteins. The composition of the fruit undergoes continuous modifications throughout the ripening process, attributed to the synthesis of novel chemicals or the degradation of existing ones (Prasanna *et al*., 2010).

A typical fruit originates from the ovary subsequent to pollination and fertilization. Throughout fruit development, the cells within the ovary wall undergo a continual series of divisions and expansions, ultimately determining the fruit's final size prior to ripening. Generally, fruit developmental stages are broadly categorized into four phases

- Phase 1: Development of ovary followed by fertilization and fruit set, wherein the floral meristem differentiates into the ovary, then after fertilisation cell division occurs to attain fruit set.
- Phase 2: During this stage rapid cell division occurs leading to seed formation, and early embryo development.
- Phase 3: This phase entails cell expansion and embryo maturation, characterized by the attainment of the fruit's ultimate size, characterized by increased cell volume due to cell expansion.

- Phase IV marks the culmination of fruit growth, initiating the ripening or maturation process (Stikic *et al.*, 2015).

Fruits typically undergo a spectrum of biochemical and physiological transformations during the process of maturation. These alterations encompass the degradation of chlorophyll, the synthesis of pigments like anthocyanins, betalains, xanthophylls and carotenoids, enhancement of flavour and aroma profiles, adjustments in acid and sugar compositions, and softening of tissue. The rate and extent of these changes occurring in fruit depends upon the type of fruit (Nath *et al.*, 2014).

However, based on their respiratory pattern and ethylene biosynthesis during ripening, harvested fruits can be further categorized as either climacteric or non-climacteric types. Ethylene and respiratory activity play pivotal roles in facilitating fruit ripening of climacteric fruits. Ethylene has been known to have a regulatory impact on fruit tissue softening (Johnston *et al.*, 2009; Yin *et al.*, 2008) and lignification (Wang *et al.*, 2013) during ripening. Climacteric fruits exhibit a surge in respiration accompanied by rapid ethylene production at the onset of ripening; this ethylene, functioning as a plant hormone, expedites the ripening process (Lelievre *et al.*, 1997). Conversely, such changes are absent in non-climacteric fruits, which mature at a comparatively slower pace (Nath *et al.*, 2014). Consequently, climacteric fruits ripen even when harvested prematurely, unlike non-climacteric fruits, which remain unripe after detachment from the parent plant. In climacteric fruits, although respiration rate and ethylene synthesis are minimal at maturity, they surge dramatically to a climacteric peak upon the onset of ripening. In contrast, non-climacteric fruits lack the ability to continue ripening post-detachment. Fruits such as strawberries, grapes, citrus fruits, pineapples, and cherries progress through maturation and ripening without experiencing a burst of ethylene production or an increase in respiration, unlike climacteric fruits. Moreover, these fruits produce minimal endogenous ethylene and are unresponsive to external ethylene treatment. While ethylene is generally dispensable for ripening, non-climacteric fruits possess the capacity to generate a certain level of endogenous ethylene and respond to exogenous ethylene. Non-climacteric fruits exhibit a gradual decline in both respiration and ethylene production throughout ripening (Rehman and Gamage, 1999). Unlike climacteric fruits, the ripening process in non-climacteric fruits is regulated by abscisic acid (ABA) (Bai *et al.*, 2020). The significance of ABA has recently been highlighted in numerous non-climacteric fruits. In highbush blueberries, ABA biosynthesis and signal transduction are intricately linked to anthocyanin biosynthesis, accumulation, and pericarp coloring (Chung *et al.*, 2019). Similar findings have been observed in sweet cherries (Fresno and Munne-Bosch, 2021). ABA has been shown to promote ripening and coloring in winter jujube, elevating anthocyanin content and the expression of anthocyanin-related genes (Kou *et al.*, 2019).

Climacteric fruits respond positively to exogenous ethylene application for ripening (e.g., mangoes, bananas, tomatoes), whereas non-climacteric fruits do not, except for chlorophyll degradation in certain fruits such as citrus and pineapples, termed as 'degreening' (Noichinda, 2017). In climacteric fruits, ethylene assumes a critical role in ripening, initiating the maturation process with a burst of ethylene production. Thus, in climacteric fruits, ethylene production follows an autocatalytic pattern, wherein an initial concentration of ethylene induces further ethylene synthesis. Once ethylene production initiates, the fruit autonomously augments the production of ethylene, thereby expediting the ripening process characterised by significant changes in colour, texture, and aroma. Depending on fruit development stage, plants producing climacteric fruits exhibit two systems of ethylene production.

- System 1 operates in immature climacteric fruits and vegetative tissues, wherein ethylene inhibits its own biosynthesis via negative feedback.
- In contrast, mature climacteric fruits and senescing petals exhibit System 2, wherein ethylene stimulates its own biosynthesis, facilitating autocatalytic ethylene production (Shakya and Lal, 2018).

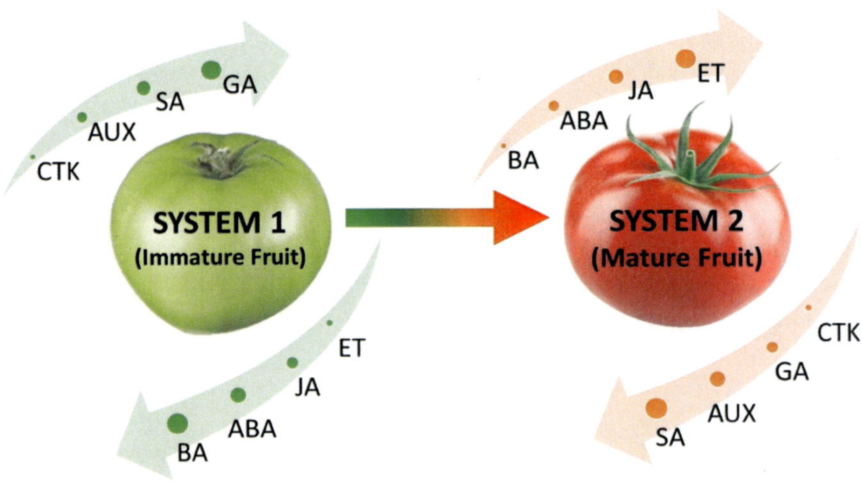

Fig. 1: Hormonal system transition during ripening of climacteric fruits. Ascending and descending arrows respectively illustrate the increment and decline in various hormones. Where. ET, ethylene; ABA, Abscisic acid; JA, Jasmonic acid; BS, Brassinosteroids; AUX, auxins; GA,Gibberellins; SA, Salicylic acid, CTK, Cytokinins.

System 2 is characterized by higher ethylene production than system 1. Furthermore, system 2-derived ethylene is indispensable for climacteric fruit ripening. Apart from ethylene biosynthesis, ethylene signalling also holds significance in this process. Exposure to exogenous ethylene induces rapid autocatalytic ethylene production in mature climacteric fruits, hastening the onset of the climacteric rise in ethylene and associated ripening changes (Fig. 1). However, ethylene treatment of pre-climacteric fruits results in a gradual increase in respiration rate due to ethylene concentration rise, without triggering further autocatalytic endogenous ethylene production or ripening (Liu *et al.* 2015).

Physical changes during ripening

Onset of fruit ripening is marked by alterations in skin hue, transformation of starch into sugars, softening of the fruit, diminishing astringency, and the emergence of distinctive flavours. Additional transformations may involve variations in the ratio of pulp to peel, leading to easier peeling, adjustments in the overall concentration of soluble solids, pH of the pulp, and total titratable acidity (Fig. 2). There may also be fluctuations in the moisture content of both the peel and pulp, as well as changes in the levels of dry matter. Additionally, alterations in respiration rate and ethylene production occur. These modifications collectively influence the ultimate taste and texture of the fruit when consumed. Farmers and consumers commonly associate fruit ripening with changes in colour of the skin and texture of fruit. Also, fruit skin color, along with size, plays a crucial role in consumer preferences and purchasing decisions at retail outlets (Adi *et al.*, 2019).

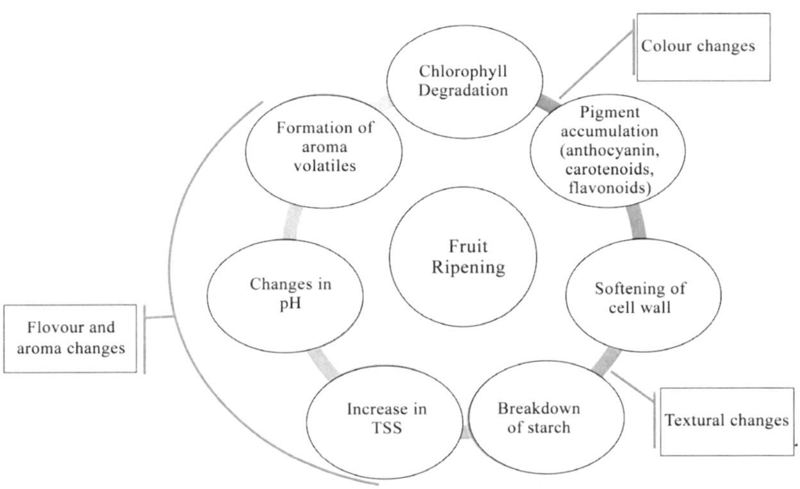

Fig. 2: Physiological changes during fruit ripening.

Throughout on-tree ripening of climatric fruits, significant physic-chemical alterations occur, serving as benchmarks for determining ripeness levels, including physiological maturity, consumption readiness, and over-ripeness. Physiological maturity denotes the stage at which fruits achieve their maximum physical development and are deemed harvestable. At the consumption ripeness or organoleptic stage, fruits exhibit optimal flavour and aroma development, accompanied by complete colouration changes such as yellow or red in the epicarp. Subsequently, beyond ripeness stage, fruits undergo complete softening and experience a decline in flavor and aroma attributes, indicative of over-ripeness (Barragan-Iglesias *et al.*, 2018).

Colour: In agricultural practices, alterations in colour serve as a pivotal indicator for assessing the ripeness and readiness for harvest of numerous fruits and vegetables. Fruit colour is thus the key criteria that is observed primarily to assess the ripeness of fruit during different stages of growth, storage as well as retail. As fruits and vegetables slowly progress towards maturity, a discernible shift in their colouration occurs, primarily attributable to the accumulation of diverse pigments. For instance, in Maradol papayas, the optimum stage of ripeness is characterised by the development of orange-red colour. During ripening chlorophyll degrades and pigments are synthesized that gives the fruit its distinct colour. The final fruit colour at optimum ripeness depends upon the chlorophyll to pigment (carotenoids/anthocyanins and betalains) ratio (Hu *et al.* 2019). The progressive loss of green pigmentation during the maturation and ripening stages is primarily attributed to the degradation of chlorophyll molecules, catalyzed by the chlorophyllase enzyme. (Schweiggert *et al.*, 2011, Kying *et al.*, 2013). Additional factors contributing and influencing the characteristic colour transition in fruits include alterations in pH levels (resulting in the release of organic acids from vacuoles), oxidative stress, and the synthesis of secondary metabolites, such as carotenoids, anthocyanins, and other pigments, which collectively obscure the expression of chlorophyll pigment. (Wills *et al.*, 1998). Chlorophyll molecules undergo hydrolysis to chlorophyllide and phytol through the action of the enzyme chlorophyllase, followed by the replacement of the magnesium atom with hydrogen, leading to the formation of brownish pheophorbide (Kanayama and Kochetov, 2015). The loss of green colouration is a widely recognized maturity index in fruits and vegetables, with exceptions including avocado, kiwifruit, granny smith apple, and feijoa.

The manifestation of characteristic fruit colour is often associated with pigments like ß-carotene responsible for hues ranging from yellow to orange, anthocyanins associated with shades of reds and blues, lycopene primarily contributing to red pigmentation and other pigments like xanthophyll, violaxanthin and lutein contribute to deep yellow to orange hues (Wills *et al.*, 2001). For some fruits like papayas, ripening is allowed and preferred on-tree so that optimum fruits colour and sweetness develops (Paull *et al.*, 2008).

The alteration of pigmentation during fruit ripening is intricately regulated, involving plant growth hormones, various transcription factors, gene families, enzymes of pigment biosynthetic pathways, and environmental cues (Wang *et al.*, 2020). Additionally, signaling molecules such as nitric oxide, melatonin, hydrogen sulfide, and sucrose have been recognized for their roles in pigment accumulation throughout different ripening stages of fruits (Duran-Soria *et al.*, 2020). Fruit ripening induces changes in colour attributed to pigment biosynthesis, degradation, and sequestration, facilitated by the development of new structures such as fibrils in peppers and plastoglobules in other fruits (Berry *et al.*, 2019; Simkin *et al.*, 2007). Furthermore, pigments undergo biochemical modifications post-synthesis to enhance their stability, resulting in the production of novel cleaved pigment products potentially possessing bioactive properties.

In fruits such as tomato and pepper, the degradation of chlorophyll is associated with a well-coordinated transformation of chloroplasts (green plastids) into chromoplasts (non-green pastids), which accumulate carotenoids, resulting in a visible shift in colour from green to yellow to orange to red (Liu *et al.*, 2015). During maturation, chromoplasts, responsible for the yellow, orange, and red hues of fruits, emerge through the differentiation of other plastid forms. The process is accompanied by the breakdown of thylakoid membranes and simultaneous accumulation of carotenoids like lycopene in the membranes. The synthesis of membranes serving as sites for carotenoid biosynthesis occurs, along with an increase in the number and size of plastoglobules (Barsan *et al.*, 2012). In papayas, chromoplasts arise from leucoplastids, as no intermediary structures such as amyloplasts or chloroplasts are encountered (Schweiggert *et al.*, 2011). The origin of chromoplasts in mangoes is highly complex, involving dynamic interconversions of plastids, although a sequential pattern could not be firmly established (Vasquez-Caicedo *et al.*, 2006).

In case of fruits and vegetables, like pomegranate, cherry, plums, turnip, blackberries, blueberries, strawberries, red pears, onions, red cabbage, red apples, prunes, eggplant, cranberries, and grapes, ripening is accompanied by the accumulation of anthocyanins leading to colour changes from green to purple or red (Patel and Rao, 2014). Anthocyanins, termed "chameleons" or the "colour diversity hub," exhibit varying expressions of colour in response to biotic or abiotic factors such as temperature drops, salinity, nitrogen deficiency, and phosphorus scarcity, typically manifesting as purple colouration on stems, leaves, and other plant parts in response to low light stress. Anthocyanins display alterations in hue due to environmental influences (Riaz *et al.*, 2016). Fruits exhibit diverse anthocyanin pigmentation in both flesh and/or skin. In some fruits, anthocyanin biosynthesis is initiated post chlorophyll and carotenoid degradation as ripening progresses (Zhao *et al.*, 2015).

In ripened avocados (*Persea americana* Mill.), colour changes from green to purple to black are attributed to a decline in chlorophyll and an increase in glycosylated anthocyanins, specifically cyanidin 3-O-glucoside. Cyanidin-3-rutinoside and cyanidin-3-glucoside have also been implicated in ripening-induced colour changes in cherries (*Prunus avium*) (Goncalves *et al.*, 2007). A study on pomegranates reported the development of red colouration alongside an increase in anthocyanin pigment and degradation of chlorophylls and carotenoids. The major anthocyanins identified in pomegranate peel were cyanidin 3,5-diglucoside and cyanidin-3-O-glucoside (Cox *et al.*, 2004).

Texture: Fruit texture gradually evolves during ripening which softens the fruit and enhances its edibility. Texture is a crucial sensory attribute for determining eating quality of food. It encompasses qualities like crispness, firmness, softening, mealiness, crunchiness, and grittiness (Toivonen and Brummell, 2008) and can be objectively assessed through techniques such as texture profile analysis (TPA), which evaluates parameters like firmness, adhesiveness, cohesiveness, springiness, chewiness, resilience, and juiciness (Nadulski and Grochowicz, 2001). Among these metrics, firmness stands out as a key parameter in fruit texture evaluation, bridging sensory perceptions with quantitative measurements (Cha¨ıb *et al.*, 2007). Thus, fruit texture can be defined as the deformation or fragmentation when force is applied, perceptible both through touch and instrumentally via destructive or non-destructive methods. Furthermore, texture is also referred to as the sensory manifestation of fruit/vegetable structure, encompassing cell type, stress response, and moisture/oil content (Meilgaard *et al.*, 2006). Certain fruits like mango, papaya, avocado, sapota and banana undergo substantial textural softening from a "stone-hard" to a "soft pulpy" stage, while apple and citrus fruits exhibit gradual textural changes during ripening. Thus, fruit softening stands as a pivotal transformation facilitating fruit edibility. This alteration in texture arises from the breakdown of the fruit's cell wall, exerting a direct influence on turgor pressure and fruit morphology (Brummell, 2006; Harker *et al.*, 2019). In citrus fruits, softening primarily arises from alterations in turgor pressure, a process linked to post-harvest dehydration. Starch, a prevalent polysaccharide in certain fruits like mango and banana, undergoes enzymatic hydrolysis, resulting in significant cell structure loosening and sweetness development (Prasanna *et al.*, 2010).

The primary constituents of the fruit cell wall comprise pectin and cellulose, alongside minor quantities of hemicellulose and non-cellulosic polysaccharides. During fruit ripening, the breakdown and solubilisation of pectic substances occur. Immature fruit in its initial stages of ripening contains highly esterified protopectins, which are typically insoluble in water and contribute to the fruit's rigid structure (Badui, 2013). Chemical composition of cell wall constantly evolves during ripening. Initially, fruit softening initiates with the dissolution of the middle lamellae, a pectin-rich layer acting as an adhesive between cells. For instance, in kiwifruit early stages of ripening does not involve degradation of pectin but solubilisation which enhances the hydrophilic nature of the cell wall causing it to swell. Further, galactose molecules are removed from pectic substances while some pectic substances undergo de-esterification. As softening continues hemicelluloses and pectin depolymerise and the cell wall weakens with the dissolution of lamella between plant cells (Atkinson *et al.*, 2012).

Pectic substances play a crucial role in providing mechanical support to plant cells and upholding their structural integrity. Within the middle lamella, pectin is intricately cross-linked with calcium salts, a process initiated during the early stages of cell development. The middle lamella is highly susceptible to heat, and its dissolution leads to the detachment of plant cells. De-esterified pectins within the middle lamella interact with calcium ions, and their removal results in cell separation (Chylinska *et al.*, 2017; Prasanna *et al.*, 2007; Brummell, 2006). During the ripening process, insoluble protopectin undergoes conversion into highly soluble polyuronides, leading to the softening of fruit texture. This transformation involves the degradation of tightly bound high molecular weight parent pectin into soluble forms loosely bound to cell walls (Gowda and Huddar, 2001). Subsequently, there occurs a breakdown and solubilization of cell wall polymers, accompanied by alterations in polysaccharide linkages and a decrease in intercellular adhesion, collectively resulting in both fruit softening and tissue disintegration (Goulao and Oliveira, 2008). As softening progresses, neutral sugars such as galactose and arabinose (major constituents of neutral protein) as well as acidic pectin (rhamnogalacturonan) are lost from the cell wall. Thus, enzymatic actions sequentially mediated by protopectinase, pectin methylesterase, pectate lyase, polygalacturonase, cellulase, and β-galactosidase play a crucial role in degrading pectic substances and ultimately resulting in accelerated firmness loss with advancing ripening stages (Gayosso-García Sancho *et al.*, 2010;

Prasanna *et al.*, 2007). While ethylene exposure can be beneficial for certain fruits and vegetables such as apricots, avocados, melons, pears, and tomatoes, promoting softening, prolonged exposure may lead to excessive softening as fruits progress into senescence. Pectins and other polysaccharides are broken down during softening, predisposing those to mechanical injuries.

Major biochemical changes during ripening

Flavour and aroma: Globally, the fruit market holds significant importance, with consumer satisfaction contingent upon the quality of organoleptic attributes. These attributes, encompassing sweetness, acidity, aroma, color, and firmness, are intricately linked to specific metabolic pathways typically coordinated during fruit ripening (Baldwin, 2011). Flavor composition, a complex determinant of quality, involves the interplay of sugars, acids, and volatile compounds. In addition to the four basic flavors recognized by humans in fruits and vegetables, aroma significantly influences consumer acceptance (Lewinsohn *et al.*, 2001). The development of these qualities is influenced by various factors, including variety, growing conditions, harvest stage, and storage conditions (Baldwin, 2011).

Accumulation of flavor compounds during fruit development has been extensively investigated across various plant species. Changes in the composition and concentration of sugars, organic acids, and aroma volatiles are commonly observed throughout various stages of fruit development during ripening. Sugars typically reach their peak levels during fruit maturation or ripening (Kader, 2008). Conversely, organic acids tend to accumulate during the early stages of fruit development, serving as respiratory substrates during ripening, and subsequently decline to minimal levels in ripe fruit (Grimplet *et al.*, 2005; Pei *et al.*, 2016). Sweetness, a critical quality parameter in many fruits, is influenced by the total sugar content and the relative proportions of different sugars present during ripening. Several fruits, including watermelon, have been reported to accumulate glucose, fructose, and sucrose as they mature. In certain developing fruits such as mango and banana, which serve as sink tissues, transported sugars or sugar alcohols are converted into starch. Meanwhile, sugars are stored either as reducing sugars or as sucrose in fruits acting as storage organs, such as tomatoes and strawberries. In some cases, such as in olives, sugars may undergo conversion into lipids. Pre-harvest variables such as sunlight exposure, water availability, fertilization practices, and chemical treatments exert influence on crop development, thereby impacting the internal quality attributes of the harvested produce, including flavor. Intense rainfall preceding harvest has been observed to diminish flavor compounds in tomatoes (*Solanum lycopersicum* L.). Studies on grapevine aroma potential indicate that vines experiencing mild water deficit and moderate nitrogen supply exhibit the highest aroma potential. Conversely, severe water deficit stress and nitrogen deficiency appear to constrain aroma potential (des Gachons *et al*., 2005).

Organic acids also play a pivotal role in contributing to fruit flavour as important intermediate metabolites. The ratio of organic acid to sugar content is crucial in determining quality parameters at the time of fruit harvest. The content of organic acids is predominantly regulated by primary metabolic pathways such as glycolysis and the tricarboxylic acid (TCA) cycle. While citrate and malate are the principal organic acids, minor ones include oxalate, succinate, isocitrate, fumarate, and aconitat (Bhatla, *et al*., 2018). Throughout fruit development, the levels of organic acids typically exhibit an inverse relationship with sugar levels. Consequently, as fruits mature and ripen, sugars accumulate primarily through sugar import or starch degradation, while the organic acids that accumulated in young fruits undergo substantial reduction (Beauvoit *et al*., 2018). Malate and citrate are recognized as the predominant organic acids present in both climacteric and non-climacteric ripe fruits. Notably, the accumulation and degradation of malate occur differently in climacteric and non-climacteric fruits. While some climacteric fruits utilize malate as a substrate during the respiratory burst, non-climacteric fruits continue to accumulate malate throughout ripening (Cherian *et al*., 2014). Interestingly, citrate levels undergo significant reduction during the ripening process, followed by decreases in malate as a respiratory substrate after the climacteric peak in papaya fruits (Silva *et al*., 2018). Similarly, during the ripening of non-climacteric orange and lemon fruits, there is a decline in titratable acidity, primarily attributable to the catabolism of citrate (Agrumes *et al*., 2018). The metabolism and gradual accumulation of different organic acids over the period of fruit ripening are subject to both genetic and environmental influences (Etienne *et al*., 2013).

Volatile compounds, comprising a diverse group of metabolites, play biologically significant roles in mediating plant interactions with other organisms and contribute substantially to the overall flavor of fruits and vegetables (Dudareva *et al*., 2004). Fruit aroma is characterised by the presence of complex combinations of different volatile chemical moieties as well as the potency, concentration and effectiveness of these individual volatiles. While different fruits may share certain aromatic traits, each fruit possesses a unique aroma profile shaped by the specific blend of volatiles present, their concentrations, and the sensitivity of receptors to individual volatile components (Tucker, 1993). Within the plant kingdom, volatile compounds serve as crucial mediators in regulating various physiological processes during plant-environment interactions (Dudareva and Pichersky, 2008). These metabolites are involved in plant defence mechanisms against microbial and herbivore threats, as well as in interplant communication. Moreover, aroma-related volatile compounds contribute to seed dispersal, similar to compounds affecting fruit colour and taste, thereby enhancing their attractiveness to organisms aiding in seed dispersal (Defilippi *et al.*, 2009). The aroma of fruits results from a complex mixture of numerous volatile compounds, including ketones, aldehydes, lactones, carbonyl compounds, terpenoids, esters and alcohols, with key aroma compounds derived from amino acids, lipids, phenols, and mono- and sesquiterpenes (Fig. 3) (Schwab *et al.*, 2008). Notably, volatile esters often make significant contributions to the aroma profiles of fruits like peach (*Prunus persica* L.) and apple (*Malus domestica* Borkh.) (Ortiz *et al.*, 2011). While in case of melons (*Cucumis melo* L), sulphur containing moieties like, *S*-methyl thiobutanoate 2-(methylthio) ethyl acetate, 3-(methylthio) propanal, 3-(methylthio) propyl acetate and3-(methylthio) ethyl propanoate impart flavor and aroma to the fruit (Obando-Ulloa *et al.*, 2008). Throughout fruit development, particularly during ripening, significant alterations occur in these volatile metabolites due to processes such as synthesis, transport, or degradation. Several studies have focused on identifying and characterizing crucial genes and enzymes involved in the biosynthesis of aroma-related volatiles (Defilippi *et al.*, 2009).

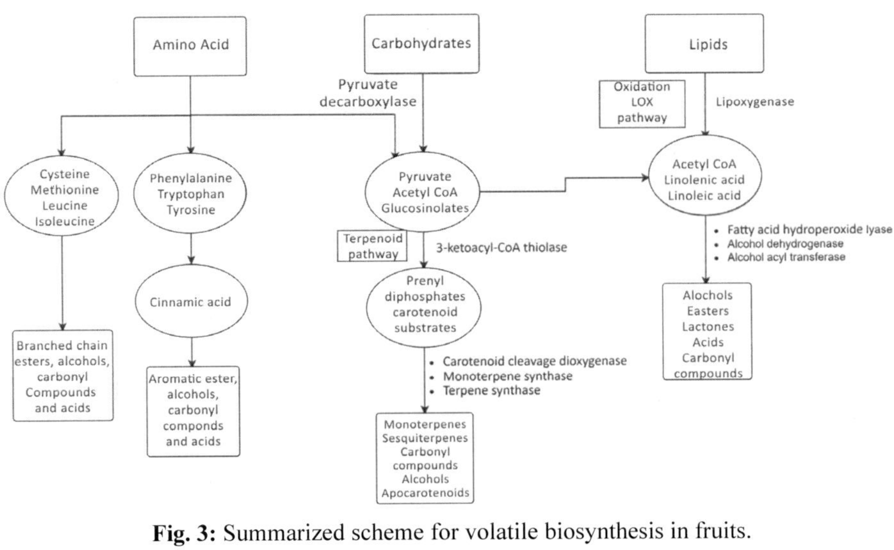

Fig. 3: Summarized scheme for volatile biosynthesis in fruits.

The volatile compositions of fruits are intricate and vary depending on factors such as cultivar, ripeness, pre- and post-harvest environmental conditions, fruit sample type (e.g., intact fruit, slices, or homogenized samples), and analytical methodologies employed. Also, volatiles can be categorized into primary or secondary compounds, denoting their presence in intact fruit tissue or production as a consequence of tissue disruption. It is essential to note that examining volatiles from either intact or disrupted fruit tissues can impact aroma profiles and the ultimate interpretation of aromas. (El Hadi *et al.*, 2013). Aroma compounds are frequently only released upon cell disruption, allowing previously compartmentalized enzymes and substrates to interact (Ishida *et al.*, 1993). Certain aroma compounds are bound to sugars as glycosides or glucosinolates. Glycosides of aroma compounds in fruits primarily comprise O-β-D-glucosides and O-diglycosides, although triglycosides have also been identified (Sarry and Günata, 2004). The proportion of glycosidically bound volatiles typically exceeds that of free volatiles, rendering them a significant potential source of flavor compounds. The odorous aglycones may be liberated from the sugar moiety during maturation, processing, and storage, or through the action of enzymes, acids, or heat (Reineccius, 2005).

Cell wall modification: The ripening process in tomatoes involves significant alterations in cell-wall modification, including changes in cell adhesion, turgor pressure, and the modification of components like pectins and hemicelluloses. Carbohydrates play a crucial role in the ripening process, as they undergo depolymerization, resulting in decreased molecular size and the activation of ripening-specific enzymes. Among the major cell wall polysaccharides undergoing modifications during ripening are starch, pectins, cellulose, and hemicelluloses. Pectins, found predominantly in the primary cell wall and middle lamella, significantly contribute to fruit texture and quality. Their degradation during ripening is a key factor in tissue softening observed in many fruits. Structurally, pectins are diverse heteropolysaccharides comprising partially methylated D-galacturonic acid residues with side chain appendages of various neutral polysaccharides. The degree of polymerization/esterification and the proportion of neutral sugar residues/side chains significantly affect their heterogeneity (Prasanna *et al.*, 2007). It was reported that the accumulation of soluble solids during fruit ripening would decrease the osmotic potential, resulting in an increase rather than a decrease in cell turgor. However, the reduction in cell turgor during ripening can be attributed to cell wall weakening, which contributes to the release of wall constraint. While cell turgor can influence fruit firmness, the decrease in firmness during ripening is more likely a comprehensive outcome of changes in wall architecture, cell expansion, and separation. Enzymes such as polygalacturonase, pectin methyl esterase, lyase, and rhamnogalacturonase play critical roles in pectin degradation and fruit tissue softening. Recent advancements in molecular biology have enhanced our understanding of fruit ripening biochemistry and facilitated genetic manipulation to regulate the ripening process. It is anticipated that further breakthroughs in these areas will yield significant societal benefits in the near future (Jia *et al.*, 2023).

Initially, it was thought that polygalacturonase (PG) was the primary enzyme responsible for tissue softening during ripening. However, further investigation revealed that the process is more intricate. Pectinmethylesterase (PME) was identified as another enzyme involved in cell-wall modification, enabling PG to access its substrates and form gel structures. While transgenic lines with reduced PME activity showed alterations in pectin composition, they did not exhibit noticeable changes in fruit softening, indicating that PME, like PG, is not solely responsible for the textural changes observed during tomato ripening (Fox and Giovannoni, 2011).

Contribution of phytohormones: Ethylene and Abscisic acid (ABA) play pivotal roles in orchestrating fruit ripening. ABA influences carotenoid production and storage in tomato fruit and also stimulates ethylene generation, acting as a stress-responsive hormone, particularly under conditions like drought and pathogen attacks. Ethylene, functioning as a gaseous phytohormone, governs a wide array of physiological and biochemical processes, from seed germination to organ aging (Fig. 4). Strong evidence supports ethylene's involvement in fruit ripening across various levels, including physiological, biochemical, and molecular mechanisms (Li *et al.*, 2022).

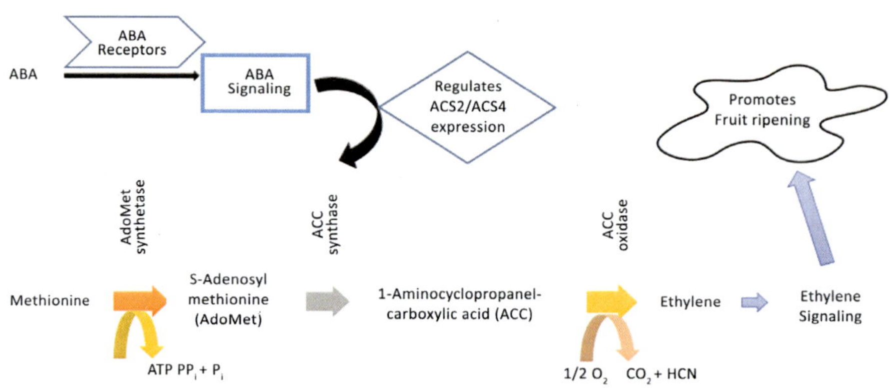

Fig. 4: Synergistic relationship between ABA and ethylene for the regulation of fruit ripening.

The production rates of ethylene and ACC (1-aminocyclopropane-1-carboxylic acid) increase concurrently with fruit ripening. This rise is associated with increased activity of the enzymes ACC synthase and ACC oxidase, along with elevated mRNA expression levels of specific genes encoding these enzymes. However, when ACC is applied to unripe fruits, ethylene production only shows a slight increase, indicating that the enhanced activity of ACC oxidase hampers full fruit ripening (Taiz and Zeiger, 2003).

Fruits that rely on ethylene for ripening typically show a notable increase in respiration prior to ripening known as climacteric fruits such as banana apple, mango, tomato. Conversely, citrus fruits and grapes, termed non-climacteric, do not display this rise in respiration or ethylene production. In climacteric fruits, changes in ethylene levels due to shifts in biosynthesis, perception, signal transduction, or gene transcription significantly impact ripening. Conversely, the ripening of non-climacteric fruits doesn't solely rely on ethylene, prompting further investigation into the underlying mechanisms (Kou *et al.*, 2021). Ethylene controls the speed of ripening, shelf life, and most maturation events in climacteric fruits. It also regulates genes responsible for volatile compound production in various fruit species, as evidenced by studies utilizing mutants in tomato, antisense RNA transgenic lines in apple, and inhibitors of ethylene receptors. Consequently, efforts to develop fruit genotypes with prolonged shelf-life often lead to a notable loss of taste and aroma, as many aroma biosynthesis genes are under ethylene regulation (Martinez-Gonzalez at al., 2017).

Different studies showed that ABA and ethylene synergistically regulate fruit ripening. Exogenous application of ABA accelerates ethylene synthesis and fruit respiration, accelerating ripening. Its regulatory role has been observed in both climacteric and non-climacteric fruits, particularly in the latter such as strawberries, grapes, and blueberries. In climacteric fruits, ABA levels rise before ethylene production, suggesting its upstream regulation of ethylene biosynthesis. Initially, ABA inhibits ethylene synthesis, but later promotes it as ripening progresses. ABA also facilitates ethylene production from ACC during ripening, through mechanisms that may be ethylene-dependent or independent (Zou *et al.*, 2022).

However, in non-climacteric fruits, ABA appears to hold greater significance. For instance, during the ripening of oranges, ABA levels rise, facilitating fruit color development. Similarly, in strawberries, another non-climacteric fruit, ABA influences growth, softening, and the production of anthocyanins and volatile compounds, whereas the involvement of ethylene is comparatively limited. Moreover, in early tomato development, a climacteric fruit, ABA levels peak before ethylene, indicating a coordinated action of both hormones in promoting fruit ripening. Various reports explained the role of ABA in promoting the accumulation of soluble sugars in apple plants *via* mechanism involving the activation of genes responsible for sugar transport, such as MdSUT2, as well as starch degradation. As a result, ABA signaling significantly enhances sweetness, which is a crucial quality trait in fruits (Duran-Soria *et al.* 2020).

Fruit color is a significant factor, which is closely linked to ABA levels in the fruit. ABA stimulates accumulation of anthocyanins, enhancing fruit coloration, and also promotes the production of phenolics (antioxidants) to enhance fruit defence mechanisms. Different studies reported that ABA treatment leads to increased levels of anthocyanins, phenyl ammonia lyase (PAL) activity, and phenolic compounds in strawberries. Similarly, in grapes, ABA treatment results in elevated levels of flavonoids, anthocyanins, total phenolic content, and antioxidants (Gupta *et al.*, 2022).

Various studies have revealed that the application of exogenous auxin influences ethylene synthesis. Treatment with indole-3-acetic acid (IAA) in bananas resulted in reduced ethylene production and delayed ripening. Similar delayed ripening effects were observed in tomatoes and other fleshy fruits following auxin treatment. Furthermore, during fruit ripening, the application of exogenous auxin changed the content of primary metabolites. These changes weakened as the ripening process progressed but did not fully recover by the end of fruit ripening (Li *et al.*, 2017).

Recent reports suggest that jasmonic acid (JA) plays a role in fruit ripening by influencing fruit color, texture, and aroma, indicating its involvement in the ripening process. JA treatment modulates the concentration of volatile organic compounds by regulating the lipoxygenase (LOX) pathway, ethylene production, and the antioxidant system. Furthermore, aroma generation during postharvest ripening of tomato fruit is facilitated by the synergistic regulation of ethylene biosynthesis by methyl jasmonate and the ascorbate-glutathione cycle (Fresno and Munne-Bosch, 2021).

Role of ROS and antioxidant defence mechanism in regulation of fruit ripening: Various studies propose that additional elements including other hormonal influences, reactive oxygen species (ROS), and specific compounds such as sucrose or carotenoids may also exert regulatory control over fruit ripening processes across different species. During fruit development and ripening, there are numerous oxidative processes occurring, resulting in the accumulation of reactive oxygen species (ROS) that can oxidize and alter cellular components, disrupting their normal functionality. To counteract the adverse effects of ROS buildup, plants employ several ROS-scavenging mechanisms to maintain proper ROS homeostasis. This includes the utilization of antioxidant defence mechanisms molecules such as ascorbic acid, carotenoids, phenolic compounds, glutathione, and tocopherols, along with enzymes such as superoxide dismutase (SOD), peroxidase (POX), glutathione peroxidase (GPX), ascorbate peroxidase (APX), and catalase (CAT). Together, these components play a crucial role in scavenging ROS in plants and fruits, ensuring cellular integrity and function (Corpas *et al.*, 2022).

Apart from their detrimental effects in plant cells, specific reactive oxygen species (ROS) such as singlet oxygen ($^1O_2$), hydrogen peroxide (H_2O_2), superoxide ($O^{2\cdot-}$), and hydroxyl radicals (•OH) are recognized as crucial signalling molecules that regulate various biological processes within fruits. These processes include fruit development and the plant/fruit response to both biotic and abiotic stresses. ROS have been observed to trigger ethylene biosynthesis by activating MAP kinase 6 (MPK6), which subsequently activates ethylene response factors (ERFs) like ERF1 and ERF6. These ERFs then stimulate the expression of genes associated with ROS, facilitating the plant's ability to counteract both biotic and abiotic stresses. Additionally, evidence suggests interactions between ROS and not only ethylene but also other hormones that regulate fundamental aspects of plant development. For example, in sunflower seed germination, ROS interact with both ethylene and abscisic acid (ABA) to regulate the process. While evidence of such interactions during fruit development is less abundant, in raspberries, the application of ABA post-fruit set alters the ascorbate/dehydroascorbate (AsA/DHA) ratio in young berries, leading to a more than twofold increase in AsA pools in ripe fruit. This effect is achieved by modifying ascorbate oxidation and recycling through the activities of specific AsA-related antioxidant enzymes (Vall-Llaura *et al.*, 2022).

Modification of carbohydrate metabolism: Although natural ripening involves catabolic pathways and the breakdown of cellular compartments, it is widely accepted as an active, biochemically regulated process. This process coordinates vital biochemical and molecular changes as fruits approach senescence, profoundly impacting fruit quality traits. During fruit ripening, starch is converted into sugars like glucose, fructose, or sucrose. The activity of invertase increases that catalyzes the breakdown of sucrose into simpler sugars like glucose and fructose, altering carbohydrate metabolism. Different enzymes such as α-amylase, β-amylase, and α-glucosidase in plastids break down starch into glucose-1-phosphate during fruit ripening, which then moves to the cytoplasm, increasing glucose and fructose levels in the fruit flesh. These sugars serve multiple metabolic roles, either fuelling glycolysis for ripening energy and contributing to the synthesis of amino acids or nucleic acids or transformed back to sucrose in a reaction catalyzed by sucrose synthase (SS) and sucrose phosphate synthase (SPS) in cytoplasm for enhancing sweetness. In glycolysis, sugar breakdown provides energy for ripening, while in respiration, pyruvate from glycolysis is converted to acetyl coenzyme A, entering the TCA cycle for overall oxidation to carbon dioxide. However, enhanced activity of SPS has been observed in different fruits such as grape berries as a contributor to increased sweetness. These changes significantly influence fruit quality traits, representing that the control of fruit ripening is vital for quality maintenance and minimising post-harvest losses (Alos *et al.*, 2019). Watrelot and Norton (2020), reported that during ripening of berries, there's a rapid increase in sugar content, which slows down approximately a week before harvest. Concurrently, the glucose to fructose ratio tends to decline, with fructose levels increasing more rapidly than glucose levels during ripening. The rise in sugar content during berry ripening is widely recognized for its association with the development of aromatic compounds and a concurrent decrease in acidity.

Reduction of acidity and tannin levels: The equilibrium between sweetness and acidity stands as a fundamental criterion in evaluating the quality of various fruits. The fruit quality is often reduced by excessive acid content, and fruit palatability can be improved by an adequate acid concentration. Processes involved in synthesis, breakdown, and transport have an impact on how sugar and acid levels change over time. The pH and titratable acidity closely influence factors such as the solubility of tartaric salts, stabilization of color, and the depolymerization and condensation reactions with tannins (Li *et al.*, 2020). Pena-Neira *et al.*, (2023) observed that all grape cultivars exhibited a substantial increase in pH accompanied by a significant decrease in overall acidity during the ripening process. Subsequently, during storage, the pH remained relatively stable, while the overall acidity of all cultivars slightly increased.

Among grape phenolic compounds, proanthocyanidins or condensed tannins are responsible for the astringent and bitter properties of the wines and are released from both grape skins and seeds. Fruit softening during maturation is attributed to the breakdown of polysaccharides in cell walls, leading to osmotic pressure and the concentration of soluble tannins within tannin cells, eventually insolubilizing them. It was suggested that during ripening, the collapse of flesh cells results in cell wall fragments adhering to tannins, reducing their astringency. Additionally, softening and collapsing of flesh cells aid in pectin solubilization during fruit ripening, forming complexes with tannins that lead to their insolubilization (Vilhena *et al.*, 2020).

Signaling pathways: In the regulation of fruit ripening, both abscisic acid (ABA) and ethylene signaling pathways play pivotal roles. In climacteric fruits, ABA interacts with ethylene, influencing the synthesis of ethylene itself. ABA's signal is detected by the PYR1–PP2C–SnRK2 module, which in turn governs the phosphorylation of ABA-responsive transcription factors (TFs), thereby stimulating the expression of genes associated with fruit development and ripening. Ethylene synthesis, on the other hand, is facilitated by ACC oxidase enzyme, utilizing SAM (S-adenosyl methionine) as a precursor. Once ethylene is produced, it binds to Ethylene Insensitive3-Like (EIL) transcription factors like EIN3, activating them (Fig. 5). These activated EIL factors can then promote the expression of ethylene response factor (ERF) TFs, which in turn modulate the genes regulated by ethylene by binding to their promoters (Gupta *et al.*, 2022).

Calcium signaling starts with increases in cytosolic Ca^{2+} levels triggered by the perception of signals like plant hormones, which play crucial roles in various plant development processes. To elicit specific physiological responses, these calcium signals, need to be interpreted by calcium-sensor proteins and relayed to downstream targets. These sensor proteins fall into two main groups: sensors-relay and sensors-responders. Sensors-relay encompass families such as calmodulin (CaM) and calmodulin-like (CML) proteins, as well as the Calcineurin B-like (CBL) family. Sensors-responders, which possess enzymatic activity directly regulated by calcium, belong to the Calcium-dependent protein kinase (CDPK) family (Fig. 5). These calcium sensors, whether directly or indirectly, modulate crucial target gene expression responsible for various mechanism involved in fruit ripening (Gao *et al.*, 2019).

Ethylene and auxin (AUX) play interconnected roles in fruit ripening and senescence. Recent studies highlight their significant influence on the ripening process of climacteric fruits. Auxin, particularly indole-3-acetic acid (IAA), contributes to fruit development and can hinder ripening processes. AUX/IAA transcription modulators and ARF (auxin response factor) proteins are indeed key components in the interaction between auxin and ethylene. They facilitate the intricate crosstalk and coordination of signaling pathways between these two hormones, influencing various cellular processes and developmental responses. Ethylene, instead of directly controlling auxin biosynthesis, modulates its transport, signaling, and metabolism by altering the expression of genes encoding key auxin response proteins and auxin carrier proteins. Ethylene regulates SlPIN7 (auxin efflux carrier protein) and SlLAX1 (auxin influx carrier protein). Auxin affects various processes such as cell wall breakdown, carotenoid pigment metabolism, energy metabolism, and the synthesis of aromatic volatiles, ultimately slowing down the ripening process (Kou *et al.*, 2021).

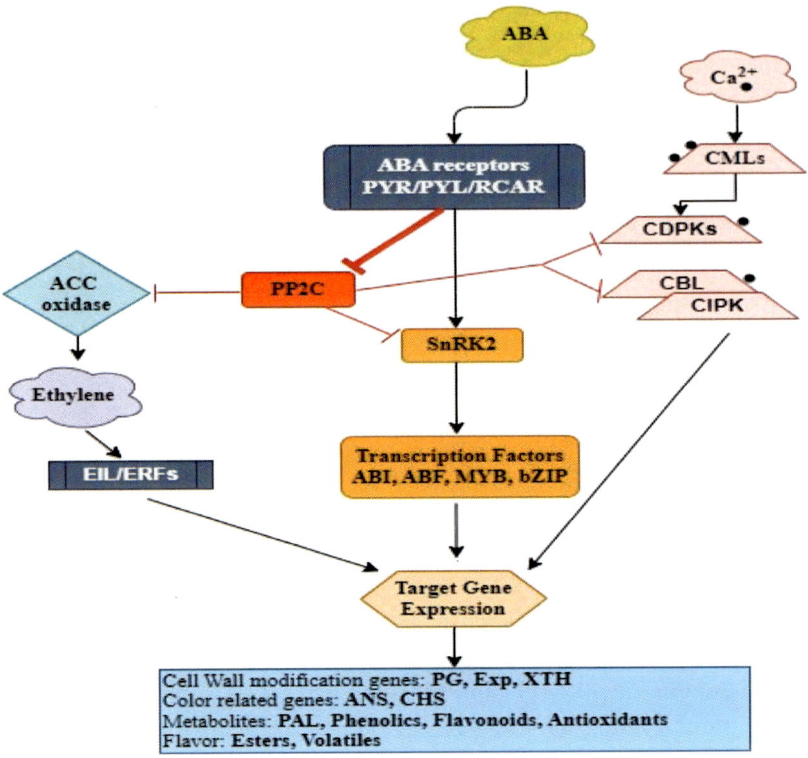

Fig. 5: Regulation of fruit ripening through various signaling pathways. EIL: Ethylene Insensitive3-Like (EIL), ERF: Ethylene response factor (ERF), PYR: Pyrabactin Resistance, PYL: PYR-LIKE, RCAR: Regulatory components of ABA Receptors, PP2C: Type 2C protein phosphatases, SnRK2: SNF1-related protein kinases 2, EXP: Expansins, PG: Polygalacturonase, XTH: Xyloglucan hydroxylase, ACC: Amino cyclopropane carboxylate, ANS: Anthocyanin synthase, CHS: Chalcone synthase. CaM: calmodulin, CML: calmodulin-like proteins, CBL: calcineurin B-like proteins, CDPK: calcium-dependent protein kinases.

In conclusion, fruits play a crucial role in our diet, offering nutritional value and economic significance. Fruit development involves a complex interplay of genes and signaling pathways, with distinct phases such as fruit set, development, and ripening. Ripening is a highly intricate process characterized by significant physiological and biochemical changes, including alterations in color, texture, aroma, and nutrient composition. The process involves various biochemical and physiological transformations, such as the breakdown of chlorophyll, synthesis of pigments, adjustments in flavor and aroma, and changes in carbohydrate content. Distinctions between climacteric and non-climacteric fruits influence ripening patterns, with physicochemical alterations marking stages from maturity to over-ripeness. Understanding fruit ripening enhances agricultural practices and food preservation strategies, bridging biology, agriculture, and nutrition for human health and sustainability. In conclusion, tomato ripening involves intricate changes in cell-wall composition mediated by enzymes like polygalacturonase and pectin methyl esterase. Hormones such as ethylene and ABA, along with other regulators like auxin and JA, orchestrate ripening processes, while ROS and antioxidant defenses play crucial roles in maintaining cellular integrity during this stage.

References

Adi A A, Oduro I and Tortoe C, 2019. Onset of fruit ripening: A review. *Journal of Agricultural Science and Technology*, 21(2):257-272. https://doi.org/10.22059/ijas.2019.282805.672693

Agrumes F, Citrusfruecht G, Agrumi I, Agrios S and Monselise S P, 2018. Citrus In *Handbook of Fruit Set and Development* (pp 87-108) CRC Press.

Alos E, Rodrigo M J and Zacarias L, 2019. Ripening and senescence In Postharvest physiology and biochemistry of fruits and vegetables (pp 131-155) Woodhead Publishing https://doiorg/101016/B978-0-12-813278-400007-5

Atkinson R G, Sutherland P W, Johnston S L, Gunaseelan K, Hallett I C, Mitra D, Brummell D A, Schröder R, Johnston J W and Schaffer R J, 2012. Down-regulation of Polygalacturonase1 alters firmness, tensile strength and water loss in apple (Malus x domestica) fruit BMC plant biology, 12, pp1-13 https://doiorg/101186/1471-2229-11-129.

Badui S, 2013. *Química de los alimentos* Pearson Educación

Bai S, Tian Y, Tan C, Bai S, Hao J and Hasi A, 2020. Genome-wide identification of microRNAs involved in the regulation of fruit ripening and climacteric stages in melon (*Cucumis melo) Horticulture research*, 7, 1-16 https://doiorg/101038/s41438-020-0302-2

Baldwin E A and Bai J, 2011. Physiology of fresh-cut fruits and vegetables *Advances in fresh-cut fruits and vegetables processing*, pp87-114.

Barragan-Iglesias E, Méndez-Lagunas L and Rodríguez-Ramirez J, 2018. Physicochemical changes during on-tree ripening of climacteric fruits *Postharvest Biology and Technology*, 144, 134-142 https://doiorg/101016/jpostharvbio201805005

Barsan C, Sanchez-Bel P, Rombaldi C V, Egea I, Rossignol M, Kuntz M, Zouine M, Latché A, Bouzayen M, Pech J C and Lelievre, J M, 2012. Characteristics of the tomato chromoplast revealed by proteomic analysis *Journal of Experimental Botany*, 63(2): 647-658 https://doiorg/101093/jxb/err304

Beauvoit B, Belouah I, Bertin N, Cakpo C B, Colombié S, Dai Z, Gautier H, Génard M, Moing A, Roch L and Vercambre G, 2018. Putting primary metabolism into perspective to obtain better fruits *Annals of botany*, 122(1), pp1-21 https://doiorg/101093/aob/mcy057

Berry H M, Roston R L and Scholl E H, 2019. Predicting the effects of structural complexity in plant plastoglobules on carotenoid bioavailability *The Journal of Physical Chemistry* B, 123(50), 10719-10729 https://doiorg/101021/acsjpcb9b09507

Bhatla S C, A Lal, M, Shakya R and A Lal M, 2018. Fruit development and ripening *Plant physiology, development and metabolism*, pp857-883 https://doiorg/101007/978-981-13-2023-1_27

Bouzayen M, Latche A, Nath P and Pech J C, 2009. Mechanism of fruit ripening In Plant Developmental Biology-Biotechnological Perspectives: Volume 1:319-339. Berlin, Heidelberg: Springer Berlin Heidelberg https://doiorg/101007/978-3-642-02301-9_16

Brummell D A, 2006. Cell wall disassembly in ripening fruit *Functional Plant Biology*, 33(2), 103-119 https://doiorg/101071/FP05177

Cha¨ıb J, Grabulos J and Guigon P, 2007. Study of different mechanical techniques for texture analysis of fruits: Application to the "pulpy" category *Journal of Food Engineering*, 81(4): 592-600 https://doiorg/101016/jjfoodeng200612007

Cherian S, Figueroa CR and Nair H, 2014. 'Movers and shakers' in the regulation of fruit ripening: a cross-dissection of climacteric versus non-climacteric fruit *Journal of experimental botany*, 65(17): 4705-4722 https://doiorg/101093/jxb/eru280

Chung S W, Yu DJ, Oh HD, Ahn J H, Huh J H and Lee H J, 2019. Transcriptional regulation of abscisic acid biosynthesis and signal transduction, and anthocyanin biosynthesis in 'Bluecrop' highbush blueberry fruit during ripening *PLoS One,* 14(7): pe0220015 https://doiorg/101371/journalpone0220015

Chylinska M, Szymanska-Chargot M, Zdunek A, Gdula K, Kruk B and Pawlaczyk-Graja I, 2017. Changes in cell wall polysaccharides and glycoproteins distribution during apple development and storage *Plant Physiology and Biochemistry*, 118: 203-212 https://doiorg/101016/jplaphy201707011

Corpas F J, Rodríguez-Ruiz M, Muñoz-Vargas M A, González-Gordo S, Reiter, R J and Palma J M, 2022. Interactions of melatonin, reactive oxygen species, and nitric oxide during fruit ripening: an update and prospective view Journal of Experimental Botany, 73(17), pp5947-5960 https://doiorg/101093/jxb/erac128

Cox S, McGhie T, White A and Woolf A, 2004. Anthocyanins and flavonoids from blueberry 'Maru' (*Vaccinium* spp) and 'Toru' (*Ribes* spp) *Journal of Agricultural and Food Chemistry*, 52(3): 402-406 https://doiorg/101021/jf0350854

Defilippi BG, Manríquez D, Luengwilai K and González-Agüero M, 2009. Aroma volatiles: biosynthesis and mechanisms of modulation during fruit ripening *Advances in Botanical Research*, 50:1-37 https://doiorg/101016/jfoodchem2019125020

Des Gachons C P, Leeuwen C V, Tominaga T, Soyer J P, Gaudillère J P and Dubourdieu D, 2005. Influence of water and nitrogen deficit on fruit ripening and aroma potential of Vitis vinifera L cv Sauvignon blanc in field conditions *Journal of the Science of Food and Agriculture*, *85*(1):73-85 https://doiorg/101002/jsfa1919

Dudareva N and Pichersky E, 2008. Metabolic engineering of plant volatiles *Current Opinion in Biotechnology*, 19(2):181-189 https://doiorg/101016/jcopbio200802011

Dudareva N, Pichersky, E and Gershenzon J, 2004. Biochemistry of plant volatiles *Plant Physiology*, 135(4):1893-1902 https://doiorg/101104/pp104049981

Duran-Soria O, Pott D M, Osorio S and Vallarino J G, 2020. Plant hormones, melatonin, and the molecular mechanisms underlying fruit ripening *Frontiers in Plant Science*, 11: 165 https://doiorg/103389/fpls202000165

Duran-Soria S, Pott D M, Osorio S and Vallarino J G, 2020. Sugar signaling during fruit ripening *Frontiers in Plant Science*, 11: 564917 https://doiorg/103389/fpls2020564917

El Hadi, MAM Zhang F J, Wu F F, Zhou CH and Tao J, 2013. Advances in fruit aroma volatile research *Molecules*, 18(7): pp8200-8229 https://doiorg/103390/molecules18078200

Etienne A, Génard M, Bancel D, Benoit S and Bugaud C, 2013. A model approach revealed the relationship between banana pulp acidity and composition during growth and post harvest ripening *Scientia Horticulturae* 162:125-134 https://doiorg/101016/jscienta201308011

Fox E and Giovannoni J, 2011. Regulating the ripening process In: Moo-Young, M (Ed), Comprehensive Biotechnology, second ed, vol 4 Elsevier, pp 339–348 https://doiorg/101016/B978-0-12-809633-809199-8

Fresno, DH and Munne-Bosch S, 2021. Differential tissue-specific jasmonic acid, salicylic acid, and abscisic acid dynamics in sweet cherry development and their implications in fruit-microbe interactions Frontiers in plant science, 12, p640601 https://doiorg/103389/fpls2021640601

Gao Q, Xiong T, Li X, Chen, W and Zhu X, 2019. Calcium and calcium sensors in fruit development and ripening Scientia Horticulturae, 253, pp412-421 https://doiorg/101016/jscienta201904069

Gayosso-García Sancho L E, Prado-Olivarez J, Loarca-Piña G, Oomah B D and Gutiérrez-Dorado R, 2010. Purification and characterization of a β-galactosidase from mango (Mangifera indica cv Ataulfo) that transglycosylates prebiotic carbohydrates *Journal of Agricultural and Food Chemistry*, 58(2): 1185-1191 https://doiorg/101021/jf903383r

Goncalves B, Landbo A K, Let M B and Silva A P, 2007. Rosa damascena flower essential oil, a rich source of beta-damascenone and its contribution to the aroma of different floral honeys *Journal of Agricultural and Food Chemistry*, 55(9): 3609-3616 https://doiorg/101021/jf063093k

Goulao L F and Oliveira C M, 2008. Cell wall modifications during fruit ripening: When a fruit is not the fruit *Trends in Food Science & Technology*, 19(1): 4-25 https://doiorg/101016/jtifs200707002

Gowda L R and Huddar A G, 2001. Characterization of polyuronides solubilized from mango (*Mangifera indica* L cv Alphonso) peel during ripening *Carbohydrate Research*, 330(2), 237-245 https://doiorg/101016/S0008-6215(00)00321-7

Grimplet J, Romieu C, Audergon J M, Marty I, Albagnac G, Lambert P, Bouchet J P and Terrier N, 2005. Transcriptomic study of apricot fruit (*Prunus armeniaca*) ripening among 13 006 expressed sequence tags *Physiologia Plantarum*, 125(3), pp281-292 https://doiorg/101111/j1399-3054200500563x

Gupta K, Wani S H, Razzaq A, Skalicky M, Samantara K, Gupta S, Pandita D, Goel S, Grewal S, Hejnak V and Shiv A, 2022. Abscisic acid: Role in fruit development and ripening *Frontiers in Plant Science,* 13: 817500 https://doiorg/103389/fpls2022817500

Harker F R, Maindonald J, Murray S H and Gunson F A, 1997. The effect of temperature on the duration of the lag phase of ethylene production in kiwifruit *Postharvest Biology and Technology*, 12(3), 239-247 https://doiorg/101016/S0925-5214(97)01417-5

Hu X, Li Y, Li C and Zhu L, 2019. Optimal ripeness stage and factors influencing fruit coloration in Maradol papaya *Acta Horticulturae Sinica,* 46(12): 2499-2508 https://doiorg/1016420/jissn0513-353x2018-0691

Ishida B K, Baldwin E A, Buttery R G, Chui SH and Ling L C, 1993. Flavor volatilcs, sugars and color development in ripening in vitro-cultured tomato fruit and calyx *Physiologia Plantarum*, 89(4): 861-867 https://doiorg/101111/j1399-30541993tb05297x

Jia K, Wang W, Zhang Q and Jia W, 2023. Cell wall integrity signaling in fruit ripening *International Journal of Molecular Sciences,* 24(4): p4054 https://doiorg/103390/ijms24044054

Johnston J W, Gunaseelan K, Pidakala P, Wang M and Schaffer R J, 2009. Co-ordination of early and late ripening events in apples is regulated through differential sensitivities to ethylene *Journal of Experimental Botany*, 60(9): 2689-2699 https://doiorg/101093/jxb/erp128

Kader A A, 2008. Flavour quality of fruits and vegetables *Journal of the Science of Food and Agriculture*, 88(11), pp1863-1868 https://doiorg/101002/jsfa3293

Kanayama Y and Kochetov A V, 2015. The role of chlorophyllase in chlorophyll metabolism in higher plants and algae *Biochemistry* (Moscow), 80(7):869-878 https://doiorg/101134/S0006297915070095

Kou X, Feng Y, Yuan S, Zhao X, Wu C, Wang C and Xue Z, 2021. Different regulatory mechanisms of plant hormones in the ripening of climacteric and non-climacteric fruits: a review Plant Molecular Biology, pp1-21 https://doiorg/101007/s11103-021-01199-9

Kou X, He Y, Li Y, Chen X, Feng Y and Xue Z, 2019. Effect of abscisic acid (ABA) and chitosan/nano-silica/sodium alginate composite film on the color development and quality of postharvest Chinese winter jujube (*Zizyphus jujuba* Mill cv Dongzao) *Food Chemistry*, 270:385-394 https://doiorg/101016/jfoodchem201807056

Kou X, Zhou J, Wu CE, Yang S, Liu Y, Chai L and Xue Z. 2021. The interplay between ABA/ ethylene and NAC TFs in tomato fruit ripening: a review Plant Molecular Biology, 106, pp223-238 https://doiorg/101007/s11103-021-01128-w

Kying T, Stintzing F C, Carle R and Steingass C B, 2013. Color, chroma, and hue angle as objective quality attributes for the determination of mango fruit maturity *Journal of Agricultural and Food Chemistry*, 61(16):4026-4035 https://doiorg/101021/jf3048025

Lelievre J M, Latchè A, Jones B, Bouzayen M and Pech J C, 1997. Ethylene and fruit ripening *Physiologia Plantarum*, 101(4): 727-739 https://doiorg/101111/j1399-30541997tb01057x

Lewinsohn E, Schalechet F, Wilkinson J, Matsui K, Tadmor Y, Nam K H, Amar O, Lastochkin E, Larkov O, Ravid U and Hiatt W, 2001. Enhanced levels of the aroma and flavor compound S-linalool by metabolic engineering of the terpenoid pathway in tomato fruits *Plant Physiology*, 127(3):1256-1265 https://doiorg/101104/pp010293

Li BJ, Grierson D, Shi Y and Chen KS, 2022. Roles of abscisic acid in regulating ripening and quality of strawberry, a model non-climacteric fruit *Horticulture Research*, 9, puhac089 https://doiorg/101093/hr/uhac089

Li J, Khan ZU, Tao X, Mao L, Luo Z and Ying T, 2017. Effects of exogenous auxin on pigments and primary metabolite profile of postharvest tomato fruit during ripening *Scientia Horticulturae*, 219: 90-97 https://doiorg/101016/jscienta201703011

Li X, Li C, Sun J and Jackson A, 2020. Dynamic changes of enzymes involved in sugar and organic acid level modification during blueberry fruit maturation Food chemistry, 309, p125617 https://doiorg/101016/jfoodchem2019125617

Liu Y, Shao Z, Zhang M and Wang Q, 2015. The changes of chloroplast ultrastructure and chlorophyll fluorescence parameters in *Amomum villosum* var *xanthioides* T L Wu et Senjen during leaf senescence *Acta Physiologiae Plantarum*, 37(7): 1-8 https://doiorg/101007/s11738-015-1914-4

Martinez-Gonzalez M E, Balois-Morales R, Alia-Tejacal I, Cortes-Cruz MA, Palomino-Hermosillo YA and Lopez-Guzman GG, 2017. Postharvest fruits: maturation and biochemical changes Revista mexicana de ciencias agrícolas, 8(SPE19), pp4075-4087 https://doiorg/1029312/remexcav0i19674

Meilgaard M C, Civille G V and Carr B T, 2006. *Sensory evaluation techniques* CRC Press.

Nadulski R and Grochowicz J, 2001. Texture profile analysis in evaluating fruit and vegetable products *Acta Agrophysica*, 58: 123-128.

Nath P, Bouzayen M, Mattoo A K and Pech J (Eds), 2014. *Fruit ripening: physiology, signalling and genomics* CABI.

Noichinda S, Bodhipadma K and Wongs-Aree C, 2017. Antioxidant Potential and Their changes during postharvest handling of tropical fruits In *Novel Postharvest Treatments of Fresh Produce* (pp 633-662) CRC Press.

Obando-Ulloa J M, Moreno E, García-Mas J, Nicolai B, Lammertyn J, Monforte A J and Fernández-Trujillo J P, 2008. Climacteric or non-climacteric behavior in melon fruit: 1 Aroma volatiles *Postharvest Biology and Technology*, 49(1):27-37 https://doiorg/101016/jpostharvbio200711004

Ortiz A, Graell J and Lara I, 2011. Volatile ester-synthesising capacity throughout on-tree maturation of 'Golden Reinders' apples *Scientia Horticulturae*, 131:6-14 https://doiorg/101016/jscienta201109020

Patel R M and Rao M V, 2014. Anthocyanins in fruits, vegetables, and grains In *Handbook of Antioxidants for Food Preservation,* 203-216, Elsevier https://doiorg/101016/B978-0-85709-478-600015-1

Paull R E, Reyes M E and Naota M, 2008. Papaya fruit quality assessment *Journal of the Science of Food and Agriculture*, 88(7): 1071-1076 https://doiorg/101002/jsfa3187

Pena-Neira A, Cortiella MGI, Ubeda C, Pastenes C, Villalobos L, Contador L, Infante R and Gómez C, 2023. Phenolic, Polysaccharides Composition, and Texture Properties during Ripening and Storage Time of New Table Grape Cultivars in Chile Plants, 12(13): p2488 https://doiorg/103390/plants12132488

Prasanna V, Prabha T N and Tharanathan, R N, 2010. Fruit ripening phenomena—an overview *Critical Reviews in Food Science and Nutrition*, 50(1):1-17 https://doiorg/101080/10408390802565864

Prasanna V, Prabha T N and Tharanathan R N, 2007. Fruit ripening phenomena–an overview *Critical Reviews in Food Science and Nutrition*, 47(1):1-19 https://doiorg/101080/10408390600976841

Rahman M S and Gamage T V, 1999. Handbook of Food Preservation, 11.

Reineccius G, 2005. *Flavor chemistry and technology* CRC press https://doiorg/101201/9780203485347

Riaz M, Zia-Ul-Haq M and Saad B, 2016. *Anthocyanins and human health: biomolecular and therapeutic aspects,* Vol 10:978-3 Switzerland: Springer https://doiorg/101007/978-3-319-26456-1

Sarry J E and Günata Z, 2004. Plant and microbial glycoside hydrolases: volatile release from glycosidic aroma precursors *Food Chemistry*, 87(4):509-521 https://doiorg/101016/jfoodchem200401003

Schwab, W, Davidovich-Rikanati, R and Lewinsohn E, 2008 Biosynthesis of plant-derived flavor compounds *The Plant Journal*, 54(4):712-732 https://doiorg/101111/j1365-313X200803446x

Schweiggert R M, Mezger D, Schimpf F, Steingass C B, Carle R and Stintzing F C, 2011. Influence of chromoplast morphology on carotenoid bioaccessibility of carrot, mango, papaya, and tomato *Food Chemistry*, 128(2): 395-406 https://doiorg/101016/jfoodchem201103059

Schweiggert R M, Steingass C B and Carle R, 2011. Profiling of carotenoids and their biosynthetic gene expressions in different papaya (*Carica papaya* L) fruit varieties by chromatographic and Real-Time PCR analysis *Journal of Agricultural and Food Chemistry,* 59(6): 3038-3046 https://doiorg/101021/jf104412s

Shakya R A and Lal M, 2018. Fruit Development and Ripening In: *Plant Physiology, Development and Metabolism* Springer, Singapore https://doiorg/101007/978-981-13-2023-1_27

Silva W B, Silva G M C Santana D B, Salvador A R, Medeiros D B, Belghith I, da Silva N M, Cordeiro M H M and Misobutsi G P, 2018. Chitosan delays ripening and ROS production in guava (*Psidium guajava* L) fruit *Food Chemistry*, 242:232-238 https://doiorg/101016/jfoodchem201709052

Simkin A J, Kuntz M and Moreau H, 2007. Genealogy of non-photochemical quenching and chlorophyll fluorescence quenching factors in plants: A case study using the Arabidopsis protein superfamily *Photosynthesis Research*, 91(2-3):265-277 https://doiorg/101007/s11120-007-9208-8

Stikic R, Jovanović Z, Vucelic-Radovic B, Marjanović M and Savić S, 2015. Tomato: a model species for fruit growth and development studies *Botanica Serbica*, 39(2); 95-102

Taiz L and Zeiger E, 2003. Plant physiology 3rd edn https://doiorg/101093/aob/mcg079

Toivonen P M and Brummell D A, 2008. Biochemical bases of appearance and texture changes in fresh-cut fruit and vegetables *Postharvest Biology and Technology*, 48(1):1-14 https://doiorg/101016/jpostharvbio200709004

Vall-Llaura N, Fernández-Cancelo P, Nativitas-Lima I, Echeverria G, Teixidó N, Larrigaudière C, Torres R and Giné-Bordonaba J, 2022. ROS-scavenging-associated transcriptional and biochemical shifts during nectarine fruit development and ripening Plant Physiology and Biochemistry, 171, pp38-48 https://doiorg/101016/jplaphy202112022

Vasquez-Caicedo A L, Heller A, Neidhart S, Carle R and Schweiggert R M, 2006. Chromoplast morphology and beta-carotene accumulation during postharvest ripening of Mango cv 'Tommy Atkins' *Journal of Agricultural and Food Chemistry*, 54(16); 5769-5776 https://doiorg/101021/jf0613117

Vilhena NQ, Gil R, Llorca E, Moraga G and Salvador A, 2020. Physico-chemical and microstructural changes during the drying of persimmon fruit cv Rojo Brillante harvested in two maturity stages *Foods,* 9(7):870 https://doiorg/103390/foods9070870

Wang H, Qian Z, Ma S, Zhou Y, Patrick J W, Duan X, Jiang Y and Qu, H, 2013. Energy status of ripening and postharvest senescent fruit of litchi (*Litchi chinensis* Sonn) *BMC Plant biology*, 13:1-16 https://doiorg/101186/1471-2229-13-55.

Wang Z, Angenent G C, Seymour G B and de Maagd R A, 2020. Revisiting the role of master regulators in tomato ripening *Trends in Plant Science*, 25(4), 291-301 https://doiorg/101016/jtplants201911010

Watrelot, AA and Norton E L, 2020 Chemistry and reactivity of tannins in *Vitis* spp: A review Molecules, 25(9):2110 https://doiorg/103390/molecules25092110

Wills R B, Golding J B and McGlasson W B, 1998. Changes in chlorophyll, chlorophyllase and peroxidase activities in the peel of green and ripe bananas *Journal of the Science of Food and Agriculture*, 77(4):441-444 https://doiorg/101002/(SICI)1097-0010(199808)77:4<441::AID-JSFA72>30CO;2-9

Wills R B H, Warton M A, Mussa, DMDN and Chew L P, 2001. Ripening of climacteric fruits initiated at low ethylene levels *Australian Journal of Experimental Agriculture*, 41(1), pp89-92 https://doiorg/101071/EA00206

Yin X R, Chen K S, Allan A C, Wu R M, Zhang B, Lallu N and Ferguson I B, 2008. Ethylene-induced modulation of genes associated with the ethylene signalling pathway in ripening kiwifruit *Journal of Experimental Botany*, 59(8): 2097-2108 https://doiorg/101093/jxb/ern065

Zhao Z, Yuan H, Yin L and Feng X, 2015. Anthocyanins from purple sweet potato (*Ipomoea batatas* (L) Lam) and their antioxidant activity *in vitro Food Science and Technology*, 35(2): 194-198 https://doiorg/101590/1678-457x6509

Zou J, Li N, Hu N, Tang N, Cao H, Liu Y, Chen J, Jian W, Gao Y, Yang J and Li Z, 2022. Co-silencing of ABA receptors (SlRCAR) reveals interactions between ABA and ethylene signaling during tomato fruit ripening Horticulture Research, 9, puhac057 https://doiorg/101093/hr/uhac057

3

Role and Mode of Action of Calcium, Anti-oxidants and Enzymes in Post-harvest Management of Fruits

K Dhinesh Babu, R A Marathe and M Sankaran

ICAR–National Research Center on Pomegranate, Solapur-413 255 Maharashtra

Calcium, anti-oxidants and certain enzymes are reported to play vital role in extending the shelf-life of harvested produce of many horticultural crops especially fruits. They help in maintenance of cell wall structure. Being an important constituent of cell wall, in the form of calcium pectate, it helps in maintenance of cell wall structure. Antioxidants are highly beneficial in extending the shelf-life of fruits besides maintaining the quality. Similarly, some enzymes play a key role in preserving the fruit quality and prolonging its market life. As some fruits are catergorized under 'Low volume high value fruits', it is essential to extend their shel-flife for reducing the post-harvest losses of fruits and also to reap better remuneration. Thus, it becomes indispensable to review the role of Ca, anti-oxidants and enzymes which are beneficial in keeping the harvested produce afresh for a relatively longer period.

Calcium

Calcium play a vital role in thc physiological and biochemical processes of organisms and cells. Calcium is essential for structural integrity of both the cell wall and plasma membrane. Calcium treatments have known to delay softening and improve the fruit quality.

Calcium has received considerable attention in recent years due to its various beneficial role viz.,

- Delaying ripening and senescence,
- Increase in firmness,

- Increase in vitamin C,
- Increase in phenolic contents,
- Reduction of physiological disorders,
- Reduction of and storage rots,
- Reduction in respiration,
- Extending storage life of fruits.

Exogenous application of Ca has been shown to delay senescence of many fruit tissue slices, ethylene production in particular and the onset of lipid peroxidation. Calcium alters intracellular and extracellular processes which retard ripening exemplified by lower the rates of colour change, softening, CO_2 and ethylene production, increase in sugars and a reduction in total acid content.

At Anand Agricultural University, Anand, foliar spray of calcium (Calcium chloride @ 0.5, 1.0 and 1.5%; Calcium nitrate @ 0.5, 1.0 and 1.5%; Calcium sulphate @ 0.5, 1.0 and 1.5% along with control) to the trees of sapota cv. Kalipatti and ambient storage of harvested fruits at 3 weeks after the spray revealed that spraying of CaCl @ 1 per cent was found effective for increasing total soluble solids, total sugars, reducing sugars, non-reducing sugars, ascorbic acid with minimum acidity and moisture content. While, pre-harvest spraying of $CaCl_2$ @ 1.5 per cent improved calcium content of sapota fruits (Patel *et al*., 2017).

TSS: The increase in TSS during storage period up to 6 days was due to the breakdown of complex polymers in to simple substances by hydrolytic enzymes, which at later storage period got utilized during respiration (Patel *et al*., 2017). The increase in total soluble solids content during the storage period have been reported by Bhalerao *et al.* (2010) and Desai (2016) in sapota, Ramakrishna *et al.* (2001) in papaya, Wahdan *et al.* (2011), Vidya *et al.* (2014) in mango.

Total sugars: The significantly higher total sugars (16.80, 20.03, 24.83 and 17.57 per cent) was observed with foliar spray of $CaCl_2$ @ 1 per cent at 2nd, 4th, 6th and 8th day of storage period, respectively. The increase in total sugars during initial storage period might be due to the hydrolysis of starch into sugars as on complete hydrolysis of starch, no further increase occurs and subsequently a decline in total sugars is predictable (Patel *et al*., 2017).

Similarly, such increase in total sugars content during the post-harvest storage has been reported in different fruit crops by others viz., Lakshmana and Reddy,

(1999), Bhalerao *et al.* (2010) and Desai (2016) in sapota, Rajkumar *et al.*, (2006) in papaya, Wahdan *et al.* (2011) and Vidya *et al.* (2014) in mango, Bisen *et al.* (2014) in guava.

Reducing sugars: The increase of reducing sugars content by calcium application might be due to the less utilization of sugars in respiration and conversion of starch into sugars, while the subsequent decline was perhaps due to consumption of sugars for respiration during storage. Similar findings have been reported by Karemera and Habimana (2014) in mango.

Non-reducing sugars: Likewise, non-reducing sugars was found significantly maximum with $CaCl_2$ @ 1 per cent *i.e.* 9.73, 11.50 and 14.40 per cent at 2nd, 4th and 6th day of storage period, respectively. While, at 8th day of storage period the significantly maximum non-reducing sugars (8.00 %) was found with foliar spray of $Ca(NO_3)$ 2 @ 1 per cent. The increase in non-reducing sugars during storage was due to the conversion of starch into sugars. While, decrease in sugars is may be due to the consumption of sugars for respiration during storage period.

Ascorbic acid: The ascorbic acid content showed higher initially then subsequently decreases during storage period. The studies revealed that at 2nd, 4th, 6th and 8th day of storage period, and different chemicals recorded significant effect on the accumulation of ascorbic acid (mg/100 g pulp). The significantly maximum ascorbic acid (22.53, 20.83, 19.77 and18.00 mg /100g pulp) was recorded with T_2 ($CaCl_2$ @ 1%) at 2nd, 4th, 6th and 8th day of storage period, respectively. The decline in ascorbic acid content may be due to the degradation through enzymatic oxidation of L-ascorbic acid to dehydro ascorbic acid during metabolic process (Patel, 2017; Desai, 2016 and Bisen *et al.*, 2014).

Calcium content: The significantly maximum calcium content was recorded with $CaCl_2$ @ 1.5 per cent *i.e.* 9.90, 9.67, 8.40 and 8.10 mg/100 g pulp at 2nd, 4th, 6th and 8th day of storage period, respectively. The increment in calcium content might be due to the pre-harvest foliar application of $CaCl_2$ @ 1.5 per cent to sapota tree. The present investigation is in conformity with the results reported by Bisen *et al.* (2014) in guava and Wahdan *et al.* (2011) in mango.

Moisture content: A non-significant differences were noticed between treatments for moisture content of sapota fruits at 4th, 6th and 8th day of storage period. While, at 2nd day of storage period, the significantly minimum moisture content (63.78 per cent) was recorded with $CaCl_2$ @ 1 per cent. This decrease in moisture content during storage was due to the either evaporation losses or utilization of water in various senescence processes, may have also resulted in reduced firmness of the fruits. The fruits sprayed with calcium chloride retained higher moisture content probably by delaying senescence or by reducing evaporation losses by stabilizing the membrane permeability and cell wall structure (Patel *et al.*, 2017). The findings obtained in the present investigation can be compared to those obtained by Yadav *et al.* (2009) in ber.

Shelf-life of fruits: Different calcium salts have been reported for delaying prevention, sanitation and nutritional enrichment of fresh fruits. Calcium carbonate and calcium citrate are the main calcium salts added to foods in order to enhance the nutritional value (Brant, 2002). Other forms of calcium used in the food industry are calcium lactate, calcium chloride, calcium phosphate, calcium propionate and calcium gluconate, which are used more when the objective is the preservation and/or the enhancement of the product firmness. The selection of the appropriate source depends on several factors: bioavailability and solubility are the most significant, followed by flavour change and the interaction with food ingredients.

Calcium chloride: It is widely used as preservative and firming agent in the fruit industry for whole and fresh-cut commodities. Some scientists used calcium chloride as firming agent for processed strawberries. Effect of calcium chloride was known in grapefruit, showing significant improvement in the quality of the final product. The use of calcium chloride is associated with bitterness and off-flavours mainly due to the residual chlorine remaining on the surface of the product.

Calcium lactate, calcium propionate and calcium gluconate have shown some of the benefits of the use of calcium chloride, such as product firmness improvement, and avoid some of the disadvantages, such as bitterness and residual flavour. The effect of calcium lactate, calcium chloride and calcium propionate dipping in peach fruits was studied. Calcium increased in tissues with no dependence on the source used. Calcium incorporation by impregnation with two calcium sources, calcium lactate and calcium gluconate was proved useful in fresh-cut apples.

Methods for calcium application - minimally processed fruits

In general, there are two main ways of application of the calcium in fresh fruits have been reported: dipping washing (I) and impregnation (II) processes.

Dipping treatments: These are commonly used for fresh products, especially those more perishable, such as leafy vegetables. It usually consists of the soaking of the product, applying or not applying mechanical agitation, followed by the removal of excess washing solution. This treatment is gentler to the product than the impregnation techniques which can cause tissue damage and metabolic stress. The washing treatment is usually applied to the whole product or followed by other minimal procedures, including washing, peeling, cutting, shredding and / or slicing.

Dipping treatments favour the dispersion of the solution on the surface of the fruits. This method has an extra benefit since the enzymes and substrates released from the injured cells during the minimal procedure are rinsed, avoiding oxidation reactions that could lead to browning and off-flavours. Different factors (pH, immersion time, temperature, and concentration) can affect product integrity.

The dipping time ranges from 1 to 5 minutes in general. A time period of 5 min for immersion of whole peach fruits was effective. Longer periods were used with whole strawberries, lasting for 15 minutes. The effect of temperature has been shown to be of major importance in the results of dipping washing treatment. Cool, room and warm temperatures (4, 20-25, 40-60°C, respectively) have been used with different calcium solutions. The use of warm temperature (40-60°C) increased the beneficial effects of the treatment due to higher washing solution retention inside the product (Garcia *et al.*, 2020). The concentrations of the calcium salts used as washing treatments are usually within a range of 0.5-3 per cent.

Impregnation: The objective of these techniques is the modification of the composition of food material through partial water removal and impregnation with solutes, without affecting the material's integrity. The process driven forces can be osmotic gradient between the sample and solution, application of vacuum followed by atmospheric condition restoration, or both osmotic gradient and vacuum conditions. Depending on the time (minutes to days) of the process and the magnitude of the vacuum (5-200 mbar) or solution concentration (20e75 °Bx) used, the product obtained might lack the characteristics of a minimally processed product and rather be considered a dehydrated product.

Under vacuum conditions, when the porous tissue is immersed in the solution, the air is extracted from the pores; when the atmospheric pressure is restored, the solution penetrates into the intracellular spaces by capillary and pressure gradients. The partial substitution of the internal gas by the new liquid phase allows the reformulation of the food by the modification of the solid matrix, avoiding eventual stress due to long exposure to gradient solute concentration as in osmotic process. Also, to obtain product stability, impregnation techniques have been usedful to enhance the textural properties and/or fortify the product.

Impregnation provides broad applications in fruit processing. Some of the advantages are the development of reformulated products by impregnation with different sources of calcium, also improve the final product quality. Extensive work by applying vacuum impregnation in fresh-cut apple has obtained good results. Calcium lactate and calcium gluconate were selected at 5 per cent for 2-22 h. The final product obtained after the long periods used for the impregnation might not be defined as a minimally processed product. Calcium decreased the force needed to sample rupture and fracture occurred over a period of deformation, as in ductile materials. The vacuum impregnation in oranges and other fruits and vegetables using different calcium sources also proved useful.

Coating: Is also a technique that can be used to enrich minimally processed fruits with calcium. The strawberries and red raspberries were treated with chitosan-based coating formulated with calcium lactate and calcium gluconate. Adding calcium did not alter the positive effects of coating, which proved to extend the Shelf-life by decreasing the incidence of decay and weight loss, changes in colour, titratable acidity and pH.

Another technique that might have application for treating fruits and vegetables with calcium solutions is spraying over the product surface. Experiments with spray jets for low-pressure cleaning have been carried out to examine the spray cleaning effect of selected nozzles on vegetable surfaces. However, no conclusions on the application could be derived from these studies.

Effect of calcium treatments on quality, safety and nutritional issues

Role of calcium: Calcium is involved in maintaining the textural quality of produce since calcium ions form cross-links or bridges between free carboxyl groups of the pectin chains, resulting in strengthening of the cell wall (Garcia *et al.*, 2020). Calcium complexes to cell wall and middle lamella polygalacturonic acid residues and improve structural integrity. It has been suggested that 62.5 mM calcium chloride immersion treatment as a potential post-harvest treatment for whole peaches, since increased tissue firmness reduced the susceptibility to physiological disorders and reduced the risk of salt-related injures.

Complementing calcium treatments with mild heating (40-70°C, 5 min or less) is an interesting alternative. Time and temperature of the heat treatment must not compromise the fresh-like characteristics of the commodities. Heat can be applied with the calcium treatment by increasing the temperature of the calcium solution in contact with the fruit or vegetable, or at a different stage. Mild heat has been reported to activate the pectin methyl- esterase (PME).

Endogenous and added calcium can make plant tissue firmer by binding to the pectin carboxyl groups that are generated through the action of PME.

Calcium can also help to keep longer the fresh-like appearance of minimally processed fruits by controlling the development of browning. Control of the flesh browning has been observed in fruits viz., peaches and pineapple. Treatments with calcium inhibited colour changes and development of tissue translucency in honeydew chunks.

Another benefit derived from the use of calcium treatments is the incorporation of significant quantities of calcium to the fruit matrix. This is also one of the objectives and driving forces promoting the research on these types of treatments. The ability of fresh-cut apple matrix for calcium incorporation by impregnation techniques using calcium lactate and calcium gluconate was also studied. These findings explored the possibility of obtaining a product fortified with calcium. The calcium amount incorporated into the apple matrix would satisfy about 23.63 per cent (200 g of fruit) of the adequate intake.

Sensory analysis also showed that the use of calcium improved the textural characteristic of the product during storage. A combination of calcium lactate and heat-shock further increased the calcium content of the samples. Impregnation as a feasible technology can be used for making fruit and vegetable matrices suitable for fortified food development.

Importance of calcium in the industry: cost and benefits

Today's consumer has, on average, a busier lifestyle than in previous decades. Income, as measured by Gross National Product (GNP), has been rising over the past two decades, throughout the EU and the US. Both unemployment and interest rates have generally been declining throughout the 1990s, substantially increasing consumer's disposable income. This means that consumers have the ability to spend more on higher value or added-value foods, even though the current percentage spent on food from the total household budget has fallen over the past 30 years. This reflects the fact that people do not consume more food as their income rises, but rather tend to purchase higher value food. It is important to diversify the sources of calcium due to dynamic dietary changes, which transcend age, gender, ethnicity, geographic, and economic status, and creates the necessity for the consumer to have at their disposal multiple sources of calcium (dairy, non-dairy, fortified foods and supplemental).

The existing methods for application of the calcium treatment in fresh fruits and vegetables have already been studied on a significant number of commodities, showing positive results for the feasibility of this type of products. The fact that the calcium treatment is applied to keep the quality and properties of the product and not only for calcium enrichment purposes is an advantage for the implementation of this type of products. Also, the perspective of growth for the enriched food market makes this a very interesting opportunity from the industry's point of view.

Some of the purified calcium sources might result to be expensive, but the fact that the treatment is also adding value to the product is an advantage to balance the cost/ benefit rate. Other alternatives are calcium chelates, although more expensive than other salts. Calcium chelates are not corrosive to processing equipment and are more likely to penetrate deeply in plant tissues; however, problems associated with their low solubility have been reported.

Future research needs

From the point of view of the possibilities in the fortified food market, there is a wide range of opportunities to develop new products using calcium treatments, due to the high and increasing demand of fresh and healthy food. When using impregnation techniques, the behaviour of the fruit matrices is highly produce- specific. Differences in density of the solution and the commodity (usually lower) makes keeping in contact sample and solution more difficult. Approaches to solve the problem are the use of stirring, compressing or sinking rods. These methods can cause damage to the product and increase the cost of the process. More research might be necessary in this area to obtain acceptable levels of quality for fresh-like minimally processed products.

An important aspect needing further research is bioavailability of the calcium incorporated to the product. The bioavailability depends on a large number of factors, such as commodity pH or fibre content, and needs to be studied for each particular case when obtaining a fortified product. Also, the antimicrobial effect of the treatments based on calcium salts on fruits needs more study.

The markets for convenient, fresh and fortified foods are continuously growing. This extensive knowledge gathered and the further research proposed can be exploited for the development of novel fresh-like minimally processed fruits enriched in calcium that can take advantage of these three market niches.

Anti-oxidants

Salicylic acid (SA): Salicylic acid acts as natural antioxidants. Therefore, it helps in decreasing respiration rate, increasing antioxidant enzyme activity and maintaining quality attributes. Salicylic acid (SA) has a significant role in disease resistance and plant defence mechanisms. Salicylic acid including its derivatives are safe and natural alternatives for delaying the ripening and softening, retard membrane breakdown and mitigate chilling injury in several horticultural crops. This stress hormone (SA) plays a role in induction of resistance related enzymes and has been shown to suppress ethylene production, delaying pericarp browning.

It reduces stress lignification, and increases chilling tolerance in fruits". Application of salicylic acid influences the post-harvest process by increasing the shelf-life of horticultural crops including ornamentals, fruits and other crops. Post-harvest application of SA has been shown to reduce fruit rot, maintain quality, and increase the antioxidant activity of fruits. In horticultural crops, it is an endogenous plant growth regulator that delays the ripening process after harvest. It has been used on a variety of horticultural crops where it has shown the ability for extending shelf-life. It is related to the ethylene being inhibited during the ripening of fruit. The quality of horticultural goods, including their appearance, flavour, astringency, and bitterness, can be significantly altered by SA.

***Functions of salicylic acid*:** Salicylic acid plays a crucial role in the growth and development of the plant for important physiological functions, such as increasing the plants response to stress conditions (biotic and abiotic), increasing the plant's resistance to System Acquired Resistance (SAR), or both. These physiological functions are accomplished by stimulating or altering the internal endogenous signalling to withstand a variety of stresses. Additionally, it has the capacity to form conjugate bonds with certain amino acids, including proline and arginine, which improves the ability of plants to withstand adverse conditions and maintains systemic acquired resistance. Additionally, the SA plays a variety of significant physiological tasks, including promoting protein synthesis, delayed ripening of fruits, seed germination, immune response, nutrition transfer, CO_2 representation, ion absorption, blooming and stomata movement. It also advances the formation of various plant dyes by accelerating up their production and raising their levels, such as chlorophyll and carotene, and contributes to the accumulation of dry matter. In contrast to ABA's work, which is responsible for the fall of leaves, it also prevents the representation of ethylene gas. Additionally, it plays a significant part in boosting metabolic rates, which helps the plant save energy by using alternate routes and changing its amount of nucleic and amino acids.

***Physiological loss in weight (PLW)*:** Physiological loss in weight of fruits is due to loss of water as a result of ongoing respiration and transpiration even after harvest. Salicylic acid has been considered as a possible and effective treatment for senescence related enzyme activity suppression (Hanif *et al.*, 2020). This action lowers the rate of fruit respiration during storage, which in turn lowers the physiological loss in weight. Fruits that have lost water during storage not only lose weight but also lose quality and appearance due to shrivelling and withering. Post-harvest application of salicylic acid (SA@2mM) was effective in reducing the physiological loss in weight in guava i.e in Allahabad Safeda and Lalit as salicylic acid helps in suppressing the metabolic activity with respect to respiration and ethylene evolution. The application of salicylic acid @2mM showed reduction of physiological loss in weight in guava cv. Lalit under Iran conditions.

Minimum spoilage and reduction of physiological loss in weight in guava was observed by Kaur *et al*., 2019 with the application of salicylic acid (200 ppm) treated fruits under cold storage. After 28 days of storage, a minimum PLW of (5.23 per cent) was observed in fruits treated with SA 200 ppm compared with control. The studies were conducted in RRS, PAU, Bathinda, India.

Application of salicylic acid @2mM was effective in maintaining the quality of Murcott mandarin fruit during storage, in terms of reducing weight loss. Awad and Qurashi (2021) found that post-harvest dipping in 2 mM salicylic acid or 0.2 mM was effective in reducing the physiological loss in weight which improved the quality of Sensation mangoes at ambient conditions via inhibiting hydrolytic enzymes and enhancing antioxidant system of fruit. The experiment was conducted in a laboratory at King Abdulaziz University in Jeddah, Saudi Arabia. In pre- and post-harvest experiments, the physiological weight loss (PLW) was also less in SA treated peach (*Prunus persica* L.) cv. Shan -e-Punjab compared to untreated ones.

The application of SA after harvest decreased the PLW in fruit. Salicylic Acid inhibited the production of ethylene, hence preventing respiration. By closing stomata, salicylic acid may lower PLW and slow down respiration. Haider *et al*. (2020) observed that post-harvest application of salicylic acid in mandarin orange budded on 'Rough Lemon' rootstock was significantly effective to minimize the physiological loss in weight in comparison to control. The study was conducted in University of Agriculture, Faisalabad, Pakistan. Hazarika and Marak (2022) reported that fruits coated with salicylic acid @ 2 mM had considerably lower PLW after 4, 8, 12 and 16 days of storage, respectively. They also revealed that SA is responsible for reduction of weight loss in fruits and their respiration rate by closing stomata. After 40 days of storage, the least weight losses were observed in salicylic acid treatments (1.8%) in ber fruit. The salicylic acid treatments dramatically inhibited weight loss compared to that of control.

Firmness: Salicylic acid has been found effective in suppressing the wall degrading enzyme. It maintains the firmness of the fruits by decreasing the rate of ethylene production and suppressing the cell wall degrading enzymes. Post-harvest application of salicylic acid helped to maintain higher firmness in guava. SA treated fruits retained their firmness due to inhibition of cell wall and membrane degrading enzymes as polygalacturonase, lipoxygenase, cellulose, and pectin methyl esterase, as well as a lower rate of ethylene production.

Kaur *et al*. (2019) reported that post-harvest application of salicylic acid was found effective in maintaining higher firmness in guava. Fruit firmness is effectively maintained by post-harvest application of salicylic acid in guava as fruit softening was delayed by the SA immersion due to the control of soluble pectin increase and insoluble pectin decrease. The post-harvest application of salicylic acid could maintain the fruit firmness which was conducted in Kurdistan University, Sanandaj, Iran. Salicylic acid treatments slowed down fruit softening possibly by suppression of ethylene production and has also shown delay in ripening and suppress ethylene production and respiration rates during ripening at ambient conditions in mango as studied by Awad and Qurashi 2020 conducted at King Abdulaziz University in Jeddah, Iran. Salicylic acid therapy applied after harvest at a concentration of 200 ppm was the most successful in slowing the rate of firmness loss in peach fruits. The current studies were conducted on fruit from experimental peach (*Prunus persica* L.) cv. Shan-e-Punjab trees that were 6 years old in 2011 at the Regional Research Station, Punjab Agricultural University, Bathinda India.

Hanif *et al*. (2020) found that the fruit freshness of stored papaya reduced during cold storage. However, fruit treated with SA at 1.5 mmol/l, remained more firm throughout storage, indicating that this dosage was most effective in inhibiting the enzyme that breaks down cell walls. The research was carried out in the Post-harvest Research Center's lab at the Ayub Agriculture Research Institute in Faisalabad, Pakistan.

***Chilling injury*:** Post-harvest application of salicylic acid on ripening proccss helps in curbing the ethylene evolution and senescence further resulting in lower chilling injury of the fruits and less production of free radicals in guava which was studied by Madhav *et al*. (2021). The induction of heat shock proteins in salicylic acid treated guava imparts chilling tolerance. Sayyari *et al*., (2009) found that post-harvest treatment of salicylic acid @ 2 mM was the most effective for reducing chilling injury and for maintenance of ascorbic acid levels in pomegranate.

Colour: Madhav *et al*. (2021) studied that post-harvest application of SA or SSA has delayed the ripening process of guava fruits significantly, probably through inhibition of ethylene production, which was conducted in The Division of Food Science and Post-harvest Technology, IARI, New Delhi, India. Luo *et al*., (2011) found that hue value decreases less when fruit is treated with salicylic acid @2mM treatment for all fruit maturity stages. The results are due to the effect of the presence of salicylic acid which delay fruit ripening by reducing respiration rate and inhibit ethylene biosynthesis. Post-harvest application of salicylic acid delayed the development of yellow color in guava cultivars signifying a delay in the respiratory processes. The delay in ripening process was mainly due to the inhibitory effect of salicylic acid on ethylene synthesis and the barrier properties of vegetable wax formulation on gaseous exchange further resulting in suppressed activities of enzymes such as ACC synthase and ACC oxidase. Similar effect of delayed transition of color due to salicylic acid treatment has also been reported in guava fruits" by Lu *et al*. (2011) and wax coated guava by Madhav *et al*. (2021).

TSS: Application of salicylic acid @2 mM as post-harvest treatment was effective in improving the total soluble solids. The increase in TSS can be attributed to the conversion of starch to sugar during ripening of guava. Since salicylic acid might have suppressed the catabolic processes such as respiration rate and ethylene production in treated guava fruits, those fruits showed a slow increase.

Phenol: Salicylic acid, as post-harvest application, was effective in decreasing the total phenol content slightly to inhibit PPO by application of salicylic acid during ripening that reflects to increase fruit color quality as studied by Lu *et al*. (2011). The post-harvest application of salicylic acid was effective in decreasing the total phenolic content of guava within the storage period. Reduced loss of phenols in salicylic acid treated fruits may be due to delayed oxidation of phenolic substances by polyphenol oxidase (PPO). Similar findings were reported earlier by Khademi and Ershadi (2013). The apricots treated with 3.0 mM salicylic acid had less phenolic content while being stored. The formation of all significant phenolics, the production of new polyphenols, and the stimulation of phenylalanine ammonia lyase activity in apricot were all produced as a result of salicylic acid treatment. The study was conducted at Tabriz University, and the research was done on Asgar-Abad, a popular commercial cultivar of Iranian apricot. Khademi and Ershadi (2013) found that in comparison to controls, fruits treated with SA at high doses of 2 and 4 mM showed the highest levels of TPC. During the first two weeks of storage, fruits treated with SA at 1 mM were found to contain more TPC than controls, but later on, there was no discernible difference from controls. After storage, total phenols showed a downward trend, and the rate of decline was noticeably greater in SA-treated fruits than in control ones (SA at a concentration of 4 mM). Many enzymatic and non-enzymatic processes have been linked to the loss of phenolic chemicals during storage. The two main enzymes in charge of oxidising phenolic compounds are peroxidase (POD) and polyphenoloxidase (PPO).

Vitamin C: Tareen *et al.* (2012) found that post-harvest application of salicylic acid had a significant effect on maintaining higher content of ascorbic acid in peach fruits. The study was conducted on peach (*Prunus persica* L. Batsch) fruits of cv. Flordaking grafted on wild rootstock at the Post-harvest Laboratory of the Department of Horticulture, Pir Mehar Ali Shah, Arid Agriculture University Rawalpindi, Pakistan.

Salicylic acid has a significant potential for reducing post-harvest losses of horticulture crops since it is a natural and safe phenolic chemical. Following SA treatment, the main outcomes included a reduction in ethylene production and action, induction of disease resistance, prevention of oxidative stresses, induction of crop tolerance to chilling injury, reduction in respiration rate, reduction in ripening and senescence rate, prevention of cell wall degrading enzymes, and maintenance of fruit firmness. SA can be utilised as a suitable substitute for chemicals in horticulture crops post-harvest technologies to ensure food safety. Since SA, like any other post-harvest treatment, may have different effects on different crops at different circumstances, it is necessary to determine the best and safe concentration for each fruit crop and its variety.

Polyphenols are one of the most important antioxidants in plants and are often found as flavonoids, gingerols, poly- phenolic acid, and curcumin. They contain aromatic rings with -OH or $-OCH_3$ substituents that play a role in antioxidant activity. Polyphenols can also damage Reactive Oxygen Spieces (ROS) species and inhibit lipid peroxidation by binding to lipid alkoxyl radicals. Flavonoids are also known to be oxidized by peroxidase and play a role in H O degradation.

Another phenolic compound that can act as an antioxidant is curcumin. Curcumin is the main component of *Curcuma longa* L. that works as an antioxidant by the electron transfer mechanism and also protects membranes from peroxidative damage. Curcumin includes many functional groups, including the β-keto group, carbon-carbon double bond, and phenyl rings. Curcumin is degradable to ferulic acid and vanillin, two well-known antioxidants.

In addition to polyphenols, protein, vitamins and minerals can also act as antioxidants. Proteins can inhibit lipid oxidation, one way of which is by using iron-binding variants. Protein also inactivates ROS, scavenges free radicals, and reduces hydroperoxides. Vitamins and minerals are micro-nutrients needed in small amounts. One of the most well-studied vitamins with antioxidant properties is vitamin C or ascorbic acid. Vitamin C works as an antioxidant in the aqueous phase of the cell and circulatory system by donating high-energy electrons to neutralize free radicals. It also plays a role in regenerating the antioxidant form of vitamin E by reducing tocopheroxyl radicals. Other antioxidants such as albumin, hepatoglobin, and nonprotein antioxidants such as bilirubin and ubiquinol can also be found in plants.

ROS and antioxidant in post-harvest

There are several ways to slow decay in fruits and vegetables such as treatment with low temperatures, atmosphere control, hormones and coating, which are summarized in Table 1. The cooling process can reduce the metabolic rate of fruits and vegetables. The cooling process in peaches can reduce the accumulation of ROS and reduce the level of membrane lipid peroxidation; thus, it can delay the senescence of peaches. Cold storage of the peach at 5°C prompted higher activity of antioxidant enzymes, such as succinic dehydrogenase (SDH) and cytochrome C oxidase (CCO).

Table 1: Nature of storage losses in some commercially important fruits.

Post-harvest loss	**Crops**
Chilling Injury	Banana, citrus, guava, mango, papaya, pomegranate, sapota
Rind breakdown	Citrus
Loss of colour	Litchi
Desiccation and decay	Litchi, pomegranate
Internal breakdown	Pineapple
Crown dehydration	Pomegranate

Decay can also be prevented by atmospheric modification. The air composition normally consists of O_2 (20%), CO_2 (0.03 %), and N_2 (78.8 %). Modification of air composition is done by reducing oxygen levels or increasing carbon dioxide content. The low oxygen content in the air and the increase in carbon dioxide content result in a decrease in the rate of the respiratory activity of vegetables and fruits. Reduction in the rate of respiratory activity can then decrease production of ROS, because these species are produced mainly as the by-product of the respiration process. Therefore, the cooling process and modification of air composition can prevent damage caused by ROS and slow decay in fruits and vegetables. Modified Atmosphere Packaging (MAP) lowered the production rate of superoxide and hydrogen peroxide. It can also maintain high activities of superoxide dismutase (SOD), catalase (CAT), and ascorbate peroxidase (APX).

Environment friendly edible coating technology is currently being developed to overcome the problem of spoilage in post-harvest vegetables, and one of the most promising options is chitosan coating. Chitosan is a by-product of the maritime industry, obtained from marine animal waste that is biodegradable and safe to the environment. It is able to form a barrier on the surface of vegetables or fruits with the effect of reducing respiration rate and loss of water to prevent spoilage and it might decrease the production of ROS. Another characteristic of chitosan is the ability to increase antioxidant enzyme activity.

An antagonistic effect of treating fruit with yeasts has also been shown to reduce ROS and increase antioxidant activity. Harvested citrus fruits have been treated with co-dipping in *Pichia membranaefaciens* and pathogen fungi of *Penicillium italicum* and *Penicillium digitatum*. The results showed that both treatments were able to decrease disease incidence and lesion diameters. It was suggested that this effect resulted from the regulated ROS content (superoxide and hydrogen peroxide) and increased antioxidant activities (CAT and SOD). The ROS-mediated action of yeast antagonists *Metschnikowia fructicola* (strain 277) and *Candida oleophila* (strain 182) was elaborately studied in post-harvest treatment of citrus fruits and apples. The treatment of inoculating both yeasts onto citrus and apple wounds increased hydrogen peroxide content about fourfold after 18 h inoculation. Living *M. fructicola* cells were also found around the wound, indicating the yeast's ability to tolerate elevated ROS content. Thus, the use of some species of yeasts could be a solution to defeat fungi attack through the regulation of ROS content in the fruit cells.

Treatment of post-harvest commodities with hormones has also shown some beneficial effects, which are understood to be the consequence of ROS regulation and increased antioxidant capacity. The use of acibenzolar-S-methyl (ASM), a functional analog of salicylic acid, decreased lesions caused by *Trichothecium roseum* in muskmelon (*Cucumis melo* L. cv. Yujingxiang).

This effect was correlated with elevated concentrations of H_2O_2 and O_2, increased activities of NADPH oxidase (NOX), SOD, and APX, and inhibition of CAT. Further, salicylic acid treatment of cornelian cherry fruit (*Cornus mas*) significantly increased DPPH scavenging activity. This led to higher total phenols, flavonoids, anthocyanins, and ascorbic acid contents and enhanced PAL enzyme activity. Senescence in cornelian cherry fruit was also delayed by the application of sodium nitroprusside (SNP) and γ-aminobutyric acid (GABA) in the post-harvest cold storage condition. Both chemicals reduce

H_2O_2 production, thus further ameliorating fruit browning. This outcome was also a consequence of improved activity of anti- oxidant enzymes, such as SOD, CAT, APX, and GR.

The application of animal analogous compounds has gained popularity in treating post-harvest fruits. Once thought to contribute only to reducing oxidative stress in animals, sodium para-aminosalicylate is also able to inhibit ROS- mediated processes in post-harvest litchi fruit. This tuberculosis drug delays pericarp ripening during senescence. Similarly, melatonin, the famous sleep hormone in mammals, is capable of regulating ROS and delaying post-harvest senescence in peach fruit. Melatonin occurs widely and naturally in plants and can be extracted and upscaled for various applications.

Inorganic chemicals and moderate heat shock have also been described to extend shelf-life of vegetables by modulating ROS. Biofortification of 15 mM potassium in KCl of tomato (*Solanum lycopersicum* L.) improved the quality during post-harvest storage at 4°C through limited loss of weight, water, and antioxidant content. Fumigation of fresh broccoli (*Brassica oleracea*) with hydrogen sulfide (H_2S) was also able to improve activity of antioxidant enzymes, while reducing ROS production. Prior to cold storage, heat shock treatment of spinach (*Spinacia oleracea*) at 40°C delayed leaf senescence and increased ascorbic acid content.

More sophisticated means of ROS post-harvest management are genetic engineering and nanotechnology. In cassava (*Manihot esculenta* Crantz), the rapid post-harvest physiological deterioration hurts farmers almost every harvest season. This problem is caused mainly by the accumulation of phenols and their oxidation which alters taste and instills brown color in the roots. The SOD and CAT overproduced by inserting an extra copy each of Me-*Cu/ZnSOD* and Me- *CAT1* into the cassava genome, which is controlled by a vascular-specific promoter p54. The transgenic plants showed a superior characteristic in terms of enzymatic activity of SOD and CAT. Further, reduced concentrations of malondialdehyde, lipid peroxidation and H_2O_2 were observed in the transgenic line as compared to the wild variant. The use of nanocomposite packaging for post-harvest enoki mushroom (*Flammulina velutipes*) has also significantly decreased ROS accumulation and lipid peroxidation activity. The nanocomposite packaging material contains 35 per cent weight. Nano-TiO_2 in the range of 40–60 nm is well known for its photocatalytic capability in degrading ethylene, a fruit-ripening hormone.

Enzymes

Endo polygalactouronase: Apart from PG, Exo-PG is also found in fruits through it is not ripening-specific. The presence of PG and exo-PG is species dependent. Both activities are present in peach. Many fruits showed relatively low or hardly detectable levels of extractable (PG) activity, probably due to the enxyme instability, but the activity or the mRNA was detected later on i.e. apple, persimmon and strawberry. Even low activity levels may suffice to produce extensive pectin disassembly but the total quantity of polyuronide residues and polyuronide length are only slightly modified in fruits like strawberry. While the largest cell wall change during strawberry ripening is described as pectin solubilisation, this occurs in apparent absence of pectin polymer digestion. Only the strawberry hemicellulose polymers appear to be broken down during fruit ripening.

The transgenic apple trees were produced that contained additional copies of a fruit-specific apple PG gene under a constitutive promoter. The PG over expression caused various novel phenotypes which included silvery colored leaves and premature leaf shedding due to reduced cell adhesion in leaf abscission zone. Transgenic modification of apple trees by a single PG gene has provided a new outlook on the role of pectin and cell wall adhesion in organs other than fruits.

Pectin methyl esterase: Pectin Methyl Esterase (PME) catalyses the hydrolysis of galactosyluronate methyl esters that results in the de-esterification of pectins. PME, a ubiquitous enzyme in plant tissues is plentiful in many fruits viz., peach.

Pectate lyase: Clones with homology to PL are abundant in banana, a climacteric fruit and also in non-climacteric fruits such as grape berry and strawberry suggesting a role for the putative enzymes relevant to wall disassembly during ripening. Two different banana cDNAs, MWPL1 and MWPL2 which were found to be full length clones were also reported in literature. Those cDNAs showed distinct temporal and spatial expression. In ripening fruit, both cDNAs were expressed in pulp and peel, and their expression increased following ethylene treatment. A previously reported cDNA, Ban17 showed high homology to MWPL1 and its mRNA also accumulated in green bananas in response to exogenous ethylene. A single PL cDNA fragment, named VvPL1 was obtained from grape berry. Relatively high levels of transcript were found in the period following veraison but PL mRNA was also present during the first month after flowering.

In the strawberry plant, the expression of the putative PL gene occurs predominantly in the receptacle and seems to be under the control of auxins. Transgenic strawberry plant incorporating an antisense sequence of a PL gene have been obtained. These plants displayed a lower degree of in vitro cell wall swelling, a lower amount of ionically bount pectins, and a significant reduction of softening during ripening and post-harvest storage. Thus, the eliminative cleavage performed by this enzyme may play a key role in wall pectin modification in strawberry that acquire a melting texture in very short period. Endo-1,4-β-D-glucanase (EGase) activity has been associated with modifications in cell wall structure that accompany several processes, including fruit growth and ripening. The EGases are usually present as a divergent family of enzymes with several molecular forms. Different isoenzymes have been found in avocado, the latter showing particularly high activity levels when extracts of fruit proteins were assayed against carboxymethylcellulose, a cellulose like substrate often used in assays.

A role for EGases in cell wall dissolution processes was suggested for avocado fruit. In peach fruit, two different EGases were identified. One form was present in early growth of the fruit while the other one was abundant during ripening. The second activity was increased by propylene treatment.

In contrast to climacteric fruit, the regulation of strawberry fruit ripening related EGase gene expression was found to associated with auxin levels. Nevertheless, ethylene can exert marked increases in EGase mRNA from other tissues of non-climacteric species, as happens in calyx abscission zones of the Valencia orange. In addition, auxins regulate the expression of some members of the EGase gene family during growth of climacteric fruit.

Xyloglucan endotransglycosylase: XET activity was identified in different fruits among them, kiwifruit and apple. XET activity in kiwifruit was found to be tissue dependent, that is, not uniformly distributed throughout the different parts of the fruit. In grape berry, two cDNA fragments encoding XET (VvXET1 and VvXET2) have been described. Their transcripts were present at relatively high levels during ripening though they were also detected before veraison. In kiwifruit, a ripening related XET was detected at relatively high activity levels and found to be regulated by ethylene.

References

Abdel-Salam M M, 2016. Effect of foliar application of salicylic acid and micronutrients on the berries quality of "Bez El Naka" local grape cultivar. *Middle East J. Appl. Sci.* 6(1):178-188.

Aghdam M S, Asghari M, Babalar M and Sarcheshmeh MAA, 2016. Impact of salicylic acid on post-harvest physiology of fruits and vegetables. In Eco-friendly technology for post-harvest produce quality. *Academic Press.* 243-268.

Ali A, Rasool K, Ganai N A and Wani A H, 2022. Role of salicylic acid in post-harvest management of peach and plum fruits: A review.

Amanullah S, Sajid M, Qamar M B and Ahmad S, 2017. Post-harvest treatment of salicylic acid on guava to enhance the Shelf-life at ambient temperature. *International Journal of Biosciences.* 10(3):92-106.

Asghari M and Aghdam M S, 2010. Impact of salicylic acid on post-harvest physiology of horticultural crops. *Trends in Food Science and Technology*, 21(10):502-509.

Awad M A and Al-Qurashi A D, 2021. Post-harvest salicylic acid and melatonin dipping delay ripening and improve quality of 'Sensation' mangoes. *Philipp. Agric. Sci.,* 104:34- 44.

Babalar M, Asghari M, Talaei A and Khosroshahi A, 2007. Effect of pre- and post-harvest salicylic acid treatment on ethylene production, fungal decay and overall quality of Selva strawberry fruit. *Food Chemistry*, 105(2): 449- 453.

Baninaiem E, Mirzaaliandastjerdi A M, Rastegar S and Abbaszade K H, 2016. Effect of pre- and post-harvest salicylic acid treatment on quality characteristics of tomato during cold storage. *Advances in Horticultural Science*, 30(3):183-192.

Barman K, Sharma S, Kumari P and Siddiqui M W, 2016. Salicylic acid. Post-harvest management approaches for maintaining quality of fresh produce. 51-68.

Bhalerao R R, Parmar B R, Padhiar B V and Bhalerao P P, 2010. Pre- harvest spray of different sources of calcium to improve the bio chemical qualities of sapota fruits [*Manilkara achras* (Mill.) Forsberg] cv. Kalipatti. *The Asian Journal of Horticulture*, 5(1):93-95.

Bisen S, Thakur R S and Tembhare D, 2014. Effect of pre-harvest spray of calcium nitrate and gibberellic acid alone and in combination on growth, yield and post-harvest behaviour of guava fruits. *An International Quarterly Journal of Environmental Sciences*, 5:56-62.

Champa W A, Gill M I S, Mahajan B V C and Arora N K, 2015. Reharvest salicylic acid treatments to improve quality and post-harvest life of table grapes (*Vitis vinifera* L.) cv. flame seedless. *Journal of Food Science and Technology*, 52(6):3607-3616.

Chen C, Sun C, Wang Y, Gong H, Zhang A, Yang Y, *et al.*, 2023. The pre-harvest and post-harvest application of salicylic acid and its derivatives on storage of fruit and vegetables: A review. *Scientia Horticulturae*. 312:111858.

Desai V N, 2016. Influence of pre-harvest spraying treatments of chemicals and plant growth regulators on quality parameters and Shelf-life of sapota [*Manilkara achras* (Mill.) Forsberg] fruits cv. Kalipatti.M.Sc. Thesis Submitted to Anand Agricultural University, Anand.

Dokhanieh A Y, Aghdam M S, Fard J R and Hassanpour H, 2013. Post-harvest salicylic acid treatment enhances antioxidant potential of cornelian cherry fruit. *Scientia Horticulturae*. 154:31-36.

Etemadipoor R, Dastjerdi A M, Ramezanian A and Ehteshami S, 2020. Ameliorative effect of gum arabic, oleic acid and/or cinnamon essential oil on chilling injury and quality loss of guava fruit. *Scientia Horticulturae*. 266:109255.

Ezzat A, Ammar A, Szabó Z and Holb I, 2017. Salicylic acid treatment saves quality and enhances antioxidant properties of apricot fruit. *Horticultural Science*, 44(2):73- 81.

García-Pastor M E, Zapata P J, Castillo S, Martínez-Romero D, Guillén F, Valero D and Serrano M, 2020. The effects of salicylic acid and its derivatives on increasing pomegranate fruit quality and bioactive compounds at harvest and during storage. *Frontiers in Plant Science*, 668.

Haider S A, *et al.*, 2020. Effects of salicylic acid on post-harvest fruit quality of "Kinnow" mandarin under cold storage. *Scientia Horticulturae*, 259:108843.

Hajilou J and Fakhimrezaei S, 2013. Effects of post-harvest calcium chloride or salicylic acid treatments on the shelf-life and quality of apricot fruit. *The Journal of Horticultural Science and Biotechnology*, 88(5): 600-601.

Hanif A, Ahmad S, Jaskani M J and Ahmad R, 2020. Papaya treatment with putrescine maintained the overall quality and promoted the antioxidative enzyme activities of the stored fruit. *Scientia Horticulturae*, 268:109367.

Hanif A, *et al.*, 2020. Post-harvest application of salicylic acid reduced decay and enhanced storage life of papaya fruit during cold storage. *Journal of Food Measurement and Characterization*, 14(6): 3078- 3088.

Hassoon S A and Abdulsattar Abduljabbar I, 2020. Review on the role of salicylic acid in plants. *Intech* Open; DOI: 10.5772/intechopen.89107

Hazarika TK, Lalrinfeli L, Lalchhanmawia Jonathan and Mandal Debashis, 2019. Alteration of quality attributes and shelf-life in strawberry (*Fragaria× ananassa*) fruits during storage as influenced by edible coatings. *Indian Journal of Agricultural Sciences*. 89(1):28-34.

Hazarika T K and Marak T, 2022. Salicylic acid and oxalic acid in enhancing the quality and extending the Shelf-life of grape cv. Thompson seedless. *Food Science and Technology International,* 28(6):463-475.

Junmatong C, Faiyue B, Rotarayanont S, Uthaibutra J, Boonyakiat D and Saengnil K, 2015. Cold storage in salicylic acid increases enzymatic and non-enzymatic antioxidants of Nam Dok Mai No. 4 mango fruit. *Science Asia.* 41(12):12-21.

Kapoor H, Baswal A K and Vishavjeet Jakhar A G, 2022. Effect of pre-and or post-harvest application of salicylic acid and ascorbic acid on post-harvest quality of horticultural crops: A review.

Karemera and Habimana, Karemera N J U and Habimana S, 2014. Effect of pre-harvest spray of calcium chloride on post-harvest behavior of mango fruits (*Mangifera indica* L.) cv. Alphoso. Universal *Journal of Agricultural Research*, 2(3):119-125.

Kaur M, Singh G and Akshi M, 2019. Post-harvest treatment for improvement of Shelf-life in guava (*Psidium guajava* L.) cultivar 'Allahabad Safeda under ambient storage condition. *Plant Arch.* 19:3005-3010.

Khademi Z and Ershadi A, 2013. Post-harvest application of salicylic acid improves storability of peach (*Prunus persica cv. Elberta*) Fruits. International *Journal of Agriculture and Crop Sciences*, 5(6):651.

Khalil H, 2014. Effects of pre-and post-harvest salicylic acid application on quality and Shelf-life of 'Flame seedless' grapes. *European J. Hortic. Sci.* 79:8-15.

Kumar N, Tokas J, Kumar P, Singal H R and Jayanti Tokas C, 2018. Effect of salicylic acid on post-harvest quality of tomato (*Solanum lycopersicum* L.) fruit. *International Journal of Chemical Studies*. 6(1):1744-1747.

Lakshmana and Reddy T V, 1999. Effect of pre-harvest foliar application of calcium on the post-harvest behaviour of sapota cv. Kalipatti fruits. *South Indian Horticulture*, 47(1-6):147-148.

Lu X, Sun D, Li Y, Shi W and Sun G, 2011. Pre-and post-harvest salicylic acid treatments alleviate internal browning and maintain the quality of winter pineapple fruit. *Scientia Horticulturae*. 130(1): 97-101.

Luo Z, Chen C and Xie J, 2011. Effect of salicylic acid treatment on alleviating post-harvest chilling injury of 'Qingnai' plum fruit. *Post-harvest Biology and Technology*. 62(2):115-120.

Madhav J V, Sethi S, Sharma R R, Nagaraja A and Varghese E, 2021. Use of salicylic acid for alleviation of chilling injury and quality assurance of guava fruits during storage. *Indian Journal of Horticulture*, 78(4):439-444.

Madhav J V, Sethi S, Sharma R R and Rudra S G, 2016. Impact of salicylic acid on the physiological and quality attributes of guava (*Psidium guajava*) fruits during storage at low temperature. Indian *Journal of Agricultural Sciences*, 86(9):1172- 1174.

Mandal D, Pachuau L, Hazarika T K and Shukla A C, 2018. Post-harvest application of salicylic acid enhanced Shelf-life and maintained quality of local mango cv Rangkuai of Mizoram at ambient storage condition. *Environment and Ecology*. 36(4):1057-1062.

Martínez-Esplá A, Serrano M, Valero D, Martínez-Romero D, Castillo S and Zapata P J, 2017. Enhancement of antioxidant systems and storability of two plum cultivars by preharvest treatments with salicylates. *International Journal of Molecular Sciences*. 18(9):1911.

Mo Y, Gong D, Liang G, Han R, Xie J and Li W, 2008. Enhanced preservation effects of sugar apple fruits by salicylic acid treatment during post-harvest storage. *Journal of the Science of Food and Agriculture*, 88(15):2693-2699.

Patel H A, Patel M J, Riddhi Vasara, N G Patel and N K Sutariya, 2017. Effect of pre- harvest spray of calcium on bio-chemical parameters of sapota [*Manilkara achras* (Mill.) Forsberg] Fruits cv. Kalipatti. *Journal of Pharmacognosy and Phytochemistry*, 6(5): 712-715.

Rajkumar M, Karuppaiah P and Andasamy R K, 2006. Effect of pre-harvest application of calcium on storage behaviour, ripening and shelflife of papaya (*Carica papaya* L.). *International Journal of Agriculture Sciences*, 2(2):480-482.

Ramakrishna M, Haribabu K, Reddy Y N and Purushotham, 2001. Effect of pre-harvest application of calcium on physico-chemical changes during ripening and storage of papaya. *Indian Journal of Horticulture*, 58(1):228- 231.

Ranjbaran E, Sarikhani H, Wakana A and Bakhshi D, 2011. Effect of salicylic acid on storage life and post-harvest quality of grape (*Vitis vinifera* L. cv. Bidaneh Sefid).

Rasouli M, Saba M K and Ramezanian A, 2019. Inhibitory effect of salicylic acid and aloe vera gel edible coating on microbial load and chilling injury of orange fruit. *Scientia Horticulturae*, 247:27-34.

Razavi F, Hajilou J, Dehgan G, Nagshi R and Turchi M, 2014. Enhancement of post-harvest quality of peach fruit by salicylic acid treatment. *Inter. J. Biosci*, 4(1):177-184.

Sarmin RA, Khan SAKU, Fatema K and Sultana S, 2018. Effect of neem leaf and banana pulp extracts on Shelf-life and quality of mango (*Mangifera indica* L.): Effect of plant extract on Shelf-life of mango. *Journal of the Bangladesh Agricultural University*. 16(3):343-350.

Sayyari M, *et al*., 2009. Effect of salicylic acid treatment on reducing chilling injury in stored pomegranates. *Post-harvest Biology and Technology*. 53(3):152-154.

Supapvanich S and Promyou S, 2013. Efficiency of salicylic acid application on post-harvest perishable crops. In Salicylic acid Springer, *Dordrecht*. 339-355.

Tahir HM, Pervez N, Nadeem J, Khan A A and Hassan Z, 2019. Esculent coating of spider silk enhanced the preservation and Shelf-life of apricot. *Brazilian Journal of Biology*, 80:115-121.

Tareen M J, Abbasi N A and Hafiz I A, 2012. Effect of salicylic acid treatments on storage life of peach fruits cv. 'Flordaking'. *Pakistan Journal of Botany*, 44(1):119-124.

Tareen MJ, *et al*., 2012. Post-harvest application of salicylic acid enhanced antioxidant enzyme activity and maintained quality of peach cv. 'Flordaking' fruit during storage. *Scientia Horticulturae*, 142: 221-228.

Vidya A, Swamy G S K, Prakash N B, Jagadeesh R C, Jagadesh S L, Gangadharappa P M and Mukesh L C, 2014. Effect of pre-harvest spray of nutrients on the physico-chemical characters in mango (*Mangifera indica* L.) cv. Mallika. *Mysore Journal of Agricultural. Sciences*. 2014; 48(4):529-533.

Wahdan M T, Habib S E, Bassal M A and Qaoud E M, 2011. Effect of some chemicals on growth, fruiting, yield and fruit quality of "Succary Abiad" mango cv. *Journal of American Science*. 7(2):651-658.

Wang J, Allan AC, Wang WQ and Yin X R, 2022. The effects of salicylic acid on quality control of horticultural commodities. *New Zealand Journal of Crop and Horticultural Science*, 50(2-3):99-117.

Wang J, Lewis D, Shi R, McGhie T, Wang L, *et al.*, 2022. The colour variations of flowers in wild Paeonia delavayi plants are determined by four classes of plant pigments. *New Zealand Journal of Crop and Horticultural Science*, 50(1):69- 84.

Wang L and Li S, 2008. Role of salicylic acid in post-harvest physiology. Fresh Produce. 2(1):1-5.

Wang Z, Ma L, Zhang X, Xu L, Cao J and Jiang W, 2015. The effect of exogenous salicylic acid on antioxidant activity, bioactive compounds and antioxidant system in apricot fruit. *Scientia Horticulturae*. 181:113-120.

Win S T and Setha S, 2022. Enhancement of anti- inflammatory and antioxidant activities of mango fruit by pre-and post-harvest application of salicylic acid. *Horticulturae*. 2022; 8(6): 555.

Yadav I, Sharma R K, Goyal R K and Siddiqui S, 2009. Effect of pre-harvest sprays of calcium on the storage of ber (*Zizyphus mauritiana* Lamk.) cv. Umran, *Haryana Journal of Horticultural Sciences*. 38(3&4):243-246.

Zheng Y and Zhang Q, 2002. Effects of polyamines and salicylic acid on post-harvest storage of Ponkan mandarin. In XXVI International Horticultural Congress: Citrus and Other Subtropical and Tropical Fruit Crops: Issues, Advances. 632:317-320.

4

Physiological Changes, Hardening and Regulation of Ripening in Fruit Crops

Karanbir Singh Gill

Department of Fruit Science, Punjab Agricultural University
Ludhiana-141 004, Punjab

Fruits, the vivid and delightful offerings of the nature, undergo remarkable transformations as they travel from the orchard to our dining tables. These changes indicate the significance of post-harvest physiology, which is the scientific study that reveals the secrets of how fruits behave after they are picked, including the mechanisms that affect fruit quality and how long it may be stored. The next section provides definitions of the various stages and key players involved in the processes of fruit maturity, ripening, and post-harvest physiology. These topics will be further explored in the subsequent sub-sections of this chapter.

Fruit developmental physiology

The life cycle of horticultural commodities (fruits and vegetables) can be effectively categorized into three primary physiological stages after fruit set or initiation, namely growth, maturation, and senescence. Still, a lack of obvious differentiation exists among the several stages. Growth encompasses the process of cell division followed by cell enlargement, which determines the ultimate size of the product. Maturation typically begins prior to the cessation of growth and involves various activities across different commodities. The stages of growth and maturity are commonly denoted as the developmental phase. Senescence can be described as the phase in which anabolic biochemical processes transition into catabolic processes, resulting in the ageing and eventual demise of the tissue. Ripening, a term specifically used for fruit, is typically regarded as the initial phase of senescence that occurs during the later phases of development. Defining the transition from growth to senescence is rather straightforward. The maturation phase is frequently characterized as the interval between these two stages, lacking a precise biochemical or physiological definition. The process of fruit development and maturation is dependent upon its attachment to the plant, but ripening and senescence can occur either on or off the plant. Fruits are typically collected at either the mature or ripe stage, although certain fruits, like zucchini, which are commonly consumed as vegetables, may be harvested prior to reaching maturity. Vegetables are typically harvested at various stages of physiological development, spanning from pre-maturation to senescence, with the exception of the ripening period.

Hardening

A freshly harvested apple gleams with promise, but beneath its glossy skin lies vulnerability. Once detached from the tree, fruits lose their protective network of living tissues. They become isolated, exposed to environmental stressors, and become susceptible to decay. This is where hardening steps in a natural defence mechanism that fortifies fruits against adversity. Hardening, also known as conditioning or pre-storage treatment, occurs during the early post-harvest phase. It involves subjecting fruits to specific conditions usually low temperatures for a brief period with the aim to prepare them for the stresses during storage and transportation. During hardening fruits are chilled to temperatures slightly above freezing. This induces physiological changes that enhance their resilience. The duration of low temperature exposure depends upon the fruit type *e.g.* for apples, it might be a few days; for tomatoes, just a few hours. Hardening walks a tightrope between preservation and ripening. While it toughens fruits, it must not hinder their eventual ripening process. If hardening persists too long, it delays ripening, affecting flavor and texture. Striking the right balance is crucial. Hardening triggers a cascade of physiological changes or responses within the fruit.

Cell membrane adjustments: As temperatures drop, cell membranes undergo alterations. Lipids within the membranes rearrange, making them more robust. This fortification shields cells from chilling injury and desiccation.

Enzyme activation: Certain enzymes become more active during hardening. These enzymes—such as peroxidase and superoxide dismutase—play critical roles in antioxidant defense and stress tolerance.

Sugar accumulation: Fruits accumulate soluble sugars during hardening. These sugars act as natural cryoprotectants, preventing ice crystal formation within cells.

Phenolic compounds: Phenolics, abundant in fruits, increase during hardening. They bolster the fruit's defense against pathogens and oxidative stress.

Ripening physiology

The process of ripening is a significant occurrence in the fruit's life cycle, wherein a physiologically mature yet non-consumable plant organ transforms into a visually appealing aromatic and alluring delicacy. Ripening involves cascade of transformations and changes, most of which are independent of each other. From a developmental biology viewpoint, one may suggest that the seed is shielded by the surrounding fruit tissue throughout its growth and maturity stages. Ripening might be hypothesized as the physiological mechanism by which the breakdown of the flesh facilitates the liberation of the fully developed seed for the purpose of germination. The ripening process can trigger seed release by attracting mammals, birds, and insects by alterations in colour, texture, taste, and fragrance characteristics. After consuming the flesh, lignified pits and seeds surrounded by a fibrous core were discarded, while smaller seeds were excreted by the animal's digestive tract thereby scatter the seeds and expand the plant's territory, may enhance the likelihood of survival for the succeeding generation.

Fruit ripening process involves a series of physiological and biochemical changes, governed by a network of regulatory mechanisms, which are genetically programmed and irreversible. After being harvested, climacteric fruits including apples, bananas, and avocados experience an increase in the production of ethylene, which then initiates a series of ripening processes. Climacteric fruits possess the characteristic ability to undergo further ripening even after they have been harvested. In contrast, non-climacteric fruits like strawberries and grapes do not exhibit a surge in ethylene production and must be harvested when ripe, as they do not ripen post-harvest. Ethylene, regulates ripening of climacteric fruits, acting in concert with other compounds. For instance, gases like nitric oxide and volatiles such as methyl jasmonate play roles in modulating the ripening process, either individually or through interactions with ethylene. Understanding these interactions is crucial for manipulating ripening and extending the shelf life of fruits. The ripening process is also governed by a complex network of genetic expressions and enzymatic activities.

Recent studies have shown that thousands of genes are differentially expressed during the ripening process, highlighting the complexity of this physiological change. Enzymes like ACC synthase and ACC oxidase, which are involved in ethylene biosynthesis, are encoded by multigene families whose expressions are finely tuned by developmental and environmental signals. The textural changes in fruits, a hallmark of ripening, are primarily due to enzymatic actions that alter the structure and composition of cell walls. A battery of cell-wall-modifying enzymes is involved in these changes, which are critical for the fruit to become softer and more palatable. Additionally, modifications in the fruit's internal structure, such as changes in carbohydrate composition and water accumulation, contribute to its overall size increase during ripening. These changes encompass a wide range of processes:

Starch conversion to sugars: This process enhances the sweetness of the fruit, making it more appealing in taste.

Cell wall degradation and textural changes: Key enzymes such as polygalacturonase and cellulase break down the cell wall, leading to the softening of the fruit, which is crucial for palatability.

Pigment accumulation: The colour of the fruit changes due to the breakdown of chlorophyll and the accumulation of pigments like carotenoids and anthocyanins, which contribute to the visual appeal and nutritional value of ripe fruits.

Volatile compound production: These compounds are responsible for the fruit's aroma and flavour, significantly influencing consumer preference.

The regulation of fruit ripening is also influenced by other phytohormones, transcription factors, and epigenetic modifications. Abscisic acid (ABA), for instance, has been identified as a key regulator in the ripening of non-climacteric fruits, although its role varies among different species. The ripening process is further modulated by environmental factors, such as temperature and light, which can affect the production and action of ripening hormones. At the molecular level, thousands of genes are involved in controlling the ripening process, orchestrating changes in fruit color, texture, flavor, and aroma. These genetic regulators work in concert with phytohormones to ensure the proper development of the fruit's edible qualities. For example, the degradation of chlorophyll and the synthesis of pigments like anthocyanins and carotenoids contribute to the fruit's colour change, while modifications in the cell wall structure lead to the softening of the fruit flesh.

Change type	**Description**	**Key compounds/ enzymes involved**
Starch to sugar	Enhances sweetness	-
Cell wall degradation	Softens fruit, making it palatable	Polygalacturonase, cellulase
Pigment accumulation	Changes color, adds nutritional value	Carotenoids, anthocyanins
Volatile compound production	Produces aroma, enhances flavor	-
Structural changes	Alters texture and size	Cell-wall-modifying enzymes

Post-harvest technology has mostly concentrated on studying the alterations in ethylene and respiration as part of its efforts to understand the process of fruit ripening.

Physiology of respiration

Respiration refers to the process of oxidatively breaking down complex materials found in cells, such as starch, sugars, and organic acids, into simpler molecules like carbon dioxide and water. This mechanism also produces energy and other compounds that can be used by the cell for synthetic reactions. Respiration can occur under two conditions: with the presence of oxygen, called aerobic respiration, or without oxygen, known as anaerobic respiration or fermentation. The respiration rate of produce is a dependable measure of the metabolic activity of the tissue and can be used to ascertain the likely duration of preservation for the produce. When assessing the respiration rate of a fruit or vegetable during its growth, maturation, ripening, and ageing phases, a clear and identifiable respiratory pattern becomes apparent. This pattern can be determined by measuring either the amount of oxygen consumed or the amount of carbon dioxide produced. The respiration rate per unit weight is greatest in the young fruit or vegetable and gradually decreases as it ages. (Fig. 1)

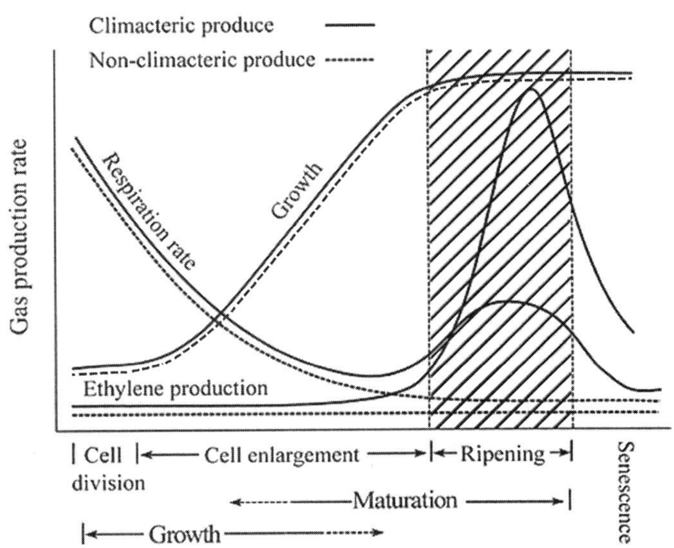

Fig. 1: Growth, respiration and ethylene production patterns of climacteric and non-climacteric plant organs.

A distinct group of fruits, such as tomato, mango, banana, and apple, deviates from the usual respiratory pattern by undergoing a significant increase in respiration throughout the ripening process. An increase in respiration is called a respiratory climacteric, and fruits belonging to this group are known as climacteric fruits. The respiratory climacteric, first documented in 1925 in apples and later found in many other fruits, has notable variations in its intensity and duration among different fruit species, as depicted in (Fig. 2).

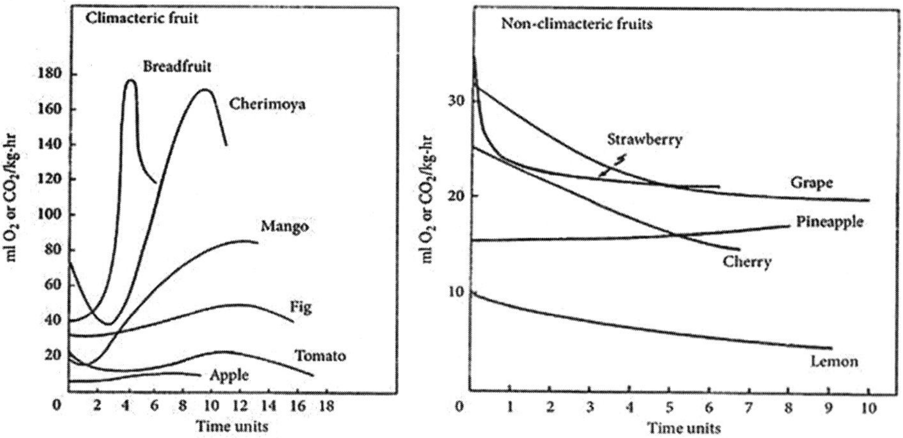

Fig. 2: Respiratory pattern during ripening of climacteric and non-climacteric fruits.

The respiratory climacteric commences simultaneously with the fruit reaching its peak size, and it is during this phase that all other distinctive changes associated with ripening take place. The respiratory climacteric and ripening processes can occur whether the fruit is still connected to the plant or has been disconnected, except for avocados, which only ripen when they are separated. Non-climacteric fruits such as citrus, pineapple, and strawberries. Non-climacteric fruits experience most of the ripening changes, although they do so at a slower pace compared to climacteric fruits. However, the majority of vegetables exhibit a non-climacteric respiratory rhythm. The categorization of fruit into two groups based on their respiratory rhythm is a subjective classification, and there is some overlap between the groups. However, this classification has encouraged extensive research to uncover the biochemical regulation of the respiratory climacteric.

Ethylene in climacteric and non-climacteric fruits

Fruit	Ethylene (μL/L)
Climacteric	
Apple	25–2500
Pear	80
Peach	0.9–20.7

Avocado	28.9–74.2
Mango	0.04–3.0
Passionfruit	466–530
Plum	0.14–0.23
Non-climacteric	
Lemon	0.11–0.17
Lime	0.30–1.96
Orange	0.13–0.32
Pineapple	0.16–0.40

SP Burg and EA Burg (1962), *Plant Physiology* 37: 179–89.

Ethylene production is strongly linked to fruit ripening in various species and serves as the plant hormone that governs and synchronizes the various elements of the ripening process; colour, aroma development, and textural changes are all influenced by ethylene. Fruit usually produces minimal amounts of ethylene until it ripens, at which point there is a sudden increase in production. Climacteric and non-climacteric fruits can be distinguished based on their reaction to ethylene and the way they produce ethylene during ripening. All fruits create modest amounts of ethylene during their development. During ripening, climacteric fruits generate significantly higher levels of ethylene compared to non-climacteric fruits. The contrast between the two types of fruit is highlighted by the varying levels of ethylene concentration detected during different stages of growth and ripening. Climacteric fruits have a significant variation in internal ethylene concentration, while non-climacteric fruits experience minimal fluctuations in ethylene levels during their development and ripening. Applying ethylene at a concentration of 0.1 microliters per litre for one day is typically enough to accelerate the ripening of climacteric fruit. However, the extent of the ripening process is not significantly affected by the concentration of ethylene applied. Applied ethylene only results in a temporary rise in the respiration of non-climacteric fruits, with the extent of the increase being influenced by the concentration of ethylene. Additionally, non-climacteric fruits may experience many increases in respiration in response to ethylene, unlike climacteric fruits which only have a single respiration increase.

Although, all fruits were formerly categorized under this classification, current research has revealed that the differentiation is not as straightforward. Some species, such as pepper and melon, have different cultivars or genotypes that display traits of either climacteric or non-climacteric behaviour. Research has demonstrated that certain ripening changes in climacteric fruit can happen without the presence of ethylene, but some non-climacteric fruit need ethylene for ripening changes to occur. Advancements in ethylene measurement technology have revealed that certain non-climacteric fruits produce higher levels of ethylene during ripening than was previously measurable. Both climacteric and non-climacteric species exhibit ethylene-dependent and ethylene-independent pathways, indicating that ethylene regulation is widespread. Climacteric and non-climacteric behaviours can be seen as opposite ends of a spectrum of responses, with sensitivity to ethylene being a crucial factor. Both unripe climacteric and non-climacteric fruit show an increased respiration rate when exposed to external ethylene, supporting this sensitivity hypothesis.

Physical and biochemical changes during fruit ripening

Colour: This is the primary visible transformation that takes place in numerous fruits and is frequently the key factor consumers rely on to assess the ripeness of the fruit. The most recurrent alteration is the disappearance of the green hue. Chlorophyll causes the green colour, while the loss of green colour occurs owing to chlorophyll structural deterioration. The main factors contributing to this deterioration include fluctuations in pH (mostly caused by the release of organic acids from the vacuole), oxidative mechanisms, and chlorophyllases (Fig. 3).

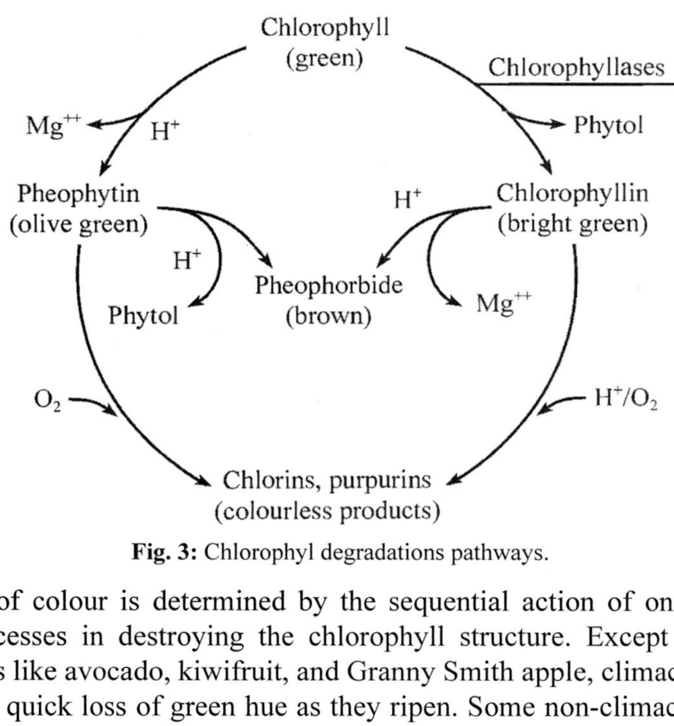

Fig. 3: Chlorophyl degradations pathways.

The loss of colour is determined by the sequential action of one or all of these processes in destroying the chlorophyll structure. Except for a few exceptions like avocado, kiwifruit, and Granny Smith apple, climacteric fruits undergo a quick loss of green hue as they ripen. Some non-climacteric fruits lose their green hue when they reach peak ripeness, such as citrus fruits in cooler temperate regions (not in tropical regions). Chlorophyll disappearance is commonly linked to the production and/or exposure of pigments that vary from yellow to red. Several of these pigments are carotenoids, which are unsaturated hydrocarbons typically consisting of around 40 carbon atoms and may have one or more oxygen atoms in the molecule.

Carotenoids are resilient compounds that remain in the tissue even after significant ageing. Carotenoids can be synthesised during the plant's development but are masked by chlorophyll. Carotenoid pigments become visible once chlorophyll breaks down. Carotenoid synthesis and chlorophyll degradation occur concurrently in specific fruits and vegetables. The banana peel follows the earlier system, whilst the tomato symbolizes the one that follows. Carotenoids are categorized into two groups: the hydrocarbon carotenes, such as lycopene (red) and b-carotene (orange); or the oxygen-containing xanthophylls, such as lutein (yellow). In addition to their visually appealing coloration, carotenoids serve to safeguard the plant's photosynthetic system against excessive light energy and are crucial for the nutritional needs of humans and animals.

Anthocyanins are responsible for the red-purple colour of fruits and vegetables and are produced through the phenylpropanoid pathway. Anthocyanin pigments are phenolic glucosides that are soluble in water. They are synthesized in the cytosol and stored in vacuoles in crops such as cherry and beetroot. However, they are mostly present in the outermost layers of fruits, such as apple and grape. The fundamental ring structure of these compounds can be altered through hydroxylation, methylation, or glycosylation. Additionally, their distinct coloration can be adjusted by changes in pH, the presence of metal ions, and the inclusion of co-pigments, resulting in a diverse range of visible colours. Anthocyanins have the capacity to generate vivid shades that often mask the pigments of carotenoids and chlorophyll. Anthocyanin accumulation is controlled by two groups of transcription factors, namely R2R3, MYB and basic helix loop helix. These regulatory proteins coordinate the gene expression of the entire phenylpropanoid system. The regulatory mechanism in grape and apple fruits has been thoroughly described. The gene VvMYBA2 in white berry grapes is rendered non-functional due to mutations in its coding area, while the gene VvMYBA1 is not translated because it has a retrotransposon in its promoter region. The MYB10 gene in apple fruit contains a mini-satellite repeat structure in its promoter region. This structure enhances the expression of the MYB10 gene, leading to increased levels of anthocyanin throughout the plant. As a result, the fruit develops a vibrant red colour in its flesh.

Carbohydrates: Ripening typically involves a significant quantitative change, notably the degradation of carbohydrate polymers, leading to the almost complete transformation of starch into sugars. This changes both the flavour and the consistency of the fruit. The rise in sugar content makes the fruit significantly sweeter and hence more appealing. Sugar accumulation in non-climacteric fruits is linked to achieving the best eating quality, even if the sugar comes from the sap entering the fruit rather than the fruit's starch reserves breaking down. Breaking down polymeric carbohydrates, specifically pectic compounds and hemicelluloses, reduces the strength of cell walls and the adhesive forces that hold cells together. Initially, the texture improves in taste, but gradually the plant's components break down. Protopectin is the initial form of pectic compounds that is not soluble. It is a big polymer that is cross-linked to other polymer chains with calcium bridges and is bonded to other sugars and phosphate derivatives to create an exceedingly large polymer. Protopectin is broken down into smaller, more water-soluble parts as it ripens and matures. The breakdown rate of pectic compounds is directly linked to the softening rate of fruit.

The degradation of starch during fruit ripening is a highly regulated process that involves a suite of enzymes working in concert to break down complex starch molecules into simpler sugars. This not only contributes to the sweetness of the fruit but also affects its overall texture and nutritional content. The process is controlled by a network of genetic and hormonal factors that ensure the timely and efficient conversion of starch as the fruit matures. Here's an explanation of the key steps and enzymes involved:

***Starch hydrolysis*:** Starch in unripe fruits is a storage form of glucose that provides energy for the fruit as it ripens. The degradation of starch into simpler sugars is primarily facilitated by the enzyme amylase. Initially, the glycosidic bonds in starch granules are hydrolyzed by α-amylase and β-amylase enzymes. α-Amylase attacks random internal α-1,4 glycosidic bonds, creating dextrin's, while β-amylase removes maltose units from the non-reducing ends of the starch molecule. Thus the enzyme breaks down starch molecules into maltose and smaller dextrin's, which are further hydrolyzed into glucose units.

***Debranching*:** In addition to amylases, debranching enzymes such as isoamylase and pullulanase are involved in the degradation of branched starch molecules. These enzymes specifically target α-1,6 glycosidic bonds at the branch points of amylopectin, creating a more linear structure that can be further degraded by amylases.

Maltose and glucose production: The maltose produced by β-amylase is then broken down into glucose by maltase. Additionally, glucoamylase (also known as amyloglucosidase) can hydrolyze both α-1,4 and α-1,6 bonds, releasing glucose from the non-reducing ends of dextrins.

Transcriptional regulation: Studies have identified transcription factors such as MabHLH6 in bananas that activate the promoters of starch degradation-related genes. These transcription factors bind to E-box motifs in the promoters, enhancing the expression of genes encoding enzymes like α-amylase, β-amylase, and debranching enzymes.

Ethylene's role: Ethylene, has been shown to regulate the expression of starch-degrading enzymes at both transcriptional and translational levels. It also influences the crosstalk between different hormones, which affects primary sugar metabolism during ripening.

Phosphorylation and dephosphorylation: Transitory starch degradation involves phosphorylating and dephosphorylating reactions that act on starch-related glucosyl residues. These reactions facilitate the transition from a hydro insoluble ordered state to a soluble state at the granule surface.

Sucrose synthesis: While not directly involved in starch degradation, enzymes like sucrose synthase and sucrose phosphate synthase play a role in transforming the glucose produced into sucrose. As a result the concentration of simple sugars like glucose, fructose, and sucrose in the fruit increases. This is not only due to the direct conversion of starch but also because of the import of sugars from other parts of the plant. The sucrose transported to the fruit is cleaved by the enzyme invertase into glucose and fructose. Additionally, sucrose synthase can convert sucrose into UDP-glucose and fructose, which can be used in various metabolic pathways or stored as starch. The increase in sugar content is regulated by a complex network of hormonal signals and gene expression.

The starch degradation and sugar increase pathways are tightly integrated with the fruit's overall developmental program. The regulation of these pathways involves a sophisticated interplay between metabolic enzymes, sugar transporters, and transcription factors. This ensures that the fruit develops the desired taste, texture, and nutritional qualities.

Fruit softening and textural changes

During the process of fruit ripening, several modifications take place, including softening and alterations in texture. These changes are responsible for the development of juiciness and enhance the desirability of consuming the ripe fruit. The feel of ripe fruit differs among various species, ranging from the firm and crisp apples to the soft and pliable texture of avocados. The different textures of different fruits and their softening process can be explained by both anatomical traits and changes in the cell wall that occur during ripening. After being picked while still firm, fruits such as kiwifruit and tomato have substantial modifications in their cell wall composition, resulting in a perceptible softening. Other fruits, such as apples or watermelons, retain their sharp texture and experience minimal softening. Their fragile cellular membranes remain mostly unaltered. Both types of softening occur within the pear family: The Asian pear exhibits a resilient and crunchy consistency, but the European pears tend to soften as they ripen, leading to a velvety and delicate feel.

The result of fruit flesh in the oral cavity is influenced by many textural qualities during its breaking and compression. The features that contribute to the textural feel of fruit pulp include cell size, cell adhesion, turgor and packing, wall thickness, wall composition, and the response of cells to shearing stress during chewing. A fully developed apple fruit has relatively large (0.1–0.3 mm diameter), solid, and thin-walled cells that are not densely packed (about 20 % of the fruit's volume is vacant space). When the flesh is bitten, the cells rupture and expel their sweet contents in the form of juice. Conversely, fully developed kiwifruit has minimal void space, accounting for approximately 2 per cent of the fruit's overall volume, and possesses thick cell walls that have a strong affinity for water. These cells exhibit a proclivity to disintegrate upon consumption, resulting in the formation of a moist paste. This paste is hydrated by liquid that is either retained within the cell walls or secreted by cells that have been ruptured. Avocado cells have robust and flexible walls that tend to detach. Moreover, it possesses a substantial quantity of oil, leading to a greasy consistency upon consumption.

Modifying the cell wall has a substantial impact on the texture and ripening properties of fruit. The plant's basic cell wall consists of a sturdy and inflexible network of cellulose microfibrils. The microfibrils are interconnected by an intricate network of polysaccharides, including of two main types: hemicelluloses (composed mainly of neutral sugars) and pectins (rich in galacturonic acid). In addition, there are lesser amounts of structural proteins present. The outer part of each cell wall, known as the middle lamella, consists mainly of pectins and functions as a site of attachment for neighbouring cells. Fruit softening involves alterations to various polysaccharide wall constituents, including pectin and hemicellulose, as well as changes in the bonding between specific polymers.

Wall modification has been extensively studied globally, primarily using tomato as a model, but also exploring other fruits to identify shared patterns. Cell wall components in various species, particularly kiwifruit and tomato, exhibit consistent alterations during the initial phases of ripening, as revealed by chemical investigations. Previous research utilising kiwifruit as a model observed the subsequent alterations throughout the ripening process:

- Pectin solubilization occurs without any additional degradation.
- The cell wall expands and exhibits a increased attraction to water, becoming more hydrophilic.
- Galactose depletion from pectins, particularly from a galactan that is strongly linked to the cellulose microfibrils;
- De-esterification of some pectin's.

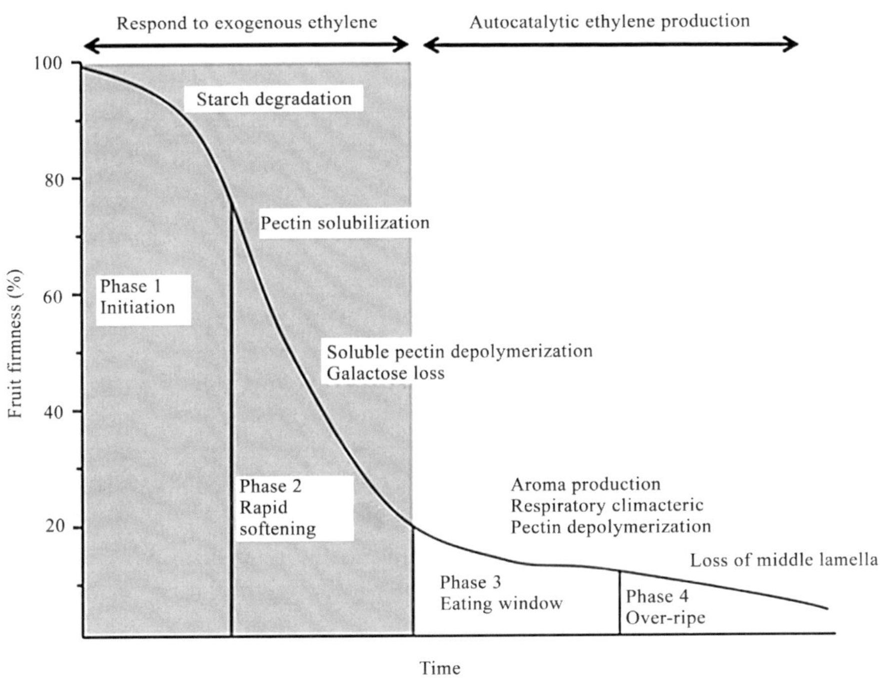

Fig. 4: Fruit softening alteration during various phases of fruit development in kiwifruit
Source: Atkinson *et al.* (2011) J. Exp. Bot. 62, 3821-3835.

These modifications persist when kiwifruit rapidly soften during the ripening process (phase 2 in Fig. 4). In the second phase fruit softness is characterised by a further rise in the solubilization of pectin, the removal of galectin and arabinan side chains from pectic polymers, and increased swelling of the cell wall. During phase 3 of softening, two further significant modifications commence, both of which seem to be controlled by ethylene:

- Depolymerisation refers to the process of reducing the size of the hemicellulosic polysaccharide xyloglucan, which leads to a decrease in the strength of the cell wall.
- The depolymerization of pectin leads to the disintegration of the middle lamella and a decrease in intercellular adhesion.

These modifications have been noted in a variety of fruits although the degree and relative timing may differ slightly among different species. These observations suggest that the process of pectin solubilization and cell wall expansion play a crucial role in regulating the softening of kiwifruit, as well as other species with a similar texture. However, fruit with a crisp texture, like apples, undergoes minimal cell wall change. As fruits reach full ripeness, the middle lamella undergoes breakdown, causing it to gradually vanish as a visible structure under a microscope. The dissolution of the middle lamella leads to a significant decrease in intercellular adhesion, causing cells to have fewer points of attachment to one another. As a result, the cells become more rounded in shape as they separate from their neighbouring cells. The primary walls also get compromised by the alterations, rendering cells susceptible to rupture when bitten or chewed, resulting in the release of cellular contents in the form of juice.

Ethylene regulates a significant number of cell wall-modifying enzymes that are synthesised during ripening. The level of polygalacturonase enzyme, which breaks down pectin's, was observed to greatly increase in ripening fruits, highlighting its crucial involvement in the process. Later on, it was shown that suppressing the activity of polygalacturonase had a limited but considerable effect on reducing the softening and enhanced shelf life of fruits like strawberries and apples. This finding confirmed that the breakdown of pectin is one of the factors contributing to the softening process. The decreased polygalacturonase activity likely resulted in lower degradation of the middle lamellae, leading to the preservation of intercellular connections and overall fruit integrity. This, in turn, contributed to the benefits observed on firmness and shelf life.

Prior research has shown that softening is not regulated by a solitary cell wall-modifying enzyme. Multiple enzymes participate in the softening process, including polygalacturonase, pectate lyase, expansin, β-galactosidase, and pectin methylesterase. Each of these enzymes has distinct roles in contributing to the softening process. The mature fruit's ultimate softening and textural qualities are a result of a well-coordinated combined action of numerous enzymes, which leads to wall swelling, reduced wall strength, and weakened middle lamellae. It has been noted that the activity of different enzymes can be interconnected. For example, polygalacturonase relies on the prior de-methylesterification of pectin by pectin methylesterase in order to make the substrate vulnerable. Additionally, the action of expansin is necessary to enhance the enzyme's access to the substrate in the cell wall.

Furthermore, in several species it was observed that a reduction in cellular turgor is a significant factor in fruit ripening and contributes to the softening process, alongside cell wall breakdown. This is mostly driven by the internal water movements occurring from the transfer of solutes from the symplast to the apoplast, as well as the loss of water from the fruit. The investigation of the non-softening DFD mutant in tomatoes revealed that its increased post-harvest firmness can be attributable to a significant reduction in water loss from the fruit. As a result, the cellular turgor in the mutant remains at greater levels compared to non-mutant tomatoes.

Flavour and aroma

Fruit ripening involves intricate molecular-level changes in biochemistry that lead to the creation of flavour and aroma. Fruits experience significant alterations in their biochemical makeup during the ripening process, resulting in the emergence of their distinct flavours and aromas. A network of metabolic activities controls the alterations, leading to the breakdown of chlorophyll, the synthesis of carotenoids and anthocyanins, and the formation of other volatile compounds. Fruit flavour is mostly influenced by the equilibrium between sugar and acid, together with the production of these volatile aroma molecules. The volatile compounds may include volatile acids, aldehydes, alcohols, esters, terpenoids, and aromatics. Advanced gas chromatography-mass spectrometry techniques has revealed that numerous ripe fruits contain more than 100 volatile compounds that contribute to their flavour and aroma. The concentration of a volatile compound is not the sole factor determining its importance in the perception of flavour and aroma. At times, minute amounts of chemicals, like parts per billion, are needed to give a fruit its unique smell. Compounds are given smell activity levels based on their importance in scent. These levels are determined by dividing the chemical concentration by the minimum concentration detectable by the human nose.

At times, certain volatile compounds are unique to a single fruit species or variety and are added to manufactured mixtures to replicate that product. Specific volatile compounds are essential in wine grapes, as one volatile component can be the main characteristic emphasized in marketing a particular type of wine. Examples include of the herbaceous methoxypyrazine note in Sauvignon Blanc, the fruity b-damascenone perfume in red wine, and the musky methyl anthranilate scent produced by *Vitis labruscana.* Noteworthy volatile compounds present in different fruits include: raspberry ketone in raspberries; ethyl butanoate and 1,8-cineole in 'Hort16A' kiwifruit; (E)-2-hexenal and hexanal in 'Hayward' kiwifruit; 2-methyl butyl acetate and a-farnesene in apples; and furaneol in strawberries.

Considerable research has focused on discovering volatile compounds and comprehending their metabolic routes through bioconversion or precursor tracing. Aldehydes, acids, and esters are derived from fatty acids and branched-chain amino acids. Lipoxygenases initiate straight-chain ester synthesis by turning linoleic acid into 13-hydroperoxide linoleic acid. Cytochrome p450 lyases convert these compounds into aldehydes such as hex-3-enal and hex-2-enal. Alcohol dehydrogenases (ADHs) transform aldehydes into corresponding alcohols, resulting in the generation of green odours. Alcohol acceptor substrates are esterified with coenzyme acid donors by alcohol acyl transferases (AATs) to create esters, which usually give off fruity and sweet characteristics. The enzyme MpAAT1 in apples produces different esters in ripe fruit and is activated by ethylene throughout the ripening process (Souleyre *et al.*, 2005). Isoleucine is transformed into branched-chain esters by the activity of branched-chain aminotransferases and pyruvate decarboxylase, leading to the formation of aldehydes. Aldehydes can be used by alcohol dehydrogenases (ADHs) and acyltransferases (AATs) to create branched chain alcohols and esters.

Sesquiterpenes and monoterpenes improve fruit taste and smell by adding floral or spicy accents. The main ripe-fruit terpene found in apples is afarnesene, which is produced by the sesquiterpene synthase AFS1. Terpenoid molecules are produced by terpene synthase enzymes using farnesyl diphosphate (FPP) and geranyl diphosphate (GPP) as substrates. Plants synthesise FPP and GPP using short chain prenyltransferases in separate processes within compartments. The MEP pathway in the plastid synthesizes GPP, whereas the mevalonate pathway in the cytoplasm yields precursors for FPP synthesis. Terpenoid structures can be modified through oxidation, hydroxylation, glycosylation, or methylation by different enzymes to increase terpenoid diversity. Ethylene controls the speed of ripening, length of storage life, and several ripening processes in climacteric fruit. Breeders have inadvertently reduced ethylene production or activity by developing genotypes with extended shelf-life. Long-lasting genotypes produced by crossing with non-ripening mutants sometimes experience a notable decrease in flavour due to the ethylene control of many aroma production genes. In the future, a key challenge is to distinguish between the reduction of ethylene and the inhibition of aroma volatile production.

Hormonal and genetic regulation of fruit ripening

The hormonal regulation of fruit ripening is a dynamic, multifaceted process, and complex process that involves a symphony of biochemical events, orchestrated by a network of plant hormones that signal various stages of fruit development and maturation. Ethylene, known as the ripening hormone, plays a pivotal role, especially in climacteric fruits—those that continue to ripen after being picked, such as apples, bananas, and tomatoes. This gaseous hormone triggers a cascade of genetic and enzymatic reactions that transform the fruit in multiple ways: it softens, sweetens, changes color, and develops its distinctive aroma and flavor. However, ethylene doesn't act alone. Abscisic acid (ABA) is another key player, often working in concert with ethylene to regulate ripening. ABA levels increase as fruits begin to ripen, promoting ethylene production and responding to environmental stressors like drought or pathogen attack, which can also influence the ripening process. Moreover, other hormones such as auxins, gibberellins, cytokinins, jasmonic acid (JA), and brassinosteroids (BR) contribute to this intricate process, each playing specific roles during different stages of fruit development and ripening. For instance, auxins and gibberellins are crucial during the early stages of fruit set and growth, promoting cell division and expansion. As the fruit matures, their levels decrease, paving the way for ripening to commence. Cytokinins, on the other hand, are involved in nutrient mobilization within the fruit, ensuring that developing seeds receive the sustenance they need. The interplay between these hormones is tightly regulated and is essential for the proper development and eventual ripening of the fruit. Traditionally, non-climacteric fruits have been categorized as a distinct group that does not exhibit the characteristic climacteric ripening pattern.

Comparative genomic data analysis conducted in tomato and hot pepper (*Capsicum* spp.) as models for climacteric and non-climacteric fruits, respectively, reveals that the expression of genes encoding transcription factors, such as non-ripening (NOR), tomato AGAMOUS-like 1 (TAGLI), and ripening inhibitor (RIN), as well as components related to the ethylene signaling pathway, are shared between both fruit categories. Furthermore, the detection of MADS-box genes in both climacteric and non-climacteric fruits implies that certain molecular regulatory mechanisms of fruit ripening are shared between these two groups. Plant hormones play a crucial role in controlling fruit development and ripening. Recent evidence suggests that the combined effects of three hormones, auxin, cytokinin, and gibberellins play a crucial role in promoting fruit growth even without fertilization. This process is known as parthenocarpy. The application of these hormones alone can initiate fruit development in various species, indicating that communication between these hormones is essential for fruit set and growth.

Ethylene:

***Climacteric fruits*:** The function of ethylene in fruit is controlled by two pathways: the biosynthesis process and the signal transduction pathway. In climacteric fruit, the production of ethylene is categorised into two systems, known as systems 1 and 2. System 1 maintains the standard levels of ethylene in vegetative tissues, and its activation is triggered by injury. System 1 regulates the generation of ethylene during the pre-climacteric stage of climacteric fruit and throughout the growth of non-climacteric fruit. System 2 is a process in which climacteric fruit undergoes auto-stimulated ethylene production, known as autocatalytic synthesis. Thus, the primary distinction between climacteric and non-climacteric fruit in relation to ethylene is presence or absence of autocatalytic ethylene biosynthesis (Fig 5). The ethylene biosynthesis occurs by utilizing methionine with the help of S-adenosyl-L-methionine (SAM) and 1-aminocyclopropane-1-carboxylic acid (ACC).

The genes that regulate the biosynthetic route have been identified and characterized. The pathway involves two key enzymes: ACC synthase (ACS), which transforms SAM into ACC, and ACC oxidase (ACO), which turns ACC into ethylene (Fig 5). The ACO and ACS genes in tomato are part of a multigene family consisting of five and nine members, respectively. The expression of these genes is regulated in a distinct manner throughout fruit growth and ripening. The LeACO1 and LeACO4 genes are elevated at the initiation and remain active throughout the ripening process. However, the LeACO3 gene is only activated temporarily during the breaker stage of fruit ripening (Fig 6).

Research demonstrated that LeACS6 and LeACS1A genes are expressed throughout the pre-climacteric stage, specifically in system 1. However, during the transition to ripening, the most active genes are LeACS4 and LeACS1A (Fig 6). Following this, LeACS4 maintains a high level of expression during the climacteric phase, while the expression of LeACS1A decreases. The increase in ethylene production during ripening leads to the activation of LeACS2 gene expression, while suppressing the expression of LeACS6 and LeACS1A genes. The precise adjustment of the ACS genes is believed to be crucial for the transition from the pre-climacteric system 1 to the climacteric system 2. It is important to note that system 1 is distinguished by the inhibitory feedback of ethylene in its own biosynthetic pathway, while the shift to system 2 is characterised by the self-sustaining synthesis. The necessity of ethylene in initiating the ripening process of climacteric fruit has been convincingly proven by suppressing the activity of ACO and ACS genes in genetically modified plants using an antisense approach. The tomato lines with suppressed ethylene exhibited significant delay in ripening, as observed in other fruits such as melon and apple. Nevertheless, climacteric fruit have alternative ripening pathways that are not dependent on ethylene. This is evident in melon fruit, where some of the processes involved in softening, sugar accumulation, and pigmentation of the flesh occur even when ethylene is repressed. The findings indicate that climacteric fruit exhibits a coexistence of climacteric (ethylene-dependent) and non-climacteric (ethylene-independent) regulation.

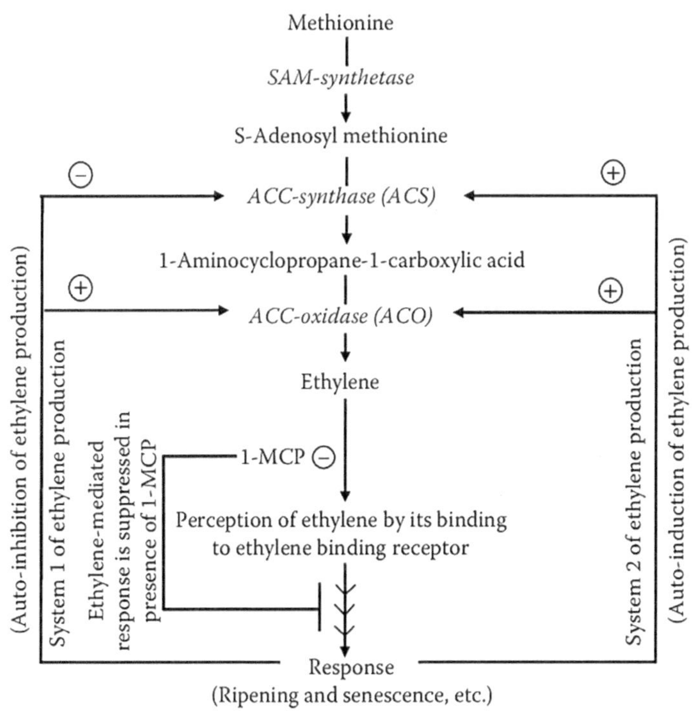

Fig. 5: The ethylene biosynthesis pathway in plants, highlighting the phenomenon of autoinhibition (self-inhibition) and autoinduction (self-stimulation) of ethylene production. These systems are denoted as systems 1 and 2 for the generation of ethylene, respectively. System 1: Ethylene suppresses its own production by reducing the expression or activity of ACS (1-aminocyclopropane-1-carboxylic acid synthase). It is important to observe that the activity of ACO is increased during system 1, but, the absence of any increase in the activity of ACS prevents autoinduction from occurring. System 2: Ethylene enhances its own synthesis by concurrently increasing the expression or activity of both enzymes (ACS and ACO), resulting in a sudden increase in ethylene production (Paul, V. *et al*., 2012).

While the maturation of non-climacteric fruit does not involve any significant alteration in ethylene production, certain ethylene-dependent processes do occur in this category of fruit. The ethylene production pathway in grape berries is activated at the beginning of the ripening process, known as the veraison stage. Exogenous ethylene treatments enhance the prolonged activation of genes associated with anthocyanin synthesis. Ethylene signals seem to have a role in controlling vascular flux, acid content, and some stages of aroma volatile generation. In citrus fruits, which are typically non-climacteric, there is a suggested presence of an autocatalytic system for ethylene production, similar to that found in climacteric fruits. It is widely recognised that citrus fruits have the capability to respond to external ethylene by undergoing chlorophyll degradation. Furthermore, in all non-climacteric fruits, the use of external ethylene speeds up the process of ageing by causing the degradation of cell membranes. Although the function of ethylene in the ripening of climacteric fruits is becoming increasingly comprehended, our understanding of the primary signalling pathways implicated in non-climacteric ripening is still relatively limited.

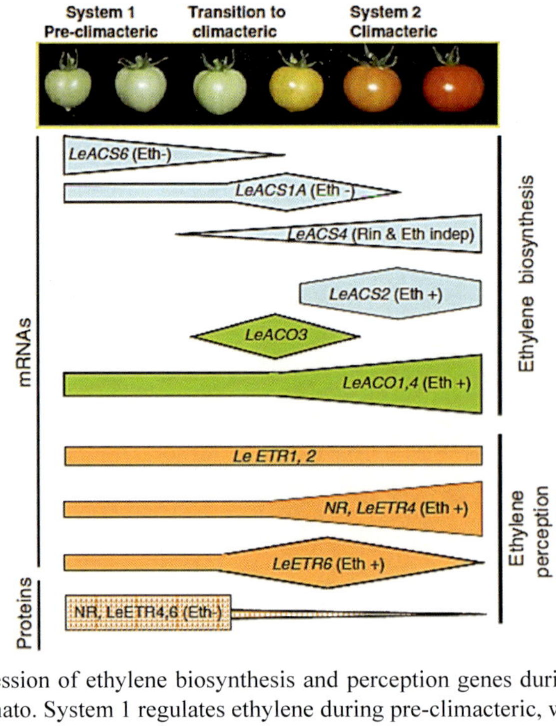

Fig. 6: The expression of ethylene biosynthesis and perception genes during the transition to climacteric in tomato. System 1 regulates ethylene during pre-climacteric, while System 2 with the autocatalytic production during the climacteric phase. LeACS refers to L. esculentum ACC synthase, while LeACO stands for L. esculentum ACC oxidase. LeETR and NR are ethylene receptors. Eth+ and Eth– denote the activation and inhibition, of gene or protein expression, respectively.

Non-climacteric fruits: Recently, scientists are aggressively working to unravel the role of ethylene in the ripening of non-climacteric fruits. Reports indicate that ethylene plays a substantial role in the ripening process of several grape varieties, with ethylene levels reaching their highest point before véraison. The application of a specific inhibitor of ethylene receptors, namely 1-methylcyclopropene (1-MCP), led to a reduction in the size of berries and the development of ripening-related traits, such as the accumulation of anthocyanins in 'Cabernet Sauvignon' berries. Administering ethylene led to a growth in berry size, which was correlated with heightened levels of transcripts related to aquaporins, polygalacturonases, xyloglucan endotransglucosylases (XTHs), cellulose synthases, and expansins. Before véraison, the use of 1-MCP reduced the ABA levels in 'Muscat Hamburg' grapes, suggesting a relationship between ethylene and ABA in the ripening process. While the concentration of ethylene in F. × ananassa is not high, its production is susceptible to changes throughout various stages of fruit development. The green fruit has a relatively high level, the white fruit has a lower level, and the level increases as the fruit matures and turns red. During the red fruit stage, there was a significant increase in the production of ethylene, a gas that plays a role in fruit ripening. This increase in ethylene production was followed by an elevated respiration rate, which was similar to that of climacteric fruit. Although categorised as a non-climacteric fruit, raspberries demonstrate the production of ethylene and an increased rate of respiration, beginning from the white fruit stage and continuing to grow until they reach full maturity. Therefore, the ethylene production pattern in raspberry fruit, which shows a consistent rise throughout the ripening process, differs greatly from that of non-climacteric fruits such as strawberry and grape. During the early phases of fruit growth, both strawberries and grapes see a significant increase in ethylene activity. This pattern is commonly found in non-climacteric fruits. There is an inverse association between the loss of firmness in raspberry fruit and the formation of ethylene. Furthermore, the container has been determined to be the main origin of ethylene. In addition, raspberry fruit in the white stage, where the drupelets are still attached to their receptacle, showed a delay in the decrease in firmness when treated with 1-MCP in a laboratory setting and stored at a temperature of 10°C. The findings suggest that ethylene may have a role in regulating the ripening process of raspberries, resulting in the softening of the fruit.

The genes responsible for the synthesis of enzymes involved in the generation of ethylene, specifically 1-aminocyclopropane-1-carboxylic acid synthase (ACS) and 1-aminocyclopropane-1-carboxylic acid oxidase (ACO), have been found to be active in non-climacteric fruit. Researchers have identified four FaACS genes and three FaACO genes during the ripening phase of strawberries. These genes display unique patterns of gene expression. The expression of VvACO genes has been observed in many grape cultivars, including as 'Cabernet Sauvignon', 'Muscat Hamburg', and 'Thompson Seedless', in conjunction with the ethylene peak that precedes véraison. During the development of strawberries, the presence of two FaACO and ethylene response sensor (FaErs1) genes has been detected. Furthermore, a significant association was found between the expression of these genes and the synthesis of ethylene in fruits that were exposed to ethylene and the auxin analogue 1-naphthaleneacetic acid (NAA). The genes RiACS1 and RiACO1 exhibited increased expression levels during the red and overripe stages of raspberry. Significantly, the levels of both transcripts were almost thrice more in the receptacle tissue than in the drupelet at the stage of overripe fruit. The expression of RiACS1 in raspberry was comparable to that observed for the FaACS1 and FaACS2 genes, which were significantly expressed in green and white strawberry receptacles. The deduced amino acid sequence of RiACO1 showed a similarity of 73.4 per cent with the protein sequence of FaACO1, which was discovered in strawberry receptacles. In addition, RiACO1 displayed a 79.3 per cent similarity with SlACO4, a gene known to be active during the maturation of tomato fruit. These past instances suggest that a similar mechanism may regulate the synthesis of ethylene in both climacteric and non-climacteric fruits.

Gene expression throughout the ripening phase of both climacteric and non-climacteric fruits has been documented in multiple studies. Alterations in the genes responsible for encoding the components involved in the ethylene signalling cascade. The genes implicated in this process include those accountable for encoding the ethylene receptor (ETR), ethylene response sensor (ERS), ethylene insensitive (EIN), and constitutive triple response (CTR1). These genes play a crucial role in transmitting the ethylene signal from the endoplasmic reticulum (ER) membrane to the EIN2s, acting as a mediator between the ETRs and EIN2s. The ethylene receptor (ETR) family comprises transmembrane proteins situated in the endoplasmic reticulum (ER) that selectively bind to ethylene. These proteins form a stable dimer with two disulfide links at the N-terminus. The ETRs serve as the initial components of ethylene signalling, functioning as inhibitory regulators of the signalling pathway by impeding the transmission of signals downstream in the absence of ethylene. Consequently, the reduction of SlETR4 resulted in the accelerated ripening of genetically modified tomatoes.

During the maturation of the grape fruit, the gene VvETR2 exhibited a significant increase in expression at the beginning of ripening, although the expression of VvETR1 remained consistent. In addition, VvERS1 and

VvEIN4 exhibited elevated expression levels in the early stages following pollination. Three genes, FaEtr1, FaErs1, and FaEtr2, have been discovered to be expressed in strawberry, coinciding with an increase in ethylene synthesis. The predominant expression of FaEtr2 in mature strawberry fruits indicates that even the minimal amount of ethylene generated during the ripening process can trigger physiological changes associated with ripening. Furthermore, the analysis of the dataset revealed that the expression pattern of two genes, namely S-adenosyl-l-methionine synthase FaSAMS1 and signaling-related gene FaCTR1, was found to be consistent with the increase in ethylene production during the process of ripe fruit-coloring. It was hypothesised that FaSAMS1 and FaCTR1 likely play a significant role in the development of characteristics during the process of fruit ripening. Suppression of both transcripts using the tobacco rattle virus-induced gene-silencing (VIGS) system not only inhibited the red coloration of fruit and changes in firmness, but it also triggered the synthesis of ethylene, thereby influencing the components involved in ethylene signalling. Significantly, the use of ethephon, a synthetic compound that releases ethylene, promoted the natural softening and red colour development in these genetically modified strawberries. This was achieved by partially restoring the production of anthocyanins, without causing a substantial impact on the firmness of the fruit. This implies that FaCTR1 has a favourable influence on the ripening of strawberry fruit. However, it is yet unclear whether ethylene plays a function in earlier stages of ripening in this non-climacteric fruit.

Abscisic acid (ABA): Abscisic acid (ABA) is a crucial phytohormone in the regulation of fruit ripening, orchestrating a range of processes that culminate in the maturation of fruit. Its role extends beyond mere participation in the ripening process; ABA is a key signal that initiates and modulates complex biochemical pathways necessary for fruit development. During the early stages of fruit growth, ABA is involved in seed formation and the initiation of fruit expansion, with its transport from the shoot to the fruit being a critical step. As the fruit matures, ABA levels rise, signaling the onset of ripening. This increase in ABA stimulates the production of ethylene, particularly in climacteric fruits, which are known for their post-harvest ripening phase characterized by a peak in respiration and ethylene synthesis.

The interplay between ABA and ethylene is intricate, with ABA regulating ethylene biosynthesis and signaling, thus influencing the ripening process. However, the molecular mechanisms underlying this interaction remain an area of active research. ABA's influence on fruit ripening is not limited to climacteric fruits; it also plays a role in non-climacteric fruits, although the relationship between ABA and ethylene in these fruits is less understood. In both types of fruits, ABA contributes to the enhancement of flavor, hastening of softening, and promotion of color development, which are hallmarks of ripe fruit. The biosynthesis and metabolism of ABA involve enzymes such as 9-cis-epoxycarotenoid dioxygenase (NCED) and ABA 8′-hydroxylase (CYP707A), which regulate the dynamic balance between ABA production and degradation. This balance is essential for maintaining appropriate levels of ABA within the fruit. Additionally, ABA conjugation and release by specific enzymes play a part in modulating ABA activity and ensuring homeostasis.

Climacteric fruits: Abscisic acid (ABA) was traditionally believed to primarily participate in the ripening of non-climacteric fruit. However, new research has demonstrated that ABA also plays a role in controlling the ripening of climacteric fruit. For instance, the administration of exogenous ABA at a concentration of 0.1 mmolL^{-1} significantly stimulated the production of ethylene and the ripening of fruit in tomato. This effect was achieved by enhancing the expression of ethylene biosynthetic genes SIACS2 and SIACO1. Mangoes treated with a concentration of 1.0 mmolL^{-1} of ABA showed a climacteric peak in ethylene production that was 36 per cent greater than the control group. Conversely, treatment with the ABA biosynthesis inhibitor NDGA (nordihydroguaiaretic acid) halted the production of ethylene and the ripening of the fruit. The application of 0.1 mmolL^{-1} NDGA to tomato plants resulted in a climacteric peak of ethylene production that was 43 per cent lower compared to the control group. During the process of ripening in fruit treated with ABA, the concentration of ACC, as well as the activity of ACS and ACO enzymes in the pulp tissues, increased by 55, 58 and 26 per cent, respectively, as compared to the control group.

The natural ABA concentration is low in immature fruit but rises during the process of fruit ripening (Fig. 7). Studies have demonstrated that the peak levels of naturally occurring ABA occur before the sudden increase in ethylene production in climacteric fruits like banana, peach, and tomato. The ABA biosynthetic gene NCED1, responsible for encoding 9-cis-epoxycarotenoid dioxygenase, exhibited significant expression exclusively during the breaker stage, characterised by the transition of fruit colour from green to yellow, coinciding with the increase in ABA concentration. The expression of the SIACS2, SIACS4, and SIACO1 genes exhibited a temporal delay following the expression of SINCED1. Several experimental data suggest that ABA treatment stimulated the production of ethylene, but ACC treatment did not affect the concentration of ABA in tomato flesh and seed. These findings suggest that ABA could potentially serve as a regulatory factor before ethylene in the process of climacteric fruit ripening.

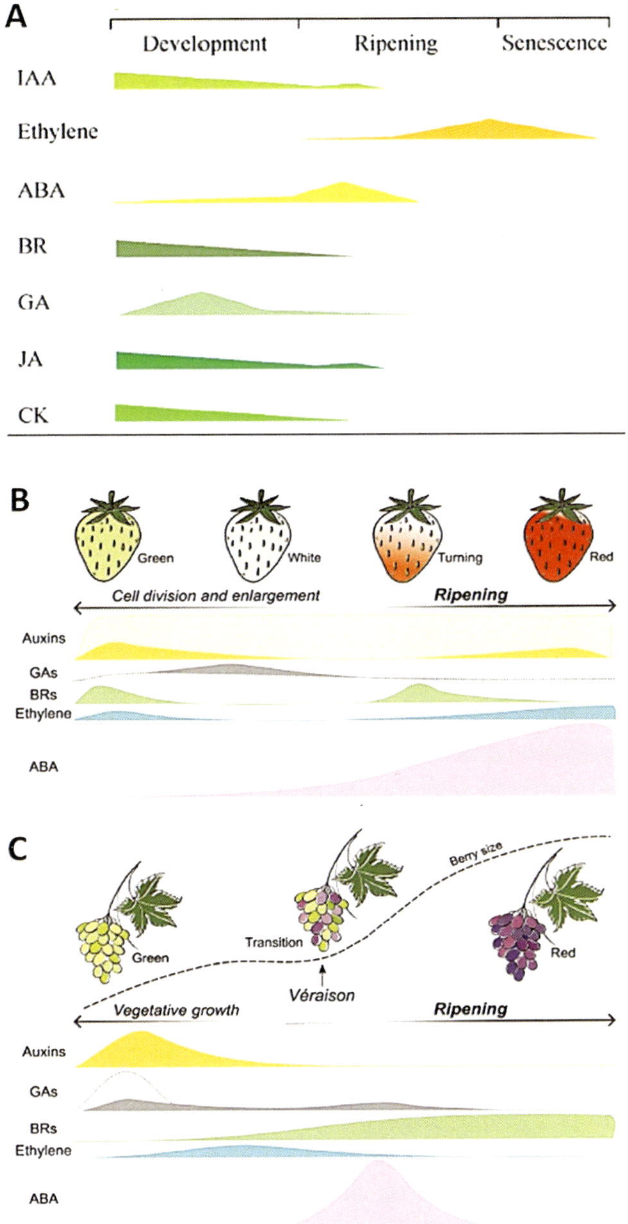

Fig. 7. Hormonal Cross-Talk In Climacteric VS Non Climacteric Fruits (A) Levels of endogenous phytohormones in climacteric fruit including apple, pear and tomato during development, maturation and senescence. (B–C) Schematic model of developmental and ripening stages and the changes in hormone content from green to ripe fruit.

Non-climacteric fruits: ABA regulates stress response, dormancy, growth of seedlings, and development of fruits in non-climacteric fruits. ABA exerts a significant influence on the ripening process of non-climacteric fruits, but ethylene lacks such a response. ABA plays a key role in fruit ripening in strawberries and grapes, which includes the development of colour, accumulation of sugar, and softening. The ABA levels in grapes and strawberries were initially found less during the early stages of fruit growth and increased significantly during the ripening process. It has been observed that ABA treatment accelerates the process of strawberry fruit softening, enhances colour and ethylene production, and stimulates the activity of the enzyme phenylalanine ammonia-lyase (PAL). This enzyme is responsible for initiating the formation of polyphenol compounds, such as flavonoids, through the phenylpropanoid pathway.

ABA treatment enhanced the anthocyanin content of strawberry fruits during ripening by activating the expression of genes related to enzymes involved in later stages of anthocyanin biosynthesis, such as anthocyanidin synthase (ANS) and glucosyltransferase. It was also reported that the ABA treatment enhanced the level of l-ascorbic acid. The expression of ABA biosynthesis genes in strawberry fruit was increased by mild salt and drought stress, resulting in improved biochemical quality attributes such as phenolics, anthocyanin, l-ascorbic acid, and enhanced antioxidant activity. Sucrose aided the fruit ripening process in strawberries in association with the stimulus of abscisic acid (ABA) production. Application of exogenous abscisic acid (ABA) enhanced the levels of anthocyanin, improved the colour, and facilitated the softening of fruits in 'Flame Seedless' grapes. This indicates that ABA controls the buildup of important metabolites that determine fruit's ripeness and also regulates the synthesis of ABA.

Genetic and biochemical studies indicate that 9-cis-epoxycarotenoid dioxygenase (NCED) is the primary enzyme responsible for ABA production in both climacteric and non-climacteric fruit. The genes -PpNCED1 and VvNCED1- showed higher levels of expression during the initial stages of fruit ripening in peach and grape. These genes were crucial for the production of ethylene in peaches. Fluridone, which is an inhibitor of ABA formation, considerably decreased ABA levels in both fruits, hence maintaining fruit firmness throughout storage. The NCED enzyme plays a vital role in the growth of strawberry fruit. The application of Nordihydroguaiaretic acid (NDGA), a compound that inhibits the enzyme NCED, resulted in a decrease in the content of abscisic acid (ABA) and the removal of red pigmentation in the fruit receptacles. The strawberry plants that were subjected to agro-infiltration with FaNCED1-RNAi exhibited comparable outcomes. An examination of fruit tissue using a sucrose-incubation test revealed elevated levels of FaNCED1 mRNA and ABA accumulation. This indicates that sucrose has the potential to affect the expression of FaNCED1 during the ripening process of strawberries.

Arabidopsis thaliana (L.) Heynh. has identified the elements of the ABA signal transduction system, which consist of the 'ABA-PYR/PYL/RCAR-PP2C-SnRK2' and 'ABA-ABAR-WRKY40-ABI5' pathways. Initially, the interaction between ABA and the pyrabactin-resistant receptor PYR1 enhances the binding with Protein Phosphatase2c (PP2C), leading to the deactivation of PP2C and the activation of SnRK2. The latter enzyme initiates ABA signalling by phosphorylating downstream targets such as ABA-response element-binding transcription factors (AREB/ABF), which in turn activate ion channels and NADPH oxidases. The putative ABA receptor ABAR initiates the second pathway. When the level of ABA is elevated, the WRKY40 transcription factor inhibits ABA signalling by interacting with ABAR. The interaction between ABAR and WRKY40 enhances the expression of ABA-responsive genes, such as transcription factors like ABA Insensitive, ABI5, and ABI4.

The ABA receptors PYR1 and ABAR, which specifically detect grape and strawberry, were recently discovered. The expression level of the PYR1-like gene (VlPYL1) in the 'Kyoho' grape was found to be highest in both fruit and leaf tissue. It exhibited an increasing trend during fruit development and subsequently decreased in mature berries. Increased expression of VlPYL1 enhanced the sensitivity of Arabidopsis plants to ABA. The upregulation of this gene resulted in an increase in the transcripts of ABA-responsive genes, such as the transcription factor ABF2 and β-glucosidases BG3, which play a critical role in ABA biosynthesis. In addition, it enhanced the quality of grape berries by elevating the levels of anthocyanins.

The initial indication of the ABA/PYR pathway in strawberries was observed through the distinct expression of genes responsible for protein phosphatase 2C (PP2C), RAS-related small GTP binding protein, and putative serine/threonine protein kinase (ARSK1) in achenes and receptacles during the ripening process. Subsequently, it was demonstrated that ABAR facilitates the maturation of strawberry fruit by forming a bond with ABA. The process of strawberry ripening is hindered by the downregulation of FaABAR, even in the presence of exogenous ABA. This indicates that the ABAR receptor is crucial for the ripening of strawberries. FaRIPK1, a gene encoding a leucine-rich repeat receptor-like kinase (LRR-RLK), is found in both the nucleus and plasma membrane. It plays a role in controlling the ripening of strawberry fruit by interacting with FaABAR. FaABAR detected the presence of ABA in order to control the activity of the transcription factor FaMYB10, which in turn controls the formation of anthocyanin during the ripening process of strawberries.

AREB/ABF, a highly renowned transcription factor in the ABA pathway, has also been investigated in the process of grape fruit ripening. ABA triggers the expression of VvABF2, often referred to as GRIP55, which progressively increases from véraison until the last stage of berry ripening in the 'Cabernet Sauvignon' grape variety. The study revealed that grape cells with an increased expression of VvABF2 had a more favourable response to ABA treatment compared to the control cells. This response resulted in changes to cell wall modification and an increase in the production of phenolic compounds. Tomato fruits that had an increased expression of VvABF2 exhibited reduced firmness, indicating that this transcription factor influences the ABA-dependent process in grape berry ripening.

Auxin:

***Climacteric fruits*:** Auxins are extensively researched regulators of growth and development in fruit. Auxins have been reported to function as regulators of fruit ripening in numerous studies. Indole-3-acetic acid (IAA), which is the most plentiful free auxin, is commonly recognised as the primary regulator in fruit. IAA concentrations are highest during fruit set and initial growth stages, but then decrease to low levels at the start of ripening in climacteric fruits like apple, mango, and tomato (Fig. 7). This decrease is associated with an increase in the conjugated form of IAA, known as IAA-Asp. Overexpressing a *Capsicum chinense* auxin-conjugating enzyme (GH3) in tomato results in a reduction in free IAA and an enhanced susceptibility to ethylene at an earlier developmental phase. The GH3 gene family consists of 15 members, but only SlGH3-1 and SlGH3-2 show increased activity during the beginning of fruit ripening. This suggests that these specific genes may play a role in the process of fruit ripening. Thus, the temporal regulation of ripening may be influenced by the ratio of IAA to IAA-Asp, rather than the actual concentration of IAA alone. Nevertheless, the mesocarp tissues of peach experience a concurrent rise in climacteric ethylene production when there is a notable elevation in IAA concentration. Additional research utilising a genomic methodology demonstrates that the hormone auxin has a significant role in the regulation of ripening in climacteric peach.

Furthermore, several genes responsible for the production of auxin (indole-3-glycerol phosphate synthase and tryptophan synthase beta subunit), its transportation, and its signaling (receptors such as TIR1, ARF, and Aux/IAA) exhibited heightened expression in the mesocarp during the ripening process. This observation provides evidence that auxin operates autonomously in the ripening of peach. The process of transmitting signals through auxin is thoroughly comprehended. In absence of auxin, the Aux/IAAs (proteins that are involved in the regulation of auxin and indole-acetic acids) bind with ARF (a protein that responds to auxin) and inhibit its activity. This inhibition stops the signaling process from progressing further. When the concentration of auxin is elevated, the receptor TIR1 (transport inhibitor response 1) identifies the auxin and facilitates the interaction between Aux/IAAs (auxin/indole-acetic acids proteins) and TIR1. Subsequently, the Aux/IAAs are eliminated by ubiquitin-mediated mechanisms, so liberating the ARF to induce the expression of genes further down the signaling pathway. ARF is the primary transcription factor in the auxin signaling pathway.

Apple MdARF5 expression is upregulated by treatment with synthetic auxin naphthaleneacetic acid (NAA). MdARF5 is a transcription factor that attaches to the regulatory regions of MdACS3a, MdACS1, and MdACO1 genes, thereby enhancing their transcription and facilitating the synthesis of ethylene. Treatment with exogenous IAA influences the ripening process in climacteric fruit. The delay in the ripening and softening of banana fruit when treated with IAA is caused by the suppression of cell wall hydrolases, specifically polygalacturonase, pectin methyl esterase, pectate lyase, and cellulase activities. Auxin therapy in tomatoes delays fruit ripening by postponing the onset of climacteric ethylene production by approximately 3 days compared to a control. Additionally, it suppresses the activity of ACO (1-aminocyclopropane-1-carboxylic acid oxidase) during the initial two weeks of storage following treatment.

On the other hand, the application of auxin or synthetic auxin stimulates the maturation process of pear (*Pyrus pyrifolia* cv. Whangkeumbae), apple, and plum. This stimulation leads to the activation of genes related to ethylene and an increase in ethylene synthesis. Auxin can promote the production of system 2 ethylene by triggering the activation of the important enzyme ACS, which controls the rate of ethylene synthesis. Thus, under these circumstances, any influence of auxin on the ripening process would be indirect and facilitated by ethylene. Significantly, the impact of auxin on ethylene production can vary according on the dosage in certain species. For instance, the application of NAA stimulated the synthesis of ethylene, leading to the acceleration of ripening processes in peach and plum fruits. Nevertheless, the use of auxin decreased the production of ethylene and postponed the ripening process in tomatoes. In addition, the application of a concentration of 100 $mmolL^{-1}$ IAA resulted in a delay in the ripening of bananas, but a concentration of 57 $mmolL^{-1}$ IAA had the opposite effect.

Non-climacteric fruits: Auxin is widely recognised for its crucial role in the growth of non-climacteric fruit. Grape blooms and young berries have been discovered to contain the highest concentrations of IAA. The levels gradually decrease to low levels during véraison during the ripening period. The negative impact of IAA on the accumulation of sugar and anthocyanin levels has been associated with a delay in the ripening process of grape berries. Achenes are widely regarded as the primary source of auxin in strawberries, playing a crucial role in the growth of the strawberry fruit. Removing achenes from immature receptacles halted the process of growth and development, resulting in the activation of genes associated with ripening. The application of auxin to strawberry fruit resulted in the downregulation of genes associated with the flavonoid pathway (such as chalcone synthase), fragrance production (such as alcohol acyltransferase), and cell wall modification (such as pectate lyase, endo-1,4-beta-glucanase, and expansins 1 and 2). Inducing higher expression of the DefH9-iaaM gene, which is involved in the production of auxin, in genetically modified raspberry and strawberry plants resulted in increased fruit size and yield. This suggests that the plant hormone IAA may have a significant impact on fruit development. There were no notable alterations observed in the firmness, total soluble solids, or titratable acidity of raspberry fruit during an in vitro IAA experiment conducted at the onset of ripening. There are still many aspects of IAA's function in raspberries and its interactions with other hormones that remain unknown.

Further investigation is required to explore the synthesis and transportation mechanisms of auxins in non-climacteric plants. IAA synthesis in Arabidopsis occurs through a two-step mechanism. The Tryptophan Aminotransferase of Arabidopsis1, also known as tryptophan Aminotransferase related (TAA1/TAR) proteins, convert tryptophan into indole-3-pyruvate. Subsequently, the YUCCA (YUC) group of flavin-containing monooxygenases catalyses the conversion of indole-3-pyruvate into IAA. The gene expression analysis of this biphasic process in grape berry development indicates that the synthesis of indole-3-acetic acid (IAA) is facilitated by the collaborative action of TAR and YUCCA proteins. During development and at the onset of ripening, there is a significant upregulation of three TAR genes and one YUCCA gene.

The genome of the wild strawberry (*Fragaria vesca* L.) contains four TAR genes. The letter "F". Based on an RNA-seq analysis of TAR-like genes in A. ananassa fruits, it was observed that three specific genes (FaTAA1, FaTAR1, and FaTAR2) were activated. During the ripening process of *F. ananassa*, specifically from the green to red stages, there was a decrease in the expression of genes associated with the production (FaTAA1 and FaTAR2), recognition (FaAux/IAAs and FaARFs), and transportation (FaPIN) of auxin. The TAA1/

TAR genes exhibited a consistent trend in the achene, with a significant decrease observed when the fruit transitioned from the green to white stage, followed by a continuing decline until reaching the red stage. The sole gene that exhibited significant expression was FaTAR2, which consistently increased from the green to red stages. Regarding the production of IAA, only five out of the nine YUCCA genes identified in *F. vesca were* found in *F. ananassa*. The achene consistently exhibited a transition from green to red stages. A few YUCCA genes were discovered in the red container, despite their low expression levels. A recent study investigated the function of the TAR and YUCCA genes by linking their promoters to a β-glucuronidase (GUS) reporter gene, which was then inserted into *F. vesca*. The investigation revealed that there were four FveYUCs and two FveTARs mostly present in the endosperm and embryo within the achenes. These studies demonstrate the crucial role of auxin production in the development and maturation of strawberry fruit's achenes and receptacle.

Throughout the process of plant growth, the AUX/LAX and PIN transport proteins facilitate the movement of auxin along polar axes. There is a scarcity of information of fruits that lack a climacteric stage. Scientists have discovered that the transportation of IAA, a plant hormone, decreases in the achene and receptacle of strawberries during the initial stages of fruit ripening and continues to slow down as the fruit ripens.

IAA conjugation to amino acids by IAA-amido synthetase (GH3) is a crucial aspect of auxin homeostasis. However, there is limited knowledge regarding the mechanisms of IAA conjugation in numerous plant species. The hormone's biologically active version is known as Free IAA. When amino acids bind to IAA, it undergoes inactivation, either through storage or degradation. Therefore, different forms of IAA are required to degrade IAA, store IAA, or inhibit auxin signalling pathways. The regulation of the pool of IAA conjugates in climacteric and non-climacteric fruits such as peach and grape is determined by the process of auxin being deactivated through conjugation and activated through conjugation hydrolysis. In peach, the presence of a protein named GH3 seems to regulate the presence of a possible IAA amidohydrolase, which is similar to Arabidopsis IAA-Leucine Resistant 1 (ILR1). This occurred due to the robust activation of both genes by NAA, which exclusively manifested during the fruit's ripening process.

Further investigation revealed that the VvGH3.1 enzyme has a role in the ripening of berries, and the addition of auxin and ABA increased the expression of the gene responsible for its production. These findings indicate that IAA undergoes inactivation through conjugation processes during the ripening of grapes. In addition to GH3, grapes exhibit reduced expression of genes encoding indole-3-acetate beta-glucosyltransferase (IAGLU) and auxin-amidohydrolase during ripening. IAGLU is responsible for auxin conjugation, whereas auxin-amidohydrolase activates auxin by breaking down conjugates. The expression of ILR1, a gene potentially associated with auxin activation by conjugation hydrolysis of IAA-amino acid hydrolase 1, increased. Transcriptomic investigation of the strawberry fruit revealed that the expression of six GH3 members was elevated in the achene during ripening. FaGH3.1 exhibited its maximum level of expression in the receptacle, but experienced a significant decrease from the green to red stages.

The only amino acid conjugates of IAA that were observed in the achene of *F. vesca* and other Fragaria species were aspartate and glutamate. These conjugates serve as a reservoir of free IAA for the growing plants. Recently, it was shown that prior to the ripening of raspberry fruits, there was an increase in their size and weight, accompanied by elevated levels of the IAA-amido synthetase RiGH3.1 transcript. During the in vitro treatment with IAA, the expression of RiGH3.1 was also enhanced. The RiGH3.5 transcript, however, was predominantly detected in flowers, and the application of IAA had minimal impact on its levels. Specifically, the deduced amino acid sequence for RiGH3.1 revealed the presence of binding motifs for IAA and aspartic acid. Given the coexistence of both motifs, it is plausible that IAA undergoes degradation during the ripening process of the raspberry.

Certain genes, such as the MADS-box genes, contribute to the growth and ripening of both climacteric and non-climacteric fruits. These findings indicate that these genes likely have a longstanding role in the ripening process of fruits. The potential MADS-box protein Ripening Inhibitor (RIN) is a transcription factor that is activated by auxin and regulates gene expression in response to ethylene. It has been suggested that the RIN MADS-box transcription factor in tomato fruit can bind to the promoter regions of genes responsible for producing cell wall hydrolases and ACS. Recent research has indicated that the presence of RIN is unnecessary for initiating the ripening process in tomatoes. Therefore, the presence of a RIN homologue could account for the expression of ripening-related genes in RIN knock-out mutants. Alternative MADS-box protein complexes capable of binding to RIN recognition sites have the potential to replace RIN.

Gibberellic acid:

Climacteric fruits: Gibberellins (GAs) are a group of complex organic compounds of class tetracyclic diterpenoid that act as hormones in higher plants. They play a crucial role in controlling several developmental processes, including seed germination, stalk growth, flower initiation, and fruit development. The primary emphasis of recent studies on GAs has been on the growth and maturation of seeds, the process of flowering, and fruit development. This focus is driven by the significant levels of GAs that are present in flowers and immature fruits. Only a select few out of several hundred plant GAs exhibit bioactivity in higher plants, including GA1, GA3, GA4, and GA7. GA1 and GA4 are present in considerable quantities, while GA3 and GA7 are found in lower quantities. Gibberellins (GAs) build up in fruit as cells divide and expand in the early stages of fruit growth, but their concentration decreases significantly during fruit ripening (Fig.7). For instance, the levels of naturally occurring GAs decrease dramatically prior to the rise in ethylene production during the transition from the mature-green (MG) to the breaker stage in tomatoes. However, the levels of SlACS2, SlACS4, and SlACO1 transcripts experience a notable increase. In addition, when the fruit-specific SlGA2ox1 gene is overexpressed, it leads to the production of a catabolic enzyme that transforms the active forms of GA (GA1 and GA4) into inactive ones (GA8 and GA34). This genetic modification causes the tomato plants to exhibit early-maturing characteristics. By administering the GA biosynthesis inhibitor prohexadione-Ca to MG tomato, the process of fruit ripening was expedited.

In addition, the application of GA3 externally can decrease the production of ethylene and inhibit the ripening process of several climacteric fruits, such as banana, persimmon, mango, and tomato. The application of GA3 to tomato plants resulted in a decrease in the transcriptional levels of SlACS2, SlACS4, and SlACO1. Additionally, the expression of SlETR3, SlETR4, and SlEIN2 exhibited a significant fall. These findings suggest that GAs suppress both the production and detection of ethylene during the ripening of fruits. The data indicate that GA functions as a suppressor of fruit ripening. Nevertheless, the precise method by which signalling genes of GAs interact with ethylene biosynthesis genes is still not fully understood.

Non-climacteric fruits: In 1969, the first report on the use of exogenous GA and its effects on the growth of strawberry receptacles was published. In the following years, numerous further researches have complemented the relationship between GA and fruit ripening, particularly strawberries. It has been found that bioactive GA1, GA3, and GA4 were present during the development of strawberry fruit. During the white stage of strawberry fruit development, the receptacle tissue was found to have high quantities of GA4, with the highest level being recorded at that time. In addition, the expression of genes encoding for GA pathway components, including the receptors, such as the FaGID1 (Gibberellin-Insensitive Dwarf1b), and FaGID1c, the Della FaRGA (Repressor of GA) and FaGAI (GA-Insensitive), and enzymes related to GA biosynthesis such as gibberellin n-oxidase (FaGA3ox), and GA hydrolysis (FaGA2ox) were also reported at high levels in the receptacle tissue during strawberry fruit development. It appears that GA had a significant part in the formation of the strawberry receptacle, as suggested by these antecedents.

The overexpression of the Gibberellin Stimulated Transcript 2 (FaGAST2) gene in various strawberry transgenic lines resulted in a reduction in fruit size. This reduction in fruit size demonstrated that the parenchymal cells were smaller than those observed in the control fruits. This finding suggests that there is a connection between the expression of FaGAST2, strawberry fruit cell elongation, and the subsequent increase in fruit size. An increase in FaGAST1 expression was observed when the expression of FaGAST2 was silenced using RNA interference techniques. This occurred without any changes in the size of the fruit cells, which suggests that both transcription factors were involved in the process of fruit development and ripening, which ultimately determined the size of the strawberry fruit cell.

The exogenous pre-bloom treatment of gibberellic acid (GA3) to grape cultivars, which is often employed to establish early ripening, also caused seedlessness and boosted berry size in seedless cultivars. This was accomplished by applying GA3 to grapes before the blooming stage. The application of GA3 to inflorescences during the pre-bloom stage of the 'Kyoho' grape plant resulted in the stimulation of flower opening, the promotion of fruit colouring, and the abortion of seeds. In cloudberry (*Rubus chamaemorus* L.), a wild berry native to the northern region, the treatment of some GAs externally resulted in the growth of parthenocarpic fruit. On the other hand, the application of prohexadione–calcium (ProCa), an inhibitor of gibberellin production, resulted in a reduction in the number of flowers in raspberry.

Brassinosteroids:

Climacteric fruits: Brassinosteroids, or BRs, are essential steroid hormones that stimulate growth and regulate male fertility, pollen production, photomorphogenesis, fruit maturation and ageing, and enhance cell enlargement and division. Furthermore, they regulate plant resilience to environmental stress at every stage of the plant's life cycle. BR production is an ongoing process that happens during the growth of tomato fruits. Recent studies have shown that BRs may also have a role in regulating the ripening process of tomato fruits. Brassinolide, the most potent brassinosteroid (BR), was administered to tomato fruit. This treatment led to the hastening of the ripening process, elevation of vitamin C and carotenoid levels, prevention of excessive accumulation of soluble solids, enhancement of respiratory activity and ethylene synthesis, and reduction in chlorophyll concentrations. However, the successful application of brassinazole, a synthetic inhibitor of brassinosteroids (BR), on tomato plants effectively blocked the expression of genes responsible for ethylene synthesis and the production of ethylene. Compared to untreated control fruit, the application of epibrassinolide (EBR), a brassinosteroid, effectively inhibited the production of ethylene and maintained the firmness of pear (*P. ussuriensis*) fruit during storage.

Extensive research has been conducted on the BR signalling system in plants. Brassinosteroid Insensitive 1 (BRI1) is the receptor that binds with BR after it has been synthesised by biosynthesis. BRI1 phosphorylates BSK1 through its interaction with BRI1-associated kinase 1, leading to the transphosphorylation of BRI1. The phosphorylated BSK1 activates BRI Suppressor 1, which in turn dephosphorylates and inhibits Brassinosteroid Insensitive 2. As a consequence, unphosphorylated Brassinazole-Resistant 1 (BZR1) is produced, and its counterpart, BRI1-Emssuppressor 1 (BES1), is transported across the membrane into the nucleus. BZR1 and BES1 are important transcription factors (TFs) in the downstream part of the BR signal transduction pathway. They specifically attach to the promoters of BR-responsive genes and control the expression of these genes.

Indeed, endogenous BR does influence the process of fruit ripening. Overexpressing the BR biosynthetic gene Dwarf in tomato leads to an augmentation in the levels of endogenous BR, which in turn causes a concomitant rise in ethylene production in the fruit. When SlBRI1 was overexpressed in tomato plants, there was a noticeable rise in the amounts of carotenoids, ascorbic acid, soluble solids, and soluble sugars during the fruit ripening phase. The increase in ethylene synthesis and fruit ripening was accompanied by an increase in the expression of the genes involved for ethylene generation, including SlACO1, SlACS2, and SlACS4. These findings suggest that BRs enhance the rate of fruit ripening by promoting the activation of genes involved in ethylene production and the synthesis of ethylene.

Furthermore, several horticultural crops have demonstrated that BRs possess the capacity to expedite the maturation of fruit. A notable example of this is that the introduction of external BRs improved the ripening of bananas, maybe due to the heightened activity of MaACS1 and MaACO13/14 genes, which subsequently stimulated the production of ethylene. Additionally, there have been reports of bioactive compounds that possess the capacity to impede the natural ripening process of fruits. The study found that the levels of naturally occurring BRs, such as typhasterol (TY), 6-deoxocastasterone (6-deoxoCS), and castasterone (CS), declined progressively as the pear (*P. ussuriensis*) fruit developed (Fig. 7). In addition, treatment with 0.2, 3, or 10 μmolL–1 EBR effectively suppressed ethylene production and preserved the firmness of the fruit, as compared to the untreated control fruit. In addition, it was noted that the activation of TF expression occurs in the presence of substantial amounts of BRs. PuBZR1 directly interacts with PuACO1 in the cytoplasm, inhibiting its enzymatic activity. Additionally, in the nucleus, PuBZR1 regulates the transcription of PuACO1 and PuACS1a. As a consequence, there is a reduction in the production of ethylene and an inhibition of the fruit ripening process. Applying 5 μmolL–1 exogenous BRs to jujube (*Zizyphus jujuba*) led to a notable decrease in ethylene production during storage. However, the addition of 10 μmolL–1 exogenous BRs stimulated the synthesis of ethylene and hastened the ripening process of the fruit. The administered dose may potentially impact the regulation of BRs on ethylene production, either positively or adversely. These data suggest that the effect of BRs on ethylene generation and fruit ripening varies among different species and is also influenced by the dosage of BRs.

Non-climacteric fruits: BRs have been associated with the growth and maturation of cucumber, grape, and strawberry fruits. Applying BRs, such as epibrassinolide and BL, to grape berries enhanced the hue and ripening of the fruit. Conversely, the injection of a BR biosynthesis inhibitor, such as brassinazole (BZ), showed a contrasting impact. BR 6-oxidase is the catalyst responsible for the conversion of 6-deoxocastasterone into the biologically active BR castasterone. BL treatment in strawberry fruit expedited the ripening process and enhanced the accumulation of FaBRI1 receptor transcript. In addition, temporary suppression of the FaBRI1 gene caused a postponement in the process of fruit ripening, resulting in a white receptacle. The results suggest that BR signalling may have a crucial function in the ripening of non-climacteric fruits.

Bud break (BR) has been hypothesised as the initial indication of grape berry ripening, potentially by regulating the level of ethylene. The gene encoding the late embryogenesis abundant (LEA) domain protein (LDP1) was discovered to be expressed in the early developmental stages of *F. chiloensis* and *F. vesca*, particularly in the receptacles, in relation to BR-responsive genes. The promoter region of the LDP1 gene harbours several BRs (brassinosteroids) and ABA (abscisic acid) response motifs. BRs and ABA have been found to regulate the transiently generated FcLDP1 promoter-GFP fusions, suggesting that these two hormones control the expression of FcLDP1 during fruit development in *F. chiloensis*.

Jasmonates:

Climacteric fruits: Most studies that have been conducted on jasmonates (JAs) in plants have focused on the physiological responses of plants to both biotic and abiotic stresses. Furthermore, it has been proven that JA plays a substantial role in the process of fruit ripening at every stage. There were larger amounts of JA present in the apple fruit during the early stages of its growth. These levels decreased as the fruit grows, but then they increased once more just prior to maturity (Fig. 7). The application of methyl jasmonate (MeJA) to apple causes an increase in the synthesis of ethylene, which in turn speeds up the process of fruit ripening. The application of MeJA was shown to have boosted the production of ethylene and hastened the ripening process in mango fruit. MeJA application resulted in a significant improvement in the ripening process of plums as well as the overall production of ethylene. Throughout the whole process of fruit ripening, the administration of MeJA resulted in an increase in the production of ethylene on tomato plants.

There is a substantial amount of information available regarding the signal transduction of JA, with the transcription factor MYC being largely acknowledged as the primary regulator. JA is found to be responsible for the stimulation of the synthesis of ethylene in apples, and the researchers found that this stimulation is dependent on MdMYC2. According to the findings, JA-activated MdMYC2 engages in direct interactions with the promoters of both MdACS1 and MdACO1, hence promoting the transcription of both of these specific genes. In addition to this, MdMYC2 establishes a connection with the promoter of MdERF3, which in turn causes the transcription of MdACS1 to become active. The ability of MdERF2 to repress both MdERF3 and MdACS1 is inhibited as a result of MdMYC2's interaction with MdERF2, which is another important finding. Jasmonates, often known as JAs, are responsible for not only controlling the process of fruit ripening but also having an effect on the colours that develop in fruit. The use of JA has the potential to considerably increase the colour development of apples; however, it may also limit the capacity of the fruit to store nutrients since it stimulates the generation of ethylene. The application of MeJA to apple trees (cv. Hanfu) three weeks prior to harvesting resulted in an improvement in the development of fruit colour without influencing the production of ethylene or the firmness of the fruit during storage.

Non climacteric fruits: Jasmonic acid (JA) and its bioactive isoleucine conjugate (JA-Ile) are significant signals for both biotic and abiotic stress responses in plants, including roles in root growth, seed germination, stamen formation, and senescence. Recently, a synchronised downregulation of JA endogenous levels, including JA-Ile, and their biosynthetic genes has been observed in strawberry fruit from flowering to ripening. Furthermore, increased amounts of JA and JA-Ile have been seen in early developing grape berries, followed by a dramatic decrease as they proceed to the mature stage. Thus, JA-Ile could be linked to early fruit development in both strawberries and grapes. It was discovered that proanthocyanidin (PA) accumulation in developing strawberry and grape berries followed a similar pattern as JA-Ile, beginning early in fruit development and reducing as the fruit ripened. An increase in PA content was seen following the administration of a pharmacological inhibitor to the major JA-Ile biosynthetic enzyme JAR1, implying that the JA pathway may be linked to PA biosynthesis in strawberry fruit

Exogenous application of JA's methyl ester (methyl jasmonate, MeJA) to *F. chiloensis* fruit altered the expression profiles of ripening-related genes, including those involved in ethylene and JA production. Furthermore, MeJA treatment boosted anthocyanin accumulation by upregulating the expression of genes involved in the anthocyanin biosynthesis pathway and flavonoid production.

In addition to increasing anthocyanin levels, MeJA increased the expression of JA biosynthesis genes in *F. chiloensis* fruits, including 13-lipoxygenase, allene oxidase synthase, and 12-oxophytodienoate reductase 3. Exogenous administration of MeJA to *F. X ananassa* fruit resulted in increased anthocyanin, JA, JA-Ile, and MeJA levels. In grape cell suspensions, JA boosts anthocyanin synthesis, while MeJA treatment increases PA concentration in two wine grape varietals. One of the features of raspberry that has received increasing attention is the molecular and metabolic processes that influence colour. Jasmonates were found to boost PAL activity and significantly increase polyphenol molecules such ellagic acid, quercetin, and myricetin.

Cytokinins:

Climacteric fruits: Studies on the impact of cytokinins (CKs) on fruit ripening have been limited to climacteric fruit for the most part. Cytokinins (CKs) play an important role in the many stages of plant growth and development. The ripening of tomatoes was shown to be progressively slowed down by increasing doses of endogenous CKs, as shown in Figure, according to the published research. When compared to the ripening fruit of a control cultivar (cv. Rutgers), the fruit of a non-ripening mutant rin (ripening inhibitor) of tomato possessed a larger levels of endogenous cytokinin, according to the findings of the researchers. The endogenous CK activity was found to be much higher in MG tomato fruit compared to mature tomato fruit, according to the findings of another investigation. Based on these observations, it appears that being exposed to a high concentration or activity of CKs may prevent fruit from ripening. A recent study found that kiwifruit fruit that had been treated with synthetic cytokinin N-(2-chloro-4-pyridyl)-N'-phenylurea inhibited the production of ethylene and maintained the softening of the fruit. This finding suggests that CKs cause tissue-dependent disruptions of proteins during the climacteric ripening process.

Non climacteric fruits: Further work is needed to understand the role of cytokinins (CKs) in the development and ripening of non-climacteric fruits. Two genes, a histidine phosphotransferase protein (AHP) and a nuclear response regulator (ARR) gene, have been identified as being connected with the CK signal transduction system. These genes were found in fruit cDNA libraries of various kinds of F. × ananassa. Several years later, researchers conducted a study comparing the transcriptomes of *F. vesca* throughout different phases of fruit development before and after fertilisation. They identified 17 genes that were expressed differently in four different fruit tissues and were associated with cytokinin (CK) manufacture, signalling, and degradation. According to reports, CKs play a crucial function in the early stages of strawberry fruit development, similar to their role in the initial development of climacteric fruits like tomatoes. Forchlorfenuron, a synthetic cytokinin also known as CPPU, has been found to contribute to the increase in grape berry weight. However, it has been discovered that the use of CPPU leads to a drop in sugar and anthocyanin content. The sole publication elucidating the function of CKs in the Rubus genus proposes that GAs may have a synergistic effect with CKs in the process of flower induction in raspberries.

Hormonal interactions

The process of fruit ripening is intricate, comprising complex alterations in physiological and metabolic characteristics, and is regulated by many hormones. Nevertheless, it seems that ethylene and ABA play a major role in regulating fruit ripening, with ethylene being notably necessary for the ripening of climacteric fruits. Other plant hormones mainly exert their effects by making small modifications to the activity of ethylene throughout the process of fruit ripening. As an illustration, the exogenous application of ABA stimulated the production of ethylene by activating the transcription of genes involved in ethylene production. Conversely, the use of the ABA biosynthetic inhibitor, fluridone, suppressed the production of ethylene in tomato fruit. Furthermore, the activation of an auxin-related GH3 gene by ABA, ethylene, and various other phytohormones during the ripening of tomatoes suggests that, in addition to ethylene, auxin can interact with ABA and other hormones.

The interaction between IAA and ethylene during ripening appears to be more intricate and captivating when compared to ABA. Firstly, it has been demonstrated that in peach, which is a climacteric fruit, there is a simultaneous increase in both ethylene production and auxin levels. In contrast, during the ripening of pear (*P. ussuriensis*), ethylene supports the reduction of IAA through the activation of PuERF-activated PuGH3.1. In apple, peach, and tomato, the genes responsible for ethylene biosynthesis (ACSs and ACOs) and the signal pathway (ETRs and ERFs) are activated by auxin. However, IAA application resulted in a decrease in ethylene production and a postponement of the ripening process in both bananas and tomatoes. In addition, ethylene can induce the upregulation of the components involved in the production and signalling of IAA during the ripening process. The data suggest that the process of fruit ripening is influenced by the coordination of numerous hormones.

Polyamines (PAs) are widely recognised for their role as cellular signals in the intricate interplay of hormonal pathways, including as ethylene, ABA, and auxin. Antagonistic interactions occur between ethylene and PAs during the ripening of fruits after they have been harvested. Both PAs and salicylic acid (SA) inhibit the expression of the ACS gene and the production of ethylene in tomato plants. Additionally, the administration of PA stimulates the synthesis of nitric oxide in olive fruit. Crucially, it has been demonstrated that nitric oxide can disrupt the impact of ethylene on fruit ripening, exerting a direct and substantial influence. The interaction between nitric oxide and SA and JA has been demonstrated to impact Post-harvest characteristics in certain climacteric fruits. However, the specific mechanism behind this interaction has yet to be fully understood.

Additionally, JA is a hormone that has been extensively researched for its ability to stimulate ethylene production and the ripening process in climacteric fruits such as apple, mango, plum, and tomato. Furthermore, the levels of endogenous JA-Me also increase when these fruits ripen, namely in apple, mango, pear, and tomato. In addition to these hormones, GAs have been reported to suppress ethylene production in various fruits including mango, peach and tomato. Currently, there is a lack of comprehensive data regarding the impact of ethylene on the biosynthesis of JA or GA, as well as the signalling pathways involved in fruit ripening. Current evidence suggests that ethylene may serve as the target for hormonal communication during the ripening of climacteric fruits. Undoubtedly, the coordination of ethylene signalling in climacteric fruit ripening is closely regulated by many phytohormones. Exploring the molecular basis of hormone regulation in fruit ripening is challenging due to the intricate nature of the ripening processes and the interference between different hormones.

Delaying fruit ripening

Ripening is an intrinsic process that triggers a sequence of biochemical transformations that result in the alteration of colour, creation of pigments, breakdown of starch, changes in texture, synthesis of volatile compounds and aroma, and ultimately the detachment of fruits from their stems. Ethylene is a naturally occurring plant hormone that has many impacts on plant growth and development, in addition to its involvement in regulating the ripening process of climacteric fruits. Plant cells possess ethylene binding receptors that feature a specific ethylene binding site. When ethylene binds to these receptors, it activates them and initiates the process of fruit ripening. When ethylene molecules bind to receptors, they trigger a sequence of events that transmit chemical messages within the cells of the fruit. These molecular interactions cause the fruits to ripen by altering the colour, flavour, scent, and composition of the fruit.

However, it has been found that ethylene functions as a rheostat rather than a trigger for the ripening of climacteric fruits. This suggests that ethylene must be consistently present in order to sustain the transcription of essential genes needed for fruit ripening. In order to extend the shelf-life of climacteric fruits, it is necessary to impede or decelerate the activity of ethylene gas, which will in turn decrease the rate at which fresh fruits ripen. Certain compounds were discovered to inhibit the basal level response of ethylene in fruits, resulting in a significant delay in the natural ripening process. These ethylene inhibitors interact with the ethylene receptor and hinder the activity of ethylene. Consequently, signals are not transmitted for the subsequent encounter, resulting in a postponement of the ripening process. However, the interaction between ethylene and receptors is a constantly changing process, and the inhibitors are unable to permanently attach to the receptor sites. Once all the chemicals that inhibit ethylene are depleted, new receptor sites are created, allowing ethylene to restore its sensitivity to them and initiate the ripening process.

Delaying fruit ripening is essential for extending shelf-life, reducing waste, and ensuring fruits reach consumers in optimal condition. Several post harvest intervention and technologies are used to achieve the objective. Fruit ripening may be regulated with the help of the following approaches:

Physical regulation of fruit ripening

Physical regulation:

Temperature: Temperature plays an important role in ripening process and may have high temperature or low temperature influence on the ripening process.

a. *High temperature*: Warm temperatures can expedite the formation of ethylene gas, which in turn hastens the ripening process of fruits. Fruits generally undergo ripening on the tree, which is a physiological process that consistently results in textural changes that indicate softening. However, the African pear fruit is unique because it remains firm on the tree even when it is mature, and only softens after it is picked and subjected to heat. Usually, the flesh of a pear is made softer by submerging it in hot water for a brief period before eating. When the pulp is subjected to moderate heat, the melting of the pulp fat causes a slight increase in volume. This expansion induces a specific level of pressure on the delicate epicarp, leading to its rupture. Usually, oils are discharged through these cracks, which are a sign of the last phase of softening. Therefore, the cracks in the pear pulp can be used as a reliable indicator of softening. Research has shown that the pear puree may be made softer by heating it to different temperatures, ranging from 35 to 85 degrees Celsius. However, the most optimal temperature range for achieving this softening effect is between 60 and 85 degrees Celsius. The pear pulp underwent solidification when exposed to temperatures above 85°C and did not revert back to its original state upon subsequent cooling. This implies that specific essential elements of the softening process are susceptible to damage from heat. Enzymes are the most likely contenders. The analysis of the pear pulp revealed that the enzymes pectin esterase, proteinase, polygalacturonase, and cellulase are most effective in softening the pear pulp while operating at temperatures between 60-85°C. Moreover, the functioning of these enzymes significantly decreased at temperatures over 85°C, which aligned with the temperature at which softening was minimal. The results strongly suggest that the softening of the pear pulp is caused by the heat-induced cell wall degrading enzymes that naturally occur in the pulp.

b. *Low temperature:* Typically, a decrease in temperature inhibits the maturation and ageing of fruit, and it also retards the process of softening or hardening that occurs during ripening. These effects occur when some enzymes involved in respiration, the initiation and control of ripening, including those associated with ethylene synthesis and signalling, as well as several enzymes specific to ripening, are suppressed. The temperature had a significant influence on the softening and breakdown of pectin in ripening durian fruits, as evidenced by the rapid softening observed when the fruit was kept at a temperature of 27 degrees Celsius. At a temperature of 27 degrees Celsius, the storage of durian fruit resulted in a notable increase in the activity of polygalacturonase. Conversely, when the fruit was kept at a temperature of 12 degrees Celsius, there was minimal softening and negligible changes in the water-soluble pectins. A temperature of 12 degrees Celsius has no impact on the hardness of durians. Based on this, it seems that a decrease in temperature inhibits the activity of enzymes involved in the degradation of pectin.

***Humidity*:** The process of fruit ripening is significantly affected by the level of relative humidity. Elevated levels of relative humidity (95-100%) can maintain the moisture content of fruit, resulting in a slower ripening process and the preservation of fruit quality. Conversely, decreased relative humidity (RH) levels facilitate evaporation of water, accelerating the ripening process. Reducing relative humidity (RH) is particularly crucial for post-harvest management since it can enhance the longevity of fruits and decrease decay. During the ripening process, there are biochemical transformations such as the conversion of starch into sugars, alterations in colour, and the creation of taste compounds. These changes are regulated by the levels of relative humidity (RH). For instance, it has been shown that the use of hexanal can effectively postpone the ripening of mangoes by a minimum of two days in typical conditions, without compromising the quality of the fruit. Comprehending these physiological and biochemical relationships is crucial for enhancing storage conditions and maximising the lifespan and quality of harvested fruits.

Several fruit ripening tests control temperature but do not control relative humidity. Nevertheless, low relative humidity (RH) clearly affects fruit ripening, as demonstrated by the inconsistencies observed in the ripening process of green bananas kept under RH levels below 80 per cent. When fruit is stored in conditions with low relative humidity (RH), it will rapidly lose water, resulting in wilting signs. Low relative humidity (RH) speeds up the ripening process in several fruits, such as avocado, mango, banana, plantains, and pear fruit. Therefore, specific RH criteria have been established for most fruits to prevent moisture loss and maintain their quality. When fruits are kept in low relative humidity (RH) conditions, they may generate ethylene due to stress, which can lead to various physiological issues. In addition to the loss of mass and wilting symptoms, water deficit stress resulting from low relative humidity (RH) has a negative impact on the quality of fruit preservation. This is due to the fact that it triggers further physiological disorders through the production of ethylene generated by stress. Japanese persimmon 'Tonewase' fruits undergo rapid softening when stored in low humidity conditions (40-60 % RH) and exhibit increased ethylene production. Conversely, when stored in high humidity conditions (>95 % RH), the fruits do not soften quickly and do not produce measurable levels of ethylene.

Storage atmosphere control: Storage atmosphere control is commonly known as controlled atmosphere (CA) or modified atmosphere (MA) storage, although the latter does not include strict regulation of the environment. The main objective of storage atmosphere control is to regulate the levels of oxygen and carbon dioxide. The ripening of many fruits can be decelerated by low oxygen levels, which can be caused by either the infusion of nitrogen gas or the fruit's natural respiration that consumes oxygen from the surroundings. For instance, 'Bosc' pears were kept in extremely low oxygen conditions (0.5% oxygen for four months, 1.0% oxygen for two months, and 1.5% oxygen for two months), with carbon dioxide levels below 0.05% during the entire storage period. These pears were stored for six, seven, and eight months, and they maintained a physiological state similar to fruit stored in normal air conditions for one to three months. This was ascertained through the synthesis of ethylene, the gradual softening of the fruit, and the decrease in the amount of juice that can be extracted during the process of ripening. High concentrations of oxygen (60 or 100%) caused the banana fruit to soften at a faster pace over a five-day period. This behaviour could be associated with the heightened production of novel ethylene binding sites that arises from high O_2 treatments.

Controlled atmosphere storage. Controlled atmosphere (CA) storage is an intricate technology that effectively maintains the quality of fruit throughout storage, mostly by decelerating the ripening process. This method involves establishing and sustaining an optimal environment around the stored fruits, typically by reducing oxygen levels and increasing carbon dioxide levels in storage facilities. CA storage has a biological impact on fruit ripening by decreasing the production and activity of ethylene, a hormone that plays a crucial role in the ripening process. By altering the environment, CA storage can significantly decrease respiration rates and other metabolic activities that contribute to the ripening process. CA storage can enhance firmness retention by reducing the activity of enzymes that degrade cell walls. It can also preserve flavour by limiting the loss of acidity and the conversion of starch into sugar.

Research has demonstrated that CA preservation can effectively reduce the respiration rate of fruits and vegetables under some circumstances. High amounts of carbon dioxide (CO_2) or low levels of oxygen (O_2) can either have no impact or actually increase the rates of respiration in certain crops. There are multiple factors that contribute to this variability. Temperature interactions can modify the crop's metabolism, causing it to become anaerobic and thereby increasing its productivity. Similarly, same principle would be applicable if the oxygen content were insufficient. Elevated carbon dioxide (CO_2) concentrations can have detrimental effects on crops, perhaps impairing their respiratory rate. These alterations in respiration rate could potentially affect the overall quality of fruits and vegetables. Crops stored in controlled atmospheres generally have an extended shelf life as a result of reduced metabolic activity. This would impede the process of ripening and decay, particularly for climacteric fruit, resulting in a lower level of maturity compared to fruits preserved in air when stored for extended periods.

Modified atmosphere packaging MAP, or modified atmosphere packaging, can extend the shelf-life of food by either removing or replacing the air surrounding the food before sealing it in vapour barrier materials. Research has shown that this method can prolong the initial freshness of food products, and it has been employed for preserving perishable food items such as vegetables, fruits, meat, and fish. This method can extend the post-harvest lifespan of the product by reducing the amount of oxygen (O_2) and adjusting the quantities of carbon dioxide (CO_2) and nitrogen (N_2). Elevated levels of oxygen promote oxidative processes, potentially impacting the product's quality. Reducing the oxygen levels in the air decelerates respiration, ethylene production, enzyme browning, and metabolic activity. Respiration initiates with the oxidation of complex substrate molecules such as starch, sugar, and organic acids. This process leads to the formation of simpler molecules, namely carbon dioxide and water, along with the release of heat. MAP has the capability to retard the ripening process, hinder the proliferation of microbes, result in packaging without any odour, and prolong the lifespan of food items. Utilising a gas blend including 4 per cent oxygen, 10 per cent carbon dioxide, and 86 per cent nitrogen has demonstrated advantageous effects on the duration that a product can be stored without spoiling. These effects encompass the inhibition of microbial growth and the maintenance of the texture and colour of mango and pineapple.

Hypobaric storage: In the 1990s, there was an increasing demand for all the fruit sold in supermarkets to be uniformly ripe. This was done to guarantee that the fruit had a shelf-life that was both satisfactory and foreseeable. This led to the development of a technique called pressure ripening, principally used for bananas but applicable to any climacteric fruit. The procedure involves directing the circulating air in the region where the fruit is ripening through the fruit boxes. This guarantees that the external ethylene gas, which is accountable for the process of ripening, makes contact with every individual fruit in the room. However, carbon dioxide, which can impede the initiation of the ripening process, must not accumulate near the fruit. Hypobaric storage, sometimes referred to as low-pressure storage, has been used for several decades to control the partial pressure of oxygen in storage facilities for fruits and vegetables. The metric unit used to quantify pressure is the millimetre of mercury (mmHg), with the atmospheric pressure at sea level being 760 mmHg. Alternative terms for this concept include "one atmosphere" and lower pressures stated as a fraction or a percentage of the standard atmospheric pressure. 76 millimetres of mercury is equivalent to 0.1 atmosphere, whereas the partial pressure of oxygen is roughly 2.09 per cent. To precisely measure the partial pressure of oxygen in the storage, it is essential to account for the presence of water vapour in the atmosphere. In order to achieve this, it is necessary to measure the relative humidity.

Research has shown that preserving bananas in hypobaric settings increases their longevity after being harvested. Under the conditions of a temperature of fourteen degrees Celsius and a pressure of 760 mm Hg, bananas could be stored for thirty days. However, if the pressure was reduced to either eighty or one hundred fifty mm Hg, the bananas would remain unripe for a longer time of one hundred twenty days. After the ripening phase, it was noted that these fruits had a really enjoyable taste, scent, and consistency. Bananas housed in a hypobaric atmosphere for three hours per day do not experience a significant increase in respiration rate, and they remain green for a period of fifteen days. The Post-harvest lifespan of pineapple was prolonged by approximately thirty to forty days when it was stored in hypobaric conditions. Storing grapefruit under low pressure conditions, known as hypobaric storage, at a temperature of 4.5 degrees Celsius and a pressure of 380 millimetres of mercury (the minimum limit of the experimental equipment), did not affect the occurrence of chilling injury. Hypobaric storage extended the post-harvest lifespan of guavas mainly by inhibiting ripening. Reports indicate that 'Hass' avocados can be stored for up to seventy days under low temperature conditions and a pressure of sixty millimetres of mercury. The pre-treatment of 'Yumyung' and 'Changhowon' peaches, which were held in a mixture containing 6 per cent oxygen and 18 per cent carbon dioxide at pressures ranging from 0 to 294.2 kPa, demonstrated that this process enhanced the preservation of firmness in both types of peaches. Nevertheless, this treatment did not provide any impact on any other assessed quality parameter throughout the ensuing three-week storage period in air at either 7 or 10 degrees Celsius.

Intelligent packaging systems: It refers to a sensor that detects the environment and characteristics of food, as well as processes information. Intelligent packaging performs many activities such as recording, detecting, sensing, tracing, and communicating information. These functions aim to prolong the product's shelf life, provide information, enhance quality, and issue warnings regarding any external or internal problems that may exist in the product. Intelligent packaging encompasses various types of indicators, including time-temperature indicators (TTIs), gas sensors, ripeness indicators, toxin indicators, and biosensors. Additionally, radio frequency identification (RFID) technology is used to directly evaluate the quality of the product inside the package.

Irradiation: Irradiation can be efficiently utilised on fruit for several purposes, including its application in quarantine treatments to eliminate pests. Nevertheless, if an incorrect dosage is administered, irradiation may have adverse consequences on the quality of the fruit. The impact of irradiation on fruit texture alterations varies according on the specific fruit crop and the dosage of irradiation used. The softening of 'Tommy Atkins' mangoes was caused by irradiating them with electron beams at doses of either 1.5 kGy or 3.1 kGy. Nevertheless, doses of up to 1.0 kGy will maintain the physical, textural, and microstructural qualities of the product and also improve its shelf life when maintained at a temperature of 12 °C for a maximum of 21 days. The softening of the fruits caused by radiation may be linked to alterations in the structural integrity of the cells, such as the presence of cracks and depressions on the surface, as well as the disintegration of the cells and their constituents. When strawberries at the 50 per cent red ripening stage were exposed to a non-lethal dosage of UV-C (4.1 kJm−2) and stored at 20 ◦C for 96 hours, it was observed that the softening of the fruit was delayed. The delay in softening was caused by a decrease in the disintegration of the cell wall, with the cell wall degrading enzymes being the specific targets of UV-C radiation.

Chemical regulations:

Salicylic acid: Salicylic acid (SA) is a widely distributed phenolic molecule in plants that plays a crucial role in regulating various plant processes and serves as a key component in the signal transduction pathway. SA also participates in both local and systemic resistance against many infections. Sodium azide (SA) was discovered to suppress the effect of ethylene on many climacteric fruits. It was stated that it effectively delayed the ripening of banana fruit. The enzyme antioxidants, catalase and peroxidase, were seen to be diminished in the presence of SA during the ripening of bananas. Sodium acetate (SA) was also discovered to significantly prolong the ripening process of strawberries, effectively preserving the fruit's quality. Succinic acid (SA) and its derivative, acetylsalicylic acid (ASA), have been found to inhibit the process of fruit softening in kiwifruit. This is most likely accomplished by either inhibiting the production or blocking the effects of ethylene. The inhibitory effect of ethylene production by salicylic acid has been shown in tomato ACC Synthase activity and apple ACC Oxidase. Research has shown that applying SA before and after harvesting can help preserve the quality of strawberry fruit during storage, hence extending its post-harvest lifespan.

The application of SA and $KMnO_4$ treatments, either separately or in conjunction, proved to be a highly efficient approach for prolonging the storage life of kiwifruit. During the extended storage time, the application of 1 mM SA followed by $KMnO_4$ was observed to be beneficial in maintaining the quality of kiwifruit without any apparent alterations. It was also proposed that kiwifruit can be preserved in polyethylene bags (specifically, LDPE-low density polyethylene, with a thickness of 13 μm) at a temperature of 0°C and a relative humidity of 85-95 per cent for a duration of 200 days, with minimal loss of quality following this process. Furthermore, SA may contribute to the regulation of fruit ripening by inhibiting the excessive formation of superoxide free radicals, which can lead to the breakdown of cell membranes and tissue ageing. This effect is in addition to its known activity on enzymes involved in ethylene biosynthesis.

Gibberellic acid: Naturally occurring plant hormone that is found throughout many tissues of plants. It is involved in physiological processes including as flowering, seed generation, and fruit development. It is also employed in particular horticultural crops to regulate the process of flowering and the growth of fruits. The research findings indicated that GA3 had an inhibitory impact on the anthocyanin synthesis process and the breakdown of chlorophylls. The application of GA3 to climacteric fruits post-harvest resulted in enhanced quality. Peaches subjected to GA3 treatment demonstrated improved firmness preservation during storage at a temperature of 2°C, accompanied by a notable decrease in both respiration rate and ethylene release. Therefore, the application of GA3 seems to be effective in reducing the susceptibility of the fruits to unexpected physical damage. Bananas are also compatible with gibberellic acid (GA3). Submerging the entire banana fruit in water-based solutions containing GA3 at concentrations ranging from 10-5 to 10-2 M has been discovered to successfully decelerate the ripening process, hence maintaining the fruit's quality and prolonging its shelf life. GA3 has been discovered to be an efficient method for delaying the ripening of mango fruit when used in post-harvest treatment. Applying GA3 to 'Alphonso' mangoes successfully inhibited the ripening process when they were stored at a temperature of 28°C. The mango fruits treated with GA3 displayed a low Brix/acid ratio and high toughness, rendering them suitable for long-distance transportation. GA3 is extensively employed in multiple countries and has obtained approval from the United States Environmental Protection Agency.

1-Methylcyclopropene (MCP): 1-Methylcyclopropene (1-MCP) is a substance that competes with ethylene to suppress its effects. It binds to the ethylene receptor and controls how tissues respond to ethylene. This prevents ethylene from triggering further signal transduction and translation processes. Application of 1.5 μL·L-1 of 1 MCP prior to storage resulted in a delay of 11 days for the climacteric peaks of respiration and ethylene production in green melon fruit. Additionally, it effectively reduced the maximum levels of respiration and ethylene production. The ripening process of both green and yellow fruit was greatly delayed by 1-MCP. The retention of firmness generated by 1-MCP was followed by a notable reduction in electrolyte leakage in the mesocarp tissue. This suggests that the softening of 'Galia' melons may be caused by membrane malfunction. Storage at a temperature of 20°C, for both the control and the samples treated with 1-MCP. The green fruit underwent a large and quick softening, while the fruit treated with 1 MCP softened at a slower pace. The firmness of control and Yellow fruit treated with 1-MCP did not show any difference immediately after the 1-MCP treatment. Subsequently, the hardness of the fruit decreased in both treatments, albeit with a substantial difference.

The widespread demonstration of 1-MCP's capacity to postpone the ripening of papaya is evident. The application of 1-MCP successfully inhibited the ripening process of 'Tainung-1' papaya fruit without any alteration in the quality of the fruit. The study discovered that it significantly slowed down the softening process of papaya and prolonged the post-harvest life of fruits when stored at moderate temperatures following treatment, hence keeping the fruit's quality. However, it is not advisable to use 1-MCP treatment on papaya during the colour break stage, since it results in a harder texture and a rather rubbery consistency when the fruit is mature. It is recommended that about 25 per cent of ripe fruits can be effectively treated with 1-MCP without any changes to their qualities.

1-MCP shown significant efficacy in retarding the ripening process of guava as well. The application of 1-MCP at a concentration of 900 nL/L for either 6 or 12 hours significantly outperformed other concentrations of 1-MCP in delaying the ripening of 'Pedro Sato' guava. 1- MCP was suggested as a means to prolong the shelf-life of climacteric bananas. Nevertheless, there were also conflicting and incongruous findings regarding Banana. Surprisingly, it was discovered that the effectiveness of 1-MCP did not have a significant impact on the length of time that bananas were immersed. Furthermore, it was discovered that immersing mature-green bananas in a water-based solution of 1-MCP after harvesting can prolong their storage life for up to 35 days. This can be achieved by placing the bananas in a 1 per cent perforated low density polyethylene bag, without causing any significant changes to the quality of the fruit. This method can be considered a viable alternative for transporting and marketing bananas over long distances, with the aim of reducing losses. The use of 1-MCP has resulted in a postponement of the ripening process in apples, leading to significant changes in the apple sector in Washington state and apple storage practices in New York.

***Aminoethoxyvinylglycine (AVG)*:** The amino acid known as AVG is responsible for controlling the biosynthesis of ethylene through processes that decrease the activity of ACC synthesis. By inhibiting ACC Synthase, it has the ability to strongly limit the conversion of SAM to ACC. As a result, it can inhibit the synthesis of ethylene, it may delay the ripening process, and it can prolong the fresh time of fruits. The findings of the experiments demonstrated that the timing of the application of AVG spraying during the marginal harvest stage in bananas had a significant impact on the inhibition of the activity of internal enzymes such ACC synthase and ACC oxidase. These enzymes were extremely important in the process of controlling the formation of ethylene. Between the 74th and 78th day after the creation of the last hand of a banana bunch, it was discovered that the most opportune time to spray with AVG is between those two days.

***Nitric oxide*:** Nitric oxide (NO) offers a compelling option for controlling fruit ripening. This is a highly diffusible free radical that was initially regarded as a potentially hazardous environmental contaminant. However, it has subsequently been demonstrated to impact various biological processes, such as stomatal movement, plant respiration, seed germination, cell differentiation, and fruit ripening. Research has demonstrated that maintaining optimal levels of nitrogen oxide (NO) can significantly increase the duration of Post-harvest storage in several horticultural crops, such as apple, banana, peach, strawberry, mango, plum, and longan. Nitric oxide (NO) decreases the generation of ethylene by blocking the activities of 1-aminocyclopropane-1-carboxylic acid synthase (ACS) and 1-aminocyclopropane-1-carboxylic acid oxidase (ACO), leading to reduced levels of 1-aminocyclopropane-1-carboxylic acid (ACC). It is believed that NO binds to ACO, creating an ACO-NO complex, which is subsequently chelated by ACC to produce an ACC-ACC oxidase-NO complex. This complex leads to a decrease in ethylene production. The banana fruits were subjected to a treatment of 100 μL/L ethylene for a duration of 24 hours. Subsequently, the fruits were sliced and treated with 5 mM sodium nitroprusside (SNP), which acts as a donor of nitric oxide (NO). All treatments failed to decrease ethylene production and pulp softening of ripening banana slices. NO suppressed ethylene production by decreasing the activity and expression of ACO. Additionally, reduced activities of polygalacturonase, pectin methyl esterase, and endo-β-1,4-glucanases resulted in a delay in softening.

Calcium: The impact of calcium treatment on reducing fruit softening and improving flesh firmness has been documented in many fruit cultivars. Calcium chloride is commonly used in the cut fruit business to enhance fruit firmness. Calcium chloride aids in the regulation of physiological problems at the cellular level by stabilising membrane systems and promoting the synthesis of Ca-pectate, which enhances the rigidity of the middle area of fruit cell walls. In addition to inhibiting cell wall degradation, it enhances the skin's resilience. Post-harvest disorder and fruit ripening can be prevented by administering calcium either prior to or following fruit harvest. An observation was made that when trees were treated with a 1.50 per cent $CaCl_2$ solution thirty days before to harvest, instead of fifteen days prior, there was a noticeable delay in the ripening of fruits in the Mallika cultivar of mango. The delay may be attributed to the greater efficacy of pre-harvest applications during the early stages of fruit growth, as opposed to applying them shortly before harvest. The capacity of $CaCl_2$ to retard the ripening process may be attributed to elevated amounts of calcium in the fruit, resulting in a reduction in the rates of respiration and ethylene generation. The fruit of cv. Mallika exhibited an extended shelf-life compared to the average when the trees were subjected to a treatment of 1.50 per cent $CaCl_2$ thirty days before harvesting (Fig. 9). The extension of shelf-life by $CaCl_2$ sprays may be attributed to the multifaceted effects of calcium on fruits when they are picked at the optimal time. One of these functions is an enhancement in fruit firmness as compared to the control group. This leads to several benefits, such as an extended duration of reservation and a reduced rate of deterioration.

Jasmonic acid (JA): Jasmonates, a type of oxylipins, are responsible for triggering a diverse array of reactions in higher plants. There is a theory that jasmonates, such as jasmonic acid (JA) and its methyl ester, methyl jasmonate (MJ), have the capacity to regulate plant growth. Jasmonates are widely distributed in the plant kingdom. Exogenous jasmonate treatments promote the ripening of climacteric fruits by boosting ethylene production, speeding up chlorophyll breakdown and b-carotene synthesis, and improving the development of aroma compounds associated with ripening. The process of lycopene biosynthesis and the activity of polygalacturonase in tomato fruit are both hindered by high levels of externally applied jasmonates. However, the concentration of naturally occurring jasmonates in fully ripe red tomatoes is rather modest.

Applying methyl jasmonate (MJ) to preclimacteric apple fruit discs for twelve hours, at doses of 1 and 10 μM, caused an increase in ethylene production, as well as an increase in the activities of ACC oxidase and ACC synthase. When exposed to a concentration of 1 μM, methyl jasmonate was observed to enhance ethylene production by a factor of three to five following the temporary treatment. At a concentration of 100 μM, the levels of MJ caused an increase in ethylene production, ACC oxidase, and ACC synthase activities, suggesting a beneficial impact on ethylene generation (data not displayed). The application of continuous MJ at doses of 1 and 10 μM was observed to enhance the activity of ACC synthase, while having a negligible effect on the activity of ACC oxidase. Administration of MJ at a dose of 1000 μM continuously was found to be extremely effective in suppressing the activity of ACC oxidase and entirely stopping the production of ethylene. However, the specific data is not provided. There is a lack of data indicating that methyl jasmonate has any impact on the activity of the two enzymes when tested in a controlled environment, implying that jasmonates may influence ACC synthase and ACC oxidase either during the process of translating genetic information into proteins or during the process of transcribing genetic information from DNA into RNA.

Polyamines: Polyamines, which are small aliphatic amines, are present in both bacterial and eukaryotic organisms. Polyamines are present in both sorts of organisms. Putrescine (Put), spermidine (Spd), and spermine (Spm) are the most abundant polyamines in plants. Physician assistants (PAs) can manifest in various manifestations, such as unstructured, conjugated forms that are capable of dissolving, and bonded forms that are incapable of dissolving. Polyphenols, when in their soluble conjugated state, undergo covalent bonding with small molecules such as phenolic compounds. Conversely, when PAs exist in the insoluble bound state, they are chemically attached to large molecules such as nucleic acids and proteins.

From a physiological perspective, PAs are growth regulators that are primarily associated with both abiotic and biotic stressors. Furthermore, they are involved in a wide range of metabolic activities, such as plant development and morphogenesis, senescence, and fruit development and ripening. Because the precursor SAM is located at a crucial intersection and functional level, polyamines and ethylene are considered to be constituents of a network. The production and catabolism of PAs are both tightly regulated processes.

Regarding plants, there are two distinct pathways that result in the production of putrescine, utilising arginine and ornithine. ODC, an abbreviation for ornithine decarboxylase, catalyses the conversion of ornithine into putrescine. The alternate approach entails the decarboxylation of arginine through the action of arginine decarboxylase (ADC), a complex process that finally transforms arginine into putrescine. Decarboxylated S-adenosylmethionine (DCSAM), derived from S-adenosylmethionine (SAM) through the action of the enzyme SAM decarboxylase (SAM-DC), can be utilised as an aminopropyl donor in the transformation of Put to Spd and Spd to Spm. This conversion is achieved through two processes that are facilitated by two distinct enzymes, specifically Spd synthase and Spm synthase, respectively. S-adenosylmethionine (SAM) plays a crucial role in metabolism as it serves as a precursor for the production of ethylene through the conversion of 1-aminocyclopropane-1-carboxylic acid (ACC). This process takes place in a reaction that is facilitated by ACC synthase, and ACC is finally converted into ethylene by ACC oxidase.

Spermidine (SPD) was administered to Kiwi fruits at concentrations of 1.0, 1.5, and 2.0 mM, while Spermine (SPM) was applied at concentrations of 0.5, 1.0, and 1.5 mM for the purpose of treatment. The kiwi fruit that received a treatment of spermine at a dose of 1.5 mM had the lowest respiration rate. The reduction in respiration activity may have been caused by the antisenescence powers of polyamines and the beneficial water activity. The untreated fruits exhibited a considerably higher level of polygalactouronase (PG) activity compared to the fruits treated with polyamines. The presence of polyamines is responsible for the increased firmness of fruits since they regulate the activity of PG, so preventing the enzyme from causing fruit deterioration. The presence of spermine at a concentration of 1.5 mM and spermidine at a concentration of 2.0 mM resulted in a six-day extension of the ripening process.

***Hexanal*:** Hexanal is classed as GRAS, meaning it is considered safe for consumption and is naturally occurring. The use of hexanal therapy resulted in promising results in terms of prolonging the storage duration of certain fruits, such as apples, bananas, cherries, peaches, and strawberries; flowers, including carnations and roses; and various freshly cut vegetables. Hexanal is typically administered as a mist or spray before harvesting or as a dip treatment after harvesting. These treatments are carried out using specialist formulations that contain antioxidants such as α tocopherol and ascorbic acid. Compared to treatment with 1-methylcyclopropene (1MCP), the use of hexanal therapy has several advantages. One of these is that it effectively delays the senescence process without impacting the development of colour and flavour.

Phospholipase D (PLD) enzyme is accountable for the breakdown of membranes, a process that can be hindered by reducing the extent of membrane degradation in fruits and vegetables. During the process of growth and development, the activity of PLD (phospholipase D) is tightly regulated. During the ripening process, PLD (phospholipase D) increases gradually in an autocatalytic manner. This rise eventually causes significant membrane breakdown and destabilisation, ultimately leading to senescence. Several diverse approaches have been developed to suppress Post-harvest loss and extend the shelf-life of various horticulture products. These strategies involve the use of hexanal. Fruits, such as tomatoes, exhibit an extended period of freshness when the activity of the PLD enzyme is suppressed. Moreover, hexanal exhibits antifungal properties that can delay the post-harvest deterioration of strawberries.

Coating: Coating fruits is widely acknowledged as a highly efficient technique for prolonging the post-harvest lifespan of fruits. Due to its antioxidant and antibacterial qualities, it was found that mangoes coated with beeswax had the longest shelf life and maintained a good quality. The starch-coated fruits exhibited superior taste, colour, fragrance, and aroma compared to other fruits. Subjecting Gala apples to hot air treatment at 38 degrees Celsius for a duration of four days is considered to be an effective method for preserving their quality. Subsequently, the apples undergo a process of being covered with a 1 per cent chitosan solution and are then stored at a temperature of 0 degrees Celsius for a duration of eight weeks. Ultimately, they are held on retail shelves at a temperature of 20 degrees Celsius for a duration of seven days. The application of chitosan coating was found to effectively slow down the ageing process of fruit and inhibit the growth of fungi in strawberries stored under specific conditions: a temperature of 10 degrees Celsius and a relative humidity of 70±5 per cent. By incorporating a low concentration of chitosan, the addition of calcium gluconate effectively inhibited fungal growth and maintained the firmness of the fruit, leading to a prolonged shelf-life of strawberries.

Ethylene management to delay fruit ripening

Various chemicals inhibit the activity of ethylene and can be utilised to decelerate the ripening process of numerous climacteric fruits. The efficacy of sodium dehydroacetate (Na-DHA) in prolonging the shelf-life of strawberries by inhibiting several ripening phases, hence extending their marketability, was identified by Watada in 1971. Indole-3-acetic acid is a highly effective chemical that can be utilised to delay the ripening process of certain fruits. Indole-3-acetic acid, when present in low concentrations of 1 and 10 μM, inhibits climacteric respiration and ethylene production, resulting in delayed ripening of avocado fruits. Emission of ethylene is a notable occurrence that takes place in several situations throughout the food chain, including as packages, transit, storage rooms, and household refrigerators. Consequently, additional techniques are necessary to reduce the presence of ethylene. The minimum concentration of ethylene in the air required to cause physiological effects is approximately 0.1 $\mu L\ L^{-1}$. Climacteric fruits, when being transported over long distances or stored for extended periods, need to be exposed to low amounts of ethylene, specifically below 1.0-0.01 $\mu L\ L^{-1}$. Ethylene scavengers, a crucial subject in the import/export of fresh produce, are commonly used to delay physiological and biochemical alterations during processing and storage.

Efficiently removing ethylene and/or mitigating its detrimental impacts is crucial for preserving the freshness and quality of fruits and vegetables. Chemical compounds with different structures and amounts can impede the generation and function of ethylene, for example by altering the activities of ACS (ACC synthase) and ACO (ACC oxidase), blocking receptor sites, and redirecting SAM through polyamine creation. An essential mechanism employed by ethylene scavengers is the utilisation of potassium permanganate ($KMnO_4$), which chemically converts ethylene into carbon dioxide (CO_2) and water through oxidation. Another method relies on the capacity of specific materials with a substantial surface area to absorb ethylene, either independently or in conjunction with an oxidising agent, such as activated carbon, clays, and zeolites. Furthermore, a material enhanced by palladium demonstrates a significant rate of adsorption of ethylene at ambient temperature. Another approach involves inhibiting ethylene activity in the horticulture plant, hence decreasing the level of ethylene released. Platinum/alumina catalysts operate at temperatures above 200°C to transform ethylene into carbon dioxide (CO_2) and water (H_2O). Another alternative is to utilise titanium dioxide (TiO_2) for the process of photocatalytic ethylene oxidation.

Potassium permanganate ($KMnO_4$) based scavengers: Although there are other ethylene removal technologies available, as previously stated, the majority of commercial ethylene control systems rely on a combination of enough ventilation and the oxidation of ethylene. This oxidation process involves breaking down ethylene's double bond using mechanisms based on $KMnO_4$. However, the implementation of ventilation in airtight facilities is impractical, such as in regulated atmospheres, packaging systems, or situations where precise ethylene control is required.

The capacity of $KMnO_4$ to decrease the levels of ethylene in the air surrounding newly harvested crops was initially discovered in apples. The oxidation of ethylene using $KMnO_4$ to generate CO_2 and H_2O is a straightforward technique for eliminating external ethylene and other contaminating gases. Consequently, the level of CO_2 rises, inhibiting the production of natural ethylene, which is essential for regulating the ripening process. While carbon dioxide (CO_2) and water are generated, there is also the possibility of the formation of partially oxidised compounds such as carboxylic acids and alcohols. Furthermore, the decrease in the effectiveness of $KMnO_4$, a potent solid oxidising agent, appears to be caused by the transformation of $KMnO_4$ into manganese oxide, resulting in a change in colour from purple to brown. Due to the fact that diffusion and natural convection are the sole factors responsible for the interaction between ethylene and $KMnO_4$, it is necessary for the oxidant to have a large surface area exposed to the surrounding atmosphere in order to efficiently decrease ethylene concentrations.

Various commercial potassium permanganate scavengers are available in sachets, filters, blankets, and other specialised trapping systems. Several commercial products containing $KMnO_4$ with alumina beads are available, such as 'Purafl,' 'Ethysorb,' 'Circul-Aire,' and 'Bloomfresh.' However, these scavengers are mostly employed in ducted air conditioning systems to eradicate organic pollutants, and there has been limited or no substantial application in commercial horticulture crops. Commercially available ethylene-scavenging systems commonly involve the insertion of tiny sachets containing the scavenger within the package. A drawback of bag-in-box active packaging methods is that they require significant quantities of scavengers, such as activated carbon and alumina, due to their limited ability to bond with ethylene. Moreover, these substances have the ability to form chemical bonds with the compounds present in the pouch, resulting in the contamination of the food. These sachets effectively reduced fruit degradation, independent of the fruit's maturity level. However, the reduction was most pronounced in mature fruits. Furthermore, the implementation of this technology has the potential to postpone the process of fruit softening and reduce weight loss.

Activated carbon ethylene removers: The application of metal catalysts on activated carbon-based absorbents is an efficient method for removing ethylene. These absorbents are not only utilised as filters, blankets, and other storage goods, but they are also included into fresh product packaging as sachets and as components of paper bags for storing fruit. The sachets include activated charcoal infused with palladium, serving as a metal catalyst. This material successfully inhibits the oxidation process that leads to the generation of ethylene in packages of bananas, kiwifruits, spinach, and broccoli.

Nano and micro particles impregnated films as ethylene absorbers: Currently, a range of packaging materials have been created using specific minerals that have the potential to absorb ethylene. Minerals such as clays, nano clays, zeolite, pumice, and Japanese oya-stone can be added to the films. The manufacturers of these films assert that the extended post-harvest lifespan of horticultural crops is primarily attributed to either the two-dimensional surface adsorption or the three-dimensional absorption of ethylene by the distributed minerals in the films. These results frequently suggest that the product has a longer shelf-life and/or less ethylene gas in the space above it, compared to typical polyethylene film bags. Mineral-filled polymeric bags can develop pores, resulting in significantly increased gas permeability and subsequently faster diffusion of ethylene. In addition, these bags are expected to create a more favourable equilibrium modified atmosphere compared to regular polyethylene bags, especially when the fresh crop has a high respiration rate. Therefore, these characteristics result in enhanced longevity after harvesting and decreased ethylene levels in the surrounding environment, without the need for any ethylene scavenging agent. Although the majority of films are not transparent and do not have the same capacity as $KMnO_4$ to absorb ethylene, these film bags are gaining popularity and can be found in many commercial markets.

Nanocomposite packages are expected to dominate a substantial portion of the food and beverage packaging business in the near future, among other emerging technologies. Nanocomposite packages have various applications. These applications encompass enhanced food barrier, mechanical, and antibacterial qualities, increased food absorption and bioavailability, and the development of novel food packaging materials that enhance the taste, flavour, colour, and texture of food. The ethylene absorbency of nanocomposites was enhanced by augmenting the content of nano-$KMnO_4$. A recent work successfully created nanocomposite films using polyolefin elastomer (POE) that incorporated impregnated nanoparticles (nanosilica and nanoclay) with $KMnO_4$. The papered nanocomposites exhibited greater ethylene scavenging capabilities and reduced water vapour permeability. Furthermore, it was observed that nanocomposites increased the longevity of bananas when stored at room temperature. Therefore, nanocomposites can be utilised in films as technologies that scavenge ethylene.

Palladium based powder as ethylene absorber: Preliminary results indicates that using palladium (Pd) based material can effectively reduce the concentration of ethylene in pre-climacteric fruits at physiologically active levels. This is the initial instance of this discovery. Palladium-based materials generally serve as adsorbers rather than catalysts. Due to its significantly increased adsorption capability in nearly 100 per cent relative humidity, Pd-promoted powdered materials outperform ethylene scavengers based on $KMnO_4$. However, in conditions of low relative humidity, the efficiency of this process increases to around sixty-fold compared to that of $KMnO_4$. This material has been proven to effectively delay the ripening process in avocados and bananas, as well as reduce the senescence of strawberries and prevent the ripening process. Moreover, as a result, the removal of this component from the environment did not cause any disruption in the ripening process. Although these fruits have already started to show signs of climacteric ripening, the removal of ethylene results in a decrease in the typical respiratory process of climacteric activity and a better retention of fruit firmness compared to the control group.

Although the respiratory enhancement during the climacteric phase has already started, it is still conceivable for an ethylene scavenger to prolong the post-harvest lifespan of agricultural products. Previous study findings indicate that palladium-promoted compounds are more efficient than $KMnO_4$ based ethylene scavengers when used in low concentrations at high relative humidity. There is a notable prospect for commercialization of the Pd-based molecule, as it can be used as an alternative or additional treatment for 1-MCP. This approach (utilising Pd-promoted material) is applied at room temperature and the choice of zeolite support makes it appropriate for most horticultural crops cultivated in conditions with high relative humidity and ambient temperatures. 1-MCP is more effective than Pd-promoted ethylene remover in slowing the ripening of avocado (*Persea americana* Mill.), cv. The reason for this is that 1-MCP has the capacity to decrease ethylene to levels that are within the range of approximately 0.01 g L^{-1}, which is considered to be physiologically active. The ripening of 'Hass' avocados is delayed in comparison to the ethylene scavenger that is activated by zinc.

Selenium (Se): Research has revealed that both plants and animals necessitate selenium (Se), a trace element. Nevertheless, an abundance of it yields detrimental consequences. There has been a recent increase in the number of research papers that specifically examine the impact of selenium on the physiological processes of fruit. Based on the results of several trials, it has been observed that pre-harvest application of selenium (Se) can effectively slow down the ripening of fruit without compromising its quality standards.

The postulated pathways for the treated fruit and leaves may involve the activation of antioxidant defence systems and the suppression of genes involved for ethylene biosynthesis, leading to the prevention of reactive oxygen species (ROS) creation. When plants are exposed to selenium, it is highly probable that it will result in an enhanced absorption of sulphur. This, in turn, leads to the synthesis of higher quantities of glutathione (GSH), a naturally occurring antioxidant that does not require the involvement of enzymes. Research on gene expression has shown that the inhibition of genes responsible for ethylene production leads to a reduction in both respiration rate and ethylene output. Furthermore, the administration of Se may have enhanced the antioxidant defence system, hence decreasing the extent of membrane damage and the generation of reactive oxygen species (ROS).

When tomatoes are picked before they are fully green, selenium (Se) has the capacity to affect the expression of ACO1, ACS2, and ACS4 in the tomato plant. Based on the results of this investigation, applying sodium selenate at a concentration of 1 mg L−1 as a foliar pretreatment effectively postponed the ripening process while preserving quality metrics. In addition, Se disrupted the processes commonly associated with fruit ripening. The data reveal that there was a greater amount of GSH compared to the control. Furthermore, se has the capacity to reduce the production of ethylene and suppress the activity of phenylalanine ammonia lyase (PAL). Foliar pretreatment with selenium not only prolongs the shelf life of pear and peach fruits, but also retards the softening of pears by facilitating the accumulation of selenium in their leaves and fruits. Se controls the ripening process by modulating the amounts of reactive oxygen species (ROS) and oxidative damage. Consequently, the utilisation of Se treatment holds promise as a viable approach to delay the ripening process and prolong the storage duration of fruits and vegetables.

Through the exploration of ethylene's role in the ripening process, it becomes clear that the hormone not only defines the biochemical trajectories of climacteric and non-climacteric fruits but also serves as a pivotal factor in post-harvest management. The distinction between these two categories of fruits, underscored by their ethylene production and sensitivity, informs a plethora of storage, handling, and treatment options aimed at prolonging shelf life and maximizing fruit quality. This elucidation not only enriches our understanding of the physiological and genetic intricacies of fruit ripening but also highlights the importance of ethylene as a target for agricultural innovation. These insights into the ethylene-dependent maturation process offer actionable strategies for delaying ripening and managing post-harvest losses, thereby addressing both economic and food security challenges. Moreover, the significant implications for future research and practice in agriculture and food science underscore the necessity for continued exploration into ethylene management techniques. By harnessing the knowledge of ethylene's multifaceted role, stakeholders across the food supply chain can better navigate the complexities of fruit ripening, ensuring that the journey from farm to table is both efficient and sustainable.

References

Atkinson R G, Gunaseelan K, Wang M Y, Luo L, Wang T, Norling C L, Johnston S L, Maddumage R, Schröder R and Schaffer R J, 2011. Dissecting the role of climacteric ethylene in kiwifruit (*Actinidia chinensis*) ripening using a 1-aminocyclopropane-1-carboxylic acid oxidase knockdown line, *Journal of Experimental Botany*, 62(11) 3821–3835, https://doi.org/10.1093/jxb/err063.

Barry, Cornelius S and James J Giovannoni, 2007. "Ethylene and Fruit Ripening." *Journal of Plant Growth Regulation* 26: 143-159.

Barua S, Rahi T, Ullah E, Ghosh D and Ahmed S, 2015. Delay in fruit ripening: a promising approach for reduction of spoilage and use of hazardous chemicals in Bangladesh. *Int. J. of Agron. and Agri. Res.* 6: 4.

Bouzayen M and Latché, A and Nath P and Pech J C, 2010. *Mechanism of Fruit Ripening - Chapter* 16. In: Plant Developmental Biology - Biotechnological Perspectives vol. 1. Springer. ISBN 978-3-642-02300-2

Brummell D A, Atkinson R G, Burdon J N, Kevin J, Patterson and Schaffer R J, 2016. Chapter 11-Fruit growth, ripening and post-harvest physiology. In Plants in Action (http://plantsinaction.science.uq.edu.au/edition1)

Burg S P and Burg E A, 1962. 'The role of ethylene in fruit ripening', *Plant Physiology* 37: 179–89.

Experimental Botany, Volume 74, Issue 20, 31 October 2023, Pages 6237–6253, https://doi.org/10.1093/jxb/erad271.

Fuentes L, Figueroa C R, Valdenegro M, 2019. Recent Advances in Hormonal Regulation and Cross-Talk during Non-Climacteric Fruit Development and Ripening. *Horticulturae*, 5(2):45.

Kevany B M, Tieman D M, Taylor M G, Dal Cin V and Klee H J, 2007. Ethylene receptor degradation controls the timing of ripening in tomato fruit. *Plant J,* 51:458–467.

Lee S Y, Lee, S J, Choi D S and Hur S J, 2015. Current topics in active and intelligent food packaging for preservation of fresh foods. *J. of the Sci. of Food and Agric.,* 95:14. 2799 –10.

María Florencia Perotti, David Posé, Carmen Martín-Pizarro, 2003. Non-climacteric fruit development and ripening regulation: 'the phytohormones show', *Journal of Experimental Botany,* 74(20): 6237-6253.

Paul V, Pandey R, and Srivastava G C, 2012. The fading distinctions between classical patterns of ripening in climacteric and non-climacteric fruit and the ubiquity of ethylene—An overview. *Journal of Food Science and Technology* 49: 1–21.

Payasi A and Sanwal G, 2010. Ripening of climacteric fruits and their control. *Journal of Food Biochemistry*, 2010, 34(4): 679–710.

Perotti M F, Posé D and Martín-Pizarro C, 2023. Non-climacteric fruit development and ripening regulation: 'the phytohormones show'. *J Exp Bot.* 74(20): 6237-6253. doi: 10.1093/jxb/erad271.

Seymour G B, Taylor J E and Tucker G A, 2017. Eds., *Biochemistry of Fruit Ripening*, Springer Science & Business Media.

Sharma M, Sitbon C, Subramanian J and Paliyath G, 2008. *Changes in nutritional quality of fruits and vegetables during storage*. In Postharvest Biology and Technology of Fruits, Vegetables, and Flowers. (Ed) Gopinadhan Paliyath, Susan Lurie, Avtar K. Handa, Dennis P. Murr ISBN: 978-0-813-80408-8 Wiley-Blackwell.

Thompson A K, Prange R K , Bancroft R D and Puttongsiri T, 2018. Controlled Atmosphere Storage of Fruits and Vegetables 3rd Ed. Oxford Shire; CAB International.

Wills RBH and Golding J B, 2016. Physiology and biochemistry CABI, (34–62) doi:10.1079/9781786391483.0034

Yinglin J I, Mingyang X U and Aide WANG, 2021. Recent advances in the regulation of climacteric fruit ripening: hormone, transcription factor and epigenetic modifications. *Front. Agr. Sci. Eng.* 8(2): 314-334.

5

Nanotechnologies for the Post-harvest Management of Fruits

K S Subramanian, Syndhiya Ranjan, G J Janavi and Jayasankar Subramanian*

Center for Agricultural Nanotechnology, Tamil Nadu Agricultural University, Coimbatore, , Tamil Nadu-641 003
**Department of Plant Agriculture, University of Guelph, Canada*

Fruits and vegetables are perishables and the post-harvest losses from the point of production at the farm gate till it reaches the consumers is estimated as 30-35 per cent which constitute a major economic drain of more than 33 billion USD equivalent to Rs 2,40,000 crores annually. In order to reduce the losses at various stages, nanotechnologies have come as potential strategies to minimize the losses while improving the marketability and profitability of perishables. One of the plants derived natural compounds "hexanal" which is known to extend the shelf-life of fruits when it is used as a pre-harvest spray or post-harvest dip as a nano-emulsion. Since, it is a highly volatile compound, efforts were undertaken to entrap the compound in the polymer-based nano-fibres (as nano stickers) or as a cyclodextrin inclusion complex (as nano pellets) in order to preserve the perishables. Further, hexanal vapour was also evaluated to preserve fruits. In this book chapter, various nanotechnologies tested, evaluated and technologies developed over the years are elucidated. Overall, the literature review suggests that use of any of the nanotechnologies delivered as pre-harvest spray, post-harvest dip, nano-stickers, nano-pellets and vapour singly or in combination found to reduce the infestation of post-harvest diseases, minimize the losses and extended shelf-life by 2-3 weeks that help the farmers to get lucrative prices due to the late arrival in the market besides significantly improve the productivity and profitability of orchards.

Introduction

Fruits and vegetables (FandV) hold immense significance in human health due to their abundance of essential nutrients and phytochemicals like vitamins, dietary fiber, anthocyanins, polyphenols, and antioxidants (Wang *et al.*, 2012; Joshi *et al.*, 2018). Increased consumption of F and V showed negative correlation of chronic diseases such as heart disease, cancer, stroke, diabetes, Alzheimer's disease, and age-related function (Pem and Jeewon, 2015). The free radicals are scavenged from the body by antioxidants which provides attractive colour to the fruits thus preventing detrimental health effects. The fibers in the fruits ferment in the colon and produces short chain fatty acids (SCFA) which helps to maintain the gut health (Lattimer and Haub, 2010). Dietary phytochemicals delay the oxidation of DNA and affect cellular signal transduction pathways controlling the cell proliferation and apoptosis (Liu, 2004). Abundant antioxidants and anti-inflammatory compounds in these plants have demonstrably prevented depression and anxiety in humans (Faezeh *et al.*, 2018; Liu *et al.*, 2020). These findings provide strong incentives for increased fruit and vegetable consumption (Wang *et al.*, 2012). However, F and V are highly susceptible to post-harvest deterioration due to their high-water content, regional and seasonal availability, and inherent physiological processes like respiration, transpiration, ethylene synthesis and membrane degradation (Huang *et al.*, 2009).

In recent years, global food production has reached unprecedented levels; nevertheless, a staggering one-third of all food intended for human consumption, equivalent to 13 billion tons, is lost or wasted post-harvest food loss emerges as a primary contributor to food insecurity for millions of families worldwide. Aspiration to achieve zero hunger by 2030 necessitates the prevention of any further loss or waste of food Mitigating post-harvest losses within food systems presents an opportunity to enhance worldwide food availability without imposing additional demand on resources or the environment. The repercussions of post-harvest losses extend beyond mere quantities, impacting nutrition, health, and finances for both consumers and farmers. Losses occur at different points in the supply chain, as commodities tend to spoil, damage, or be lost during harvesting, handling, processing, storage, and transportation. A striking difference exists in food loss patterns across the globe (i) developed nations waste over 40 per cent of food (including cereals, roots and tubers, pulses and oil crops, vegetables and fruits, fish meat, and dairy) at the trade and consumer levels, (ii) Over 40 per cent of food losses in developing countries occur at the crucial post-harvest stage and during processing, disproportionately impacting smallholder farmers who often lack adequate storage facilities which forces them into a cycle of loss and insecurity: selling their produce at low prices immediately after harvest due to abundant supply, only to repurchase the same food later at inflated prices.

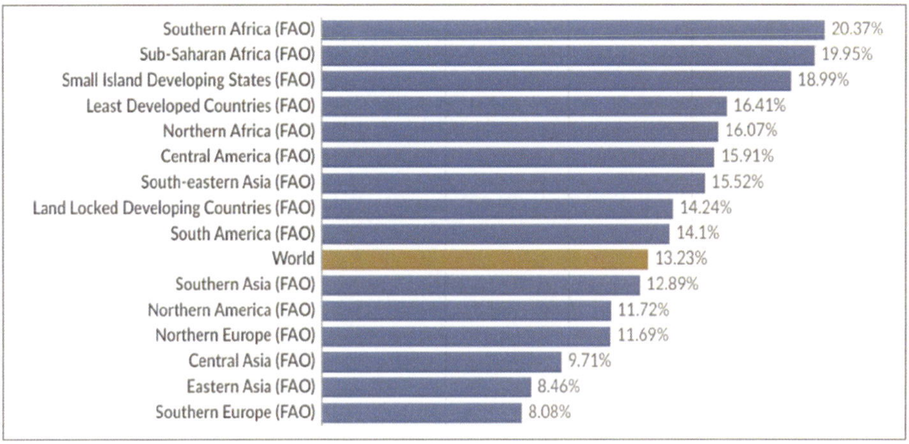

Fig 1: Share of food lost in post-harvest processes by region, 2021.
Data source: Food and Agriculture Organization of the United Nations.

The global agricultural landscape is witnessing a surge in innovation, with farmers increasingly deploying cutting-edge technologies to maximize crop yields. Intensive and extensive farming practices are being enhanced by the introduction of nano-modified stimulants and precision farming methods. Focusing on areas like agricultural efficiency, soil improvement, water security, reliable food distribution, and maintaining quality aligns perfectly with the potential of nanotechnology research to enhance food security worldwide (Ashraf *et al.*, 2021; Sastry *et al.*, 2011). Embracing these new technologies not only increases production and reduces waste, but also paves the way for sustainable living standards and a brighter future for food security.

India is the second largest producer of both fruits (108 million tonnes) and vegetables (213 million tonnes) next to China during the year 2022-2023 but the per capita availability continues to be lower due to the post-harvest losses which is estimated as 30-40 per cent. Such losses are due to several factors including inadequate infrastructure, subpar transportation, limited knowledge of post-harvest handling, market inefficiencies and technological gaps (Rajan and Sahini, 2023). These losses have profound consequences, spanning economic setbacks for farmers, compromised food security, environmental degradation, and market instability. Addressing this challenge requires a multifaceted approach, encompassing improved infrastructure, education, technology adoption, market reforms, and government initiatives. By strategically mitigating post-harvest losses, India can elevate its agricultural productivity, ensure food availability, and pave the way for a more sustainable and resilient food supply chain.

Anaerobic respiration in perishables is detrimental, leading to significant quality and flavor loss, undesirable ethanol production, and unpleasant odors. Water constitutes a major portion (80-95%) of most fruits and vegetables Excessive moisture loss, primarily through transpiration and respiration heat (Bovi *et al.*, 2016; Xanthopoulos *et al.*, 2017), leads to wilting, shriveling, and weight reduction, impacting both market value and physiological integrity (Jalali *et al.*, 2017). High in-package humidity can further promote microbial growth and shorten shelf-life. Ethylene, a plant hormone, plays a vital role in F and V development but can also trigger excessive post-harvest ripening, leading to discoloration, softening, and spoilage (Pathak *et al.*, 2017). Li *et al.* (2019) demonstrated the ripening and senescence in button mushrooms promoted by ethylene, highlighting its influence on F and V quality. Recognizing the immense health benefits of fruits and vegetables, coupled with their inherent post-harvest challenges, warrants focused research and interventions to significantly extend the shelf life and availability of these vital dietary components.

Strategies to reduce post-harvest losses

Maintaining the quality and quantity of fresh produce, particularly those with high nutritional value, can be achieved to a certain extent through various available post-harvest treatments. Careful harvesting at the ideal maturity stage is crucial for preserving the quality and longevity of fruits and vegetables. Early or late harvest compromises their shelf-life and accelerates spoilage. Extending shelf life of fruits and vegetables hinges on three key strategies: slowing down ripening (Minas *et al.*, 2012), shielding against microbial threats (Sharma *et al.*, 2009), and minimizing water loss (Hubert and Lang, 2012). A vast array of techniques has been developed to tackle these challenges. The key to develop effective solutions for reducing food loss and waste lies in recognizing the interconnectedness of different stages in the food supply chain. The choice of post-harvest technology depends on factors such as crop type, climatic conditions, affordability, and ease of use (Kitinoja and Barrett, 2015). Some major post-harvest technologies available for extending the shelf-life of fruits and vegetables include modified atmospheric packaging (Henig and Gilbert, 1975), ethylene management (Deikman *et al.*, 1997; Forney *et al.*, 2003; Lee *et al.*, 2015; Opiyo and Ying, 2006; Ebrahimi *et al.*, 2022), use of 1-Methylcyclopropene (1-MCP) (Smith *et al.*, 2009; Zewter *et al.*, 2012), irradiation (Murr *et al.*, 2008; Jeong *et al.*, 2015; Kalaiselvan *et al.*, 2018; Bisht *et al.*, 2021), heat treatment (Lu *et al.*, 2007) and edible coating (Pham *et al.*, 2023).

Emerging nanotechnologies

Conventional preservation methods often fall short due to high costs, inadequate extension of shelf-life, and the potential presence of harmful residues, thereby leaving a gap for novel approaches. Nanotechnology, leveraging its distinctive properties and capabilities, holds significant promise in overcoming these limitations and transforming the preservation of fresh produce. By transitioning materials from the macro to the nanoscale level, nanomaterials (any one of the dimension in nano scale of one-billionth a metre) acquire exceptional properties such as high surface mass ratio, enhanced hardness, superior barrier capabilities, and unique optical characteristics (Duncan 2011). These unique properties can further optimize the electrical, thermal, and optical functions to expand the application potential in agriculture, medicine, energy, chemistry, and structural materials (Subramanian *et al.*, 2018; Nair *et al.*, 2020; Jagtiani, 2022). Use nanotechnology in post-harvest technologies of fruits and vegetables has been extensively reviewed (Paliyath *et al.*, 2018). Combination of nanomaterials with traditional preservation methods can effectively compensate for the shortcomings of traditional preservation technology (Nair *et al.*, 2020; Nile *et al.*, 2020). These distinct material traits form the cornerstone of applying nanomaterials the preservation of fruits and vegetables.

This chapter intends to provide (i) a comprehensive detail of hexanal based preservation technologies to reduce post-harvest losses in fruits and vegetables and (ii) the update on other nanotechnologies that are reported to be promising to minimizing the losses of perishables during transport and storage.

Chemistry of hexanal

Hexanal is a naturally occurring C6 aldehyde that is generated via the lipoxygenase pathway following tissue damage (Vick and Zimmerman, 1987). It is alternatively known as hexaldehyde, caproaldehyde, or hexylaldehyde, with the chemical formula CH□(CH□)□CHO. The biosynthesis of hexanal is stimulated after wounding or during fruit ripening and is derived from the downstream breakdown of unsaturated fatty acids that are released during membrane phospholipid degradation. Linolenic acid is subjected to lipoxygenase-catalyzed lipid peroxidation, followed by hydroperoxide lyase-mediated degradation, ultimately yielding 3Z-hexenal. Subsequent action of alkenal oxidoreductase on 3Z-hexenal leads to the formation of hexanal (Schwab *et al.*, 2008). In general, hexanal represents a saturated fatty aldehyde wherein one of the terminal methyl groups in hexane undergoes mono-oxygenation to yield the corresponding aldehyde. The availability of this volatile components in fruits like apples, pears, and bananas, contributes to their characteristic fruity aromas (Paillard, 1990).

Hexanal - molecular formula and structure

$C_6H_{12}O$	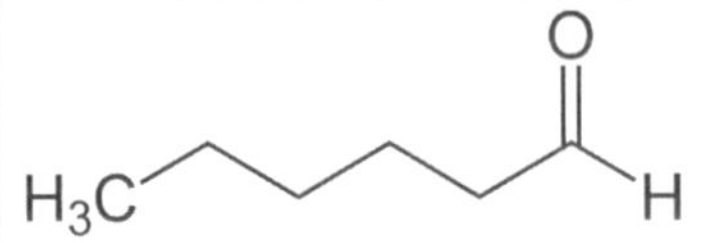

The physical properties of hexanal: the molecular weight of 100158g mol^{-1}, melting point, and boiling point of hexanal are 20°C and 130-131°C. respectively, the constant pressure heat capacity of hexanal is 2104 Jmol^{-1} K^{-1} at 285° K (Vasilev *et al.*, 1991) while its reduced pressure boiling point under 0016 bar pressure is 3012K (Weast and Grasselli, 1989). Hexanal contributes to the "green" taste of many vegetables and fruit varieties (Croteau, 1978). Hexanal has been approved as a food additive by the US Food and Drug Administration (FDA) (Song *et al.*, 1996) and now commercial products containing hexanal as an active ingredient for improving shelf life is also available. According to FDA, hexanal is referred as GRAS (generally regarded as safe) compound.

Mode of action:

Structural elements, including the cell wall and plasma membrane, undergo inherent degradation during ripening and senescence which are considered as final stages of development in fruits and vegetables during which their intrinsic characteristics are fully expressed. Harvested perishables, when subjected to abiotic stresses, experience detrimental alterations in the cell membrane, ultimately leading to a decline in quality. Seasonal cues like decreased light duration and temperature in fruits triggers ethylene biosynthesis which naturally onset ripening in fruits and senescence in leaves. At the ultrastructural level, both processes involve progressive breakdown of cellular structures, particularly the plasma membrane which disrupts compartmentalization of ions and metabolites, leading to loss of tissue integrity and ultimately cellular homeostasis.

The plasma membrane is a dynamic, fluid-phase lipid bilayer consisting of liquid crystalline and gel-phase lipids. Its components include phospholipids, sterols, and ceramide monohexosides, with predominant phospholipids like phosphatidylcholine and phosphatidylethanolamine, along with smaller amounts of phosphatidylinositol, phosphatidylglycerol, and phosphatidylserine (Yoshida and Uemura, 1986). To maintain fluidity and functionality, the plasma membrane undergoes constant modification through the turnover of its protein and lipid constituents. During ripening and senescence, a cascade of enzymatic phospholipid catabolism is induced by involving sequential action of enzymes such as phospholipase D, phosphotidate phosphatase, lipolytic acyl hydrolase

(LAH) and lipoxygenase leading to profound biophysical transformations of membranes. These include a shift from liquid crystalline to gel-phase lipids, decreased bulk fluidity and increased microviscosity, elevated phase transition temperature, and formation of nonbilayer structures. These changes, driven by accumulation of degradation products like phosphatidic acid, diacylglycerols, and free fatty acids, disrupt membrane integrity and fluidity, impacting cellular functions and contributing to the senescence process (Thompson *et al.*, 1987). Among the various enzymes involved, PLD is considered as the key enzymes as it regulates the sequence of enzyme reaction and the flow of metabolites through the catabolic pathways. Inhibiting the activity of PLD results in controlling the activity of other enzymes with the intermediates, thereby preventing the accumulation of neutral lipids and destabilization of membrane (Paliyath and Droillard, 1992). Hence, controlling the PLD action is considered critical to increase the shelf life and quality of fruits and vegetables (Paliyath *et al.,* 2009).

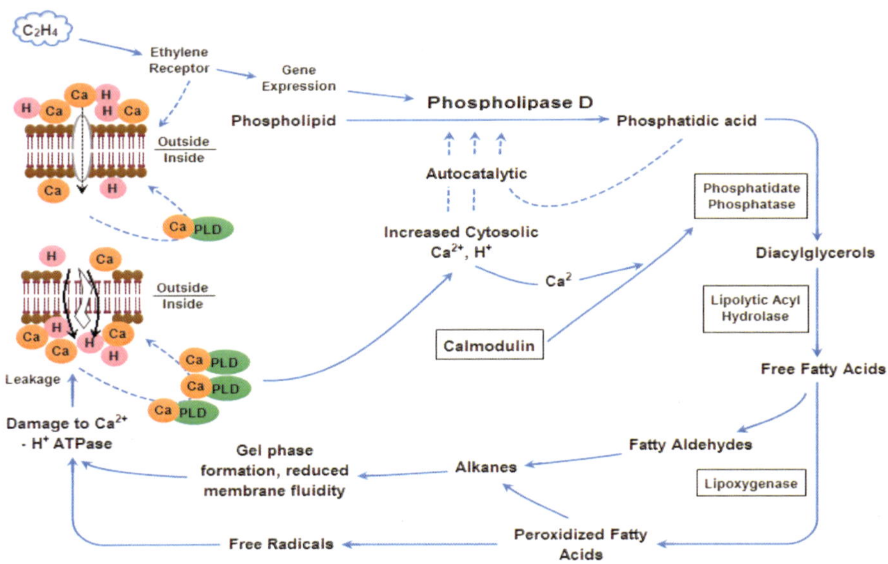

Fig. 2: Schematic representation of membrane lipid catabolism.

Unique in its mechanism of action, PLD catalyzes the release of phosphatidic acid from the active site through hydrolysis, forming phosphatidyl alcohol when encountering a primary alcohol instead of water, a process known as transphosphatidylation. Treating plant tissues with primary alcohols like methanol, ethanol, propanol, or butanol produces various products such as phosphatidylmethanol and phosphatidylethanol. The reaction is heightened by heating plant tissues in the presence of these alcohols, leading to artifact formation during phospholipid isolation. Although, incubating plant tissues with alcohols is used experimentally to assess enzyme activation, it is deemed unreliable due to PLD compartmentalization and inhibition by longer-chain primary alcohols. Investigating the impact of primary alcohols and aldehydes on PLD activity in corn kernel fractions revealed that hexanol and hexanal effectively inhibit PLD, with hexanal exhibiting greater potency. Aldehydes, lacking the primary alcohol structure for participation in the reaction, may obstruct the water-binding site. This inhibitory potential becomes particularly applicable considering hexanal's natural production during lipid peroxidation which contributes to the characteristic green flavor in wounded vegetables and its GRAS (generally regarded as safe) status. One of the significant advantages of this innovative treatment is that the treated fruits and vegetables remain in a pre-ripening state when stored at cooler temperatures (10-12°C), effectively pausing their development. Upon returning to ambient temperature, they resume ripening normally, offering several key benefits to farmers by improving fruit retention, prolonging the storage duration for packers and shipping agents, and ensuring an optimal shelf life for consumers (Paliyath *et al.*, 2018). Hexanal inhibits the phospholipsae-D in the skin of fruits besides slows down ethylene that facilitate shelf-life extension (Paliyath *et al.*, 2003, US patent # 6,514,914).

Hexanal aided physiological changes in perishables

Phospholipase D consists of an active hydrophobic site such as the HKD motifs which by a two-step ping-pong mechanism, hydrolyze phospholipid generating a phosphatidate PLD intermediate (Koonin, 1996). It has been postulated that hexanal exhibits the ability to bind to this active site of PLD, thereby mimicking an incomplete transphosphatidylation reaction and leading to the inhibition of PLD activity (Paliyath *et al.*, 2008). As a result, production of phosphatidic acid (PA) is inhibited, since the hydrolysis of phospholipids is impeded. Additionally, studies have demonstrated that hexanal serves as an effective inhibitor of the ethylene signal transduction pathway (Paliyath *et al.*, 2003; Jakubowicz *et al.*, 2010; Pak Dek *et al.*, 2018). Ethylene perception occurs through the binding of ethylene to its receptors, resulting in the deactivation of receptor activity and subsequent removal of downstream inhibition of ethylene signal transduction (Lin *et al.*, 2009). Constitutive triple response protein 1 (CTR1, MAPKKK Raf-1), a negative regulator, is modulated by PA binding to its kinase domain, inhibiting its interaction with ethylene receptor 1 (ETR1) (Testerink *et al.*, 2007). This implicates PLD in ethylene signaling, as it generates PA Additionally, PA acts as a CTR1 inhibitor (Testerink *et al.*,

2007), potentially linking PLD action to signal transduction activation and senescence-associated gene expression. The down-regulation of PA production through PLD inhibition by hexanal may consequently affect the function of ethylene signal transduction, thereby influencing the gene expression patterns related to fruit ripening and senescence PA, synthesized by PLD, serves as a crucial lipid messenger, which directly modulates the activity of numerous protein targets through direct interaction suggesting that PLD activation may indirectly regulate CTR1 activity and ethylene signaling. Inhibition of CTR1 activates downstream intermediates like ethylene-insensitive protein 2 (EIN2) and EIN3, leading to increased ethylene-related gene expression (Stepanova *et al.*, 2005). Furthermore, increased PA generation has been reported as an early response to stress (Zhang *et al.*, 2003; Bargmann and Munnik, 2006). Notably, membrane-localized PLD activity increases during fruit ripening, accompanied by a simultaneous rise in PA generation (Pinhero *et al.*, 2003), suggesting the potential involvement of PA as a regulator of ethylene signal transduction during fruit ripening through the inhibition of CTR1.

Consequently, the ripening process may be modulated by the levels of available PA through the regulation of the CTR protein, wherein lower levels of PA facilitate the existence of relatively more functional CTR, thereby downregulating signal transduction and inhibiting ripening processes. While hexanal application may directly inhibit PLD, thereby reducing membrane deterioration, its action may also be directly related to gene expression. Previous studies have demonstrated that hexanal specifically inhibits ripening-related gene expression compared to the global downregulation of gene expression caused by 1-MCP application in tomato fruits (Tiwari and Paliyath 2011). Also, hexanal does not accumulate in tissue, as it gets converted to hexanol and can get further metabolized through the respiratory cycle (Cheema *et al.*, 2014).

Approximately, there are 300 sources of hexanal including but not limited to apple, apricot, berries, and cherries. However, the levels of hexanal produced are typically low to elicit any physiological effects, such as the inhibition of PLD, possibly due to temporal or spatial separation of its biosynthesis. The vapor pressure of hexanal is very low (10 mm Hg at 20°C) which makes it highly volatile in nature limiting its application to improve the shelf-life of fruits and vegetables. Hence, it is important to develop a reliable mechanism to achieve a controlled release of hexanal. The natural waxy coating on the fruit surface may also be one of the favorable methods of application of hexanal for the absorption and storage thereby preventing the volatilization. As elaborated in the preceding section, nanotechnology presents unique strategies to mitigate hexanal's volatilization. Encapsulating/incorporating hexanal into nanostructures enhance the uptake of hexanal, facilitate the controlled release and improve absorption into the cell solution.

Hexanal technologies for preserving perishables

Hexanal-based technologies, targeting phospholipase D (PLD) inhibition, offer promising strategies for preserving various horticultural produce. Hexanal can be applied both before and after harvest of the produce. As a pre-harvest treatment, hexanal is typically applied as a spray, aiming to increase the fruit retention on tree and to prevent quality deterioration before harvest. Conversely, post-harvest treatments involve dipping the harvested produce in solutions containing hexanal or exposing them to hexanal vapor, targeting preservation and control of spoilage during storage and transportation. Each application is comprehensively discussed in the following sections.

Pre-harvest application: Hexanal's high volatility poses a limitation for its direct application. Hence two types hexanal formulation known as *Enhanced Freshness Formulation (EFF)* was developed to facilitate its practical application. The EFF is also referred as the nano-emulsion as the dimension of the micelle is in the range of 30-40 nm.

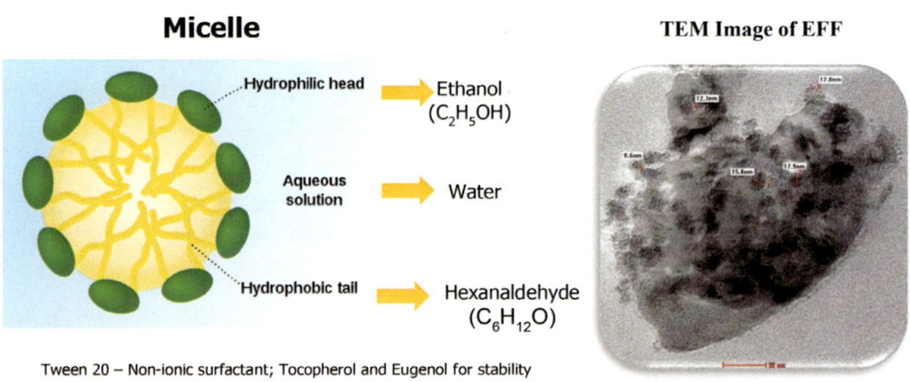

Fig. 3: Micelle of enhanced freshness formulation.

The EFF has central molecule as hexanal and encapsulated with surfactant (Tween 20) and co-surfactant (ethanol). The formulation is diluted 50 times prior to pre-harvest spray or post-harvest dip. While diluting, the mixture should be properly stirred in suitable frame for the formation of nano micelles which measures approximately 30-40 nm (Paliyath *et al.*, 2018). Nano micelles are colloidal dispersions of hexanal in water forming a hydrophobic core and hydrophilic shell. These tiny spheres have the ability to decrease the deterioration and evaporation of hexanal while enhancing its stability (Janik *et al.*, 2023). Using a mechanized sprayer, apply a very fine mist of the diluted solution containing nano-micelles directly onto the fruits, ensuring complete coverage until dripping occurs. The recommended dosage per acre (typically 50-100 liters) should be adjusted based on factors like tree size, fruit load, and plant density. A plethora of field trials have been undertaken across varying climates to assess the efficacy of EFF on a wide range of fruits such as cherries, mango, guava, and others in enhancing the post-harvest longevity. These trials consistently demonstrated a significant improvement in post-harvest shelf-life compared to untreated controls.

Fig 4: Pre-harvest spray of EFF in mango orchards in Krishnagiri, Tamil Nadu, India.

Fruit retention and disease reduction: Pre-harvest application of EFF on mango trees extended fruit retention on trees for 3- 4 weeks longer period than the usual retention period. Furthermore, EFF sprays applied on 30 and 15 days before harvest effectively reduced post-harvest diseases like anthracnose and stem end rot, while preserving other quality attributes which resulted in extended fruit freshness under both ambient (24–25 days) and cold (36–40 days) storage conditions. The treatment also significantly reduced pectin methyl esterase (PME) enzyme activity, a key factor in fruit softening (Anusuya *et al.*, 2016; Kaur *et al.*, 2020). In India, Tamil Nadu Agricultural University, Coimbatore, conducted a field experiment in mango orchards (var Alphonso) to determine the effect of pre-harvest spray of 2 per cent EFF (sprayed at the 4- and 2-weeks prior to harvest) on retention of mango fruits (Anusuya *et al.*, 2016). The data have clearly demonstrated that the fruit retention was significantly higher in EFF sprayed mango trees during the six weeks of observation in comparison to unsprayed trees (Fig. 5). Similar benefits were observed with pre-harvest EFF sprays on "Honeycrisp" apples (DeBrouwer *et al.*, 2020), sweet oranges (Samwel *et al.*, 2020) and banana (Yumbya *et al.*, 2018). These sprays not only extended fruit retention on trees but also offered secondary benefits like a 70 per cent reduction in bitter pit incidence on apples and reduced incidences of green mold, sooty mold, and blossom end rot on sweet oranges. Recent research by Sharma *et al.* (2023) demonstrated the effectiveness of EFF on Indian Jujube fruits. Pre-harvest application during both pit hardening and color break stages improved the fruits' physiochemical characteristics, preserved bioactive compounds and antioxidant enzyme activities, and extended their shelf-life up to 21 days under cold storage.

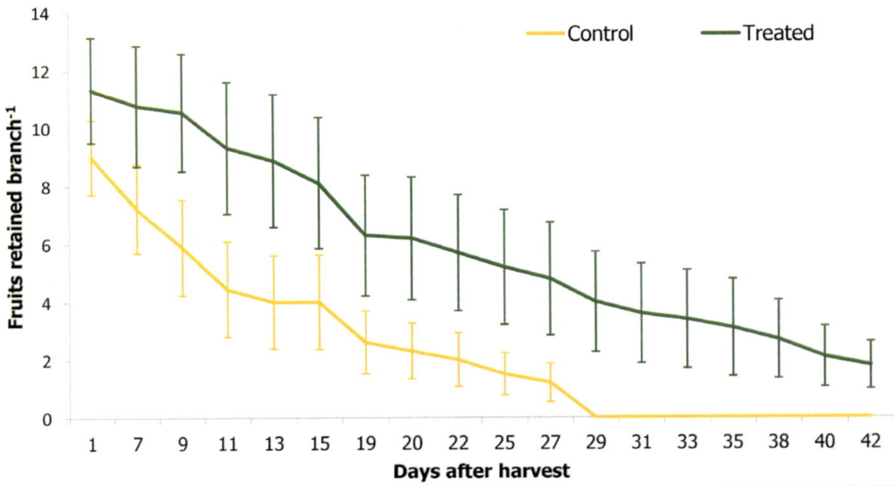

Fig. 5: Fruit retention in mango orchards (var Alphonso) after the two rounds of pre-harvest spray at 4 and 2 weeks prior to harvest (green line) in comparison to unsprayed trees (control).

In Tamil Nadu Agricultural University, Coimbatore, has experimented the effects of pre-harvest sprays in several varieties of mangoes (Banganapalli, Bangalora, Banganapalli, Neelam and Himampasand), guava and grapes under a collaborative Canadian project jointly supported by Global Affairs Canada and International Development Research Center during 2012-2018. In all the cases, there was a significant extension of shelf-life of fruits under ambient or cold storage conditions (Table 1). The data have clearly shown that EFF assisted in extending shelf-life of mango fruits 1-3 weeks and 3-6 weeks under ambient and cold storage conditions, respectively. On the other hand, grapes and guava, the shelflife hardly exceeds 5-7 days under ambient storage while 3-4 weeks under cold storage conditions.

Further, the EFF sprayed trees produced fruits with less incidence of post-harvest diseases like anthrocnose (*Colletotrichum gloeosporioides*), stem end rot (*Lasiodiplodia theobromae*) and phenolic injury associated with sooty mould (Anusha, 2014).

Table 1: Pre-harvest spray of nano-emulsion of hexanal twice (at 30 and 15 days before harvest) to extend shelf-life of fruits (mango, guava, grapes).

Fruits	Varieties	Shelf-life (Days)		Reference
		Ambient (27±2°C)	***Cold (27±2°C)***	
Mango	Banganapalli	20-24	36-40	Preethy *et al.*, 2021
	Alphonso	18-20	24-26	
	Himampasnd	15-16	18-20	
	Neelam	8-10	12-14	
	Bangalura	12-14	16-18	
Guava	Allahabad Safeda	5-7	18-20	Nandhini, 2017
Grapes	Muscat Hamburg	*4-5*	*28-30*	Tamileezham, 2017

Table 2: Effect of pre-harvest spray on the incident of post-harvest diseases in mango (var Neelam).

S.N.	Parameters	Control	Treated	SEd
1	No fruits harvested per tree	714	726	308
2	Mean fruit yield (kg tree^{-1})	143	145	443
3	Anthracnose incidence (%)	178	562	239
4	Stem end rot incidence (%)	133	533	282
5	Phenolic injury (%)	107	602	259
6	Post-harvest losses (%)	418	169	617
7	Net fruit yield (kg tree^{-1})	**832**	**1205**	**108**

Fig. 6: Mango fruits infected with anthracnose, stem end rot and phenolic injury.

***Impact on fruit color and ripening*:** The effect of EFF is particularly evident in sweet cherries. Bing cherries is a characteristic purple red color fruit (observed on control trees representing the ripening of fruits on trees) while bright red color was observed on the EFF sprayed trees. However, during post-harvest storage the characteristic color was developed as the fruits naturally show the transition from red to purple during ripening which proves the role of hexanal in delaying the ripening process. Also, cherries sprayed with EFF had a lower hue angle signifying an upregulation of carotenoid biosynthesis. Control trees displayed the characteristic progression from red to purple fruits, signifying natural ripening. In contrast, EFF-treated cherries maintained a brighter red color (indicated by a lower hue angle). This color retention suggests a delay in ripening, further supported by the observed upregulation of carotenoid biosynthesis in these cherries. While pre-harvest treatment alone offered color retention benefits, research by Sharma *et al.* (2010) recommends a combination of both pre and post-harvest treatment for optimal shelf life extension in sweet cherries.

Changes in biochemical profiles: Studies on other fruits reveal alterations in their biochemical profiles due to EFF treatment. For instance, pre-harvest spraying of hexanal formulation on strawberry cultivars at a regular interval of once per week for 3 consecutive weeks showed existence of variation in the production of total phenolics and volatile compounds which are crucial components of the ripening process. The levels of specific anthocyanins (pelargonidin-3-glucoside) and ferulic acid hexose increased, while several volatile compounds volatile compounds including esters, ketone, lactone, and acid group and levels of flavanols such as PA dimer, PA trimer, and catechin decreased. These changes highlight the influence of EFF on specific metabolic pathways which affects the organoleptic characteristics of the fruits.Cultivar-specific quality changes was observed on the hexanal treated fruits (Misran *et al.*, 2014).

Fruit quality enhancement: Greenhouse tomato sprayed with the hexanal formulation resulted in higher levels of ascorbic acid and soluble solids in fruit. The pre-harvest spraying was carried out as described for strawberries Unlike cherries, carotenoid biosynthesis remained unaffected. A clear indication of delayed ripening processes was observed by the reduced red colour intensity in hexanal formulation sprayed tomatoes. Inhibition of pectin degradation by inhibiting the downregulation of polygalacturonase and β-galactosidase genes results in increased fruit firmness of the treated fruits. In addition, the treated fruits also showed increase in ascorbic acid and the soluble solid contents (Cheema *et al.*, 2014). Similar positive effects were observed in guava fruits subjected to hexanal formulation spray prior to harvest at 4 and 2 weeks Treatment improved firmness, total soluble solids, acidity, pectin, and phenol contents. Additionally, they exhibited minimum decay incidence, reduced pectin methyl- esterase activity and maintained high organoleptic quality for up to 4 weeks of storage (Gill *et al.*, 2016).

In conclusion, pre-harvest application of EFF showed a promise as a versatile strategy across climates, improving post-harvest quality, extending shelf life, and reducing disease in diverse fruits. The specific effects of EFF vary depending on the fruit cultivar, but the overall trend suggests a delay in ripening processes and positive alterations in biochemical profiles that contribute to extend the shelf life and enhance fruit quality which significantly benefit both producers and consumers. The distinguished effect of EFF as pre-harvest spray includes the fruit retention on tree for longer period which prolongs the harvest time for the growers.

Post-harvest dip: The dip treatment is quite easier and practically possible method to reduce the post-harvest losses in fruits and vegetables. In packhouses where fruits and vegetables are assembled, stored for a while prior to transport to the wholesale and retail markets. Profound increase in the shelf-life of various F and V was observed on studying the effect of hexanal as a dip treatment. This technique involves dipping fruits harvested at their optimal maturity stage in EFF (2%) for a brief period (2-5 minutes) in an aqueous hexanal formulation similar to those used for spraying (Fig 6). Following the dip, the perishables are stored under appropriate conditions to further enhance preservation. The effectiveness of hexanal is demonstrably linked to its influence on key physiological processes.

Dipping Washing

Fig. 6: Dipping of banana and mango fruits in the EFF solution in packhouses.

Studies on tomatoes, for example, revealed that 'Rapsodie' tomatoes dipped in the formulation exhibited superior storage life at both cold and ambient temperatures. Even without refrigeration, these treated tomatoes remained 15 per cent firmer than control fruits after two weeks of storage. Further gene ontology analysis based on type, location, and function showed the gene expression changes in response to hexanal. Notably, the treatment did not negatively impact lycopene biosynthesis, a vital pigment responsible for the red color of tomatoes. However, the ethylene biosynthesis was heavily down-regulated with the inhibition in the expression of number of ACS alone (Tiwari and Paliyath, 2011). The optimal timing for the hexanal dip treatment appears to coincide with the natural rise in phospholipase D (PLD) activity that occurs during fruit ripening. Studies have shown that dipping fruits at the mature green stage, when PLD activity is on the rise, leads to the most effective results (Pinhero *et al.*, 2003). This targeted approach highlights the potential for hexanal to act as a precise tool for manipulating fruit physiology and extending shelf-life. The benefits of hexanal extend beyond tomatoes

Dipping fully mature Neelum mangoes in a EFF solution prior to storage under ambient conditions demonstrably extended their shelf-life without compromising fruit quality. This effect is attributed to the treatment's ability to significantly reduce ethylene evolution rate, oxidant content, and phospholipase D (PLD) enzyme activity – all key factors influencing ripening and senescence in fruits. Conversely, the activities of antioxidant enzymes (superoxide dismutase, catalase, ascorbate peroxidase, glutathione peroxidase) and ascorbic acid content all increased in response to hexanal treatment. This suggests that the observed delay in ripening and senescence is likely caused by the combined effects of decreased ethylene production, PLD activity, and oxidant generation, coupled with enhanced antioxidant activity and ascorbic acid content (Jincy *et al.*, 2017). Similar positive effects have been observed in other fruits like papaya (Hutchinson *et al.*, 2018) and banana (Yumbya *et al.*, 2018). Dipping mature green papayas and bananas for just 5 minutes in the hexanal formulation significantly extended their shelf life by 6 and 3 days, respectively, compared to untreated controls. Additionally, the treated fruits exhibited a remarkable 38 per cent greater firmness throughout their storage period. Beyond firmness, hexanal treatment demonstrably reduces respiration rate and fruit softening in dipped fruits and vegetables, further contributing to their extended shelf life and improved overall quality. These findings suggest that hexanal dip treatment holds promise as a versatile and effective post-harvest strategy for a diverse range of fruits and vegetables.

Table 3: Dipping of fruits/vegetables in nano-emulsion of hexanal (2 % EFF) for 5 minutes prior to packing or long-term transport on extension of shelf-life.

Fruits / Vegetables	Varieties	Shelf-life (Days)		Reference
		Ambient (27±2°C)	Cold (27±2°C)	
Mango	Alphonso	16-18	22-24	Preethy *et al.* (2021) Ajith (2016)
	Neelam	6-8	10-11	
	Bangalura	10-12	14-16	
Banana	Grand Naine	10-12	28-30	Kanmani (2016) Mutharasu (2018)
	Nendran	17-18	-	
	Ney Poovan	10-12	30-32	
	Red Banana	5-6	10-11	
Guava	Allahabad Safeda	4-5	9-10	Nandhini (2017)
Grapes	Muscat Hamburg	4-5	10-12	Tamileezham (2017)
Acid lime	Balaji	8-10	15-18	Dhanalakshmi (2018)
Tomato	PKM -1	16-18	-	Krishna Dev (2017)
Cucumber	English Long	9-10	12-14	

Fig. 7: Dipping of banana and mango fruits (var Alphonso) after 18 days of storage under ambient conditions.

The mango fruits that are treated in 2 per cent EFF also maintained flavour and colour during storage. The volatilomics was performed on nano-formulation treated (NFT) and untreated fruits to explore the molecular basis of such effect (Dar *et al.*, 2020). Untreated and treated fruits were sampled starting from 5th to 21st day after NFT at an interval of 4 days. Moderate changes in visual and digital colour parameters were evident from the intact and dissected fruits of NFT set compared to untreated fruits. Biochemical assays affirmed the phenotypic differences with significant changes in the colour imparting compounds like carotenoids and anthocyanins among them. Further, gas chromatography-mass spectrometry analysis revealed significant qualitative and quantitative variations in the different classes of compounds like lactones, furanones, esters, aldehydes and alcohols. Some of the key metabolites showed differential modulations among the NFT and untreated fruit sets indicating their potential role in various processes, which ultimately might have resulted in delayed ripening of the mango. Overall, this study has demonstrated the beneficial effect of hexanal and identified important metabolites with the enhanced shelf-life in Alphonso that could be useful for farmers and mango-based food/flavour industries.

Vapor treatment: Due to its high volatility, hexanal readily transitions from a liquid state to a vapor upon exposure to ambient conditions. Therefore, hexanal vapor treatment involves the introduction of liquid hexanal into a sealed storage environment. As the hexanal vaporizes, it permeates the atmosphere within the container, directly exposing F and V to the vapor, effectively extending their shelf-life. In addition to the enhancement of shelf-life, the treatment showed promising effect of decreasing/eradicating the growth of spoilage micro-organisms. Hexanal vapor exposure has also been linked to an improvement in apple's volatile and aroma compounds (Lanciotti *et al.*, 2003; Song *et al.*, 1996). For this treatment, any matrix (eg, Filter paper) dipped in hexanal was placed inside the container (Sharma *et al.*, 2010) or directly pure hexanal was either placed in a container or petriplates and placed inside the sealed container (Ashwini *et al.*, 2018; Dhakshinamoorthy *et al.*, 2020). By this way hexanal vapor gets absorbed on to the surface of F and V. The volume of hexanal to be added was calculated based on the size of the container, liquid to gas conversion ratio based on Avogadro's law, standard temperature and pressure of 27315 K and 101325 kPa, respectively (Dhakshinamoorthy *et al.*, 2020). Vapor treatment showed promising result in extending the shelf life of fruits such as grand naine banana (Ashwini *et al.*, 2018), and sweet cherry (Sharma *et al.*, 2010). In addition to PLD inhibition, the treatment also inhibited the growth of micro-organisms such as *Penicillium expansum* and *Botrytis cinerea* (Sholberg and Randall, 2007), *M fructicola* and *M laxa* (Baggio *et al.*, 2014) caused by *Monilinia* spp,, is one of the most important peach *(Prunus persica (L.,* *Colletotrichum gloeosporioides* and *Lasiodiplodia theobromae* (Dhakshinamoorthy *et al.*, 2020). However, commercial application faces challenges due to the high hexanal volume needed to attain the desired concentration and also its rapid release profile, potentially causing stress to the stored perishables. Hence, research is directed towards continuous application of hexanal for extended periods which is potentially achieved through impregnation of hexanal within a matrix including nanofiber and nanopellets.

Table 4: Hexanal vapour exposure of fruits (banana / mango) at 800 ppm for 3 hours extended shelf-life (days) and reduced the incidence of post-harvest diseases (%) treated (T) and untreated (C) banana (Grand Naine and Ney Poovan) and mango (Alphonso and Neelam) fruits.

Fruits	**Varieties**	**Vapour conc (ppm)**	**Shelf-life (days)**		**Disease incidence (%)**			
			Control	**Treated**	**Anthracnose**		**Stem end rot**	
Banana		800 ppm 3 hrs	C	T	C	T	C	T
	Grand Naine		7	14	752	422	802	420
	Ney Poovan		5	12	413	283	216	166
Mango	Alphonso		17	22	463	264	787	383
	Neelum		5	10	863	401	421	232

Fig 8: Banana (var Grand Naine) fruits exposed varying hexanal vapour concentrations under ambient conditions The fruits treated with 800 ppm hexanal vapour is free from post-harvest diseases (anthracnose and stem end rot) in comparison to the control.

The hexanal vapour treatment is doable and feasible in fruit storage go downs and large container lorries for long-distance transport. The technology is very powerful as extremely small quantity of hexanal transformed into vapour is sufficient to treat bulk quantities of fruits.

Nano sticker: The hexanal is a biochemical constituent and highly volatile even under room temperature It has been estimated that it completely vapourised within two hours. The formulation of hexanal which is popularly referred as the enhanced freshness formulation was found effective as a pre-harvest spray or post-harvest dip in reducing the post-harvest losses, reduction in post-harvest diseases besides extending shelf-life for 2-3 weeks. This technology is quite effective in both temperate and tropical fruits in extending shelf-life. However, this technology has application difficulty such as right time of spray and number of sprays besides trace of EFF sticking to the produce. In order to overcome these issues, the hexanal molecules have been encapsulated into suitable FDA recommended polymer and developed as electrospun nano-fibre matrix.

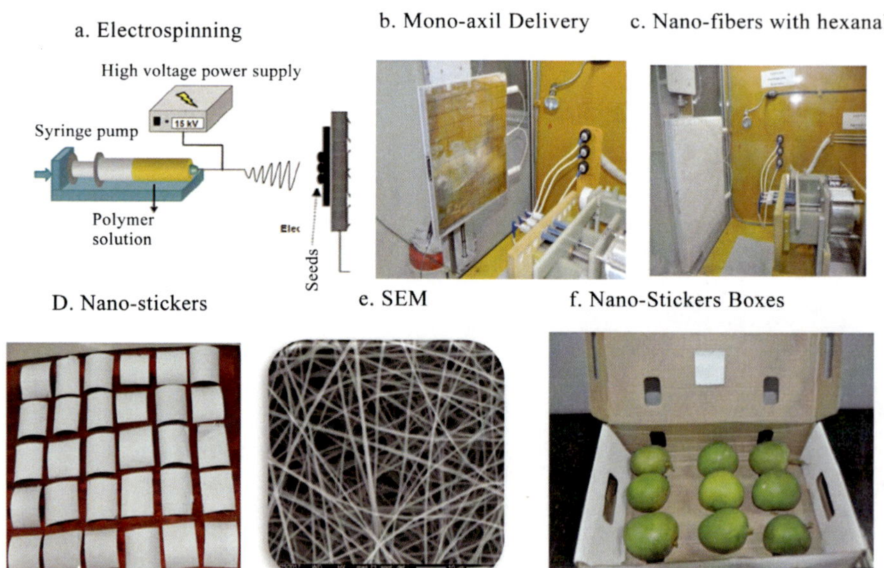

Fig. 9: Sequential steps involved in nano-sticker production a Electrospinning process, b Monoaxil production of fibres, c Nano-fibres loaded with hexanal, d Nano-stickers, e Nano-fibres viewed under Scanning Electron Microscope and f Nano-stickers affixed in the flip-top of fruit storing boxes.

Electrospinning is an elegant technique used to develop nano-fibre which is capable of retaining volatile molecules and can be served as nano-matrix to extend shelf-life of fruits. The TNAU has optimized various solution and process parameters to achieve good quality stable nano-fibres The electrospinning was performed with PVA (polyvinyl alcohol) (7%) and and β-Cyclodextrin (40%) and the polymer solution was subjected high intensity electrical voltage of 25 kV with a flow rate of 03 ml per hr and the tip to collector distance of 15cm. Within 3 hrs of electrospinning, hexanal fortified nano-fibre matrix can be developed (1 sqft) and cut into 5 cm^2 nano-matrices (Stickers). Approximately 36 nano-matrices can be developed from 1 sqft nano-fibre mat carrying 25 ml of hexanal (Patent awarded to Dr KS Subramanian in 2023 No 423341 dated 2722023). Nano-fibres possess has a surface area of 10,000 m^2 per gram and the length is is about 1000 KMs. Such an extensive surface area facilitates retention and release of hexanal vapour that assists in slowing down of the ripening process. Extensive studies have been done on nano-fibre technology on fruit preservation.

Nanofibers represent a class of one-dimensional nanomaterials fabricated from natural, synthetic, or composite polymers. These versatile structures possess the capability to encapsulate a broad range of active compounds which can be released by modulating its external stimuli (Ranjan *et al.*, 2020). Hexanal has been successfully encapsulated within nanofibers fabricated from various biodegradable polymers. These polymers include poly (vinyl alcohol) (PVA) alone (Jan *et al.*, 2022), or in composite forms with zein (Ranjan *et al.*, 2020) or nanofibrillated cellulose (Biswal and Subramanian, 2019). Employed in the form of nanostickers within the storage environment for F and V, the nanofibers act as a controlled release system. Moisture released during fruit respiration serves as a trigger mechanism for the release of hexanal vapor from the nanofibers. This innovative approach has demonstrated efficacy in extending the shelf life of perishable fruits such as peaches (Ranjan *et al.*, 2020) and mangoes (Jan *et al.*, 2022).

Table 4: Shelf-life of banana (var Grand Naine and Ney Poovan) and mango (Alphonso and Bangalora) fruits (in days) exposed to nano-stickers or nano-pellets under ambient storage conditions.

Fruits	Varieties	Shelf-life (days)			Reference
		Control	Sticker	Pellets	
Banana	Grand Naine	14	18	17	Preetha, 2017
	Ney Poovan	8	12	-	Pradeep, 2017
Mango	Alphonso	15	20	-	Shanmugapriya, 2016 Jan *et al.*, 2022
	Bangalora	12	15	15	Vivekumar, 2016 Kanithi Prasanna Kumar, 2018

Nano pellets: Nanopellets involves the process of compressing or molding a material into a definite shape. Hexanal entrapped in the metal organic frameworks (MOF) are briquette into pellets using a pelletizer which was fixed in the fruit storage boxes. Extended release of hexanal upto 10 h was observed compared to pure hexanal that vaporizes completely within 2h of exposure. Nanopellets containing hexanal significantly improved the shelf life of mango up to 15 days (Nagarajan *et al.*, 2021). The production of nano-pellets fortified with hexanal is illustrated in Fig. 10. Nano-pellets found to be effective in extending shelf-life of banana and mango as narrated in Table 4. The nano-pellet technology was awarded patent in 2019 (Patent No 392783 dated 2432022 Awarded to Dr S Marimuthu and Dr KS Subramanian).

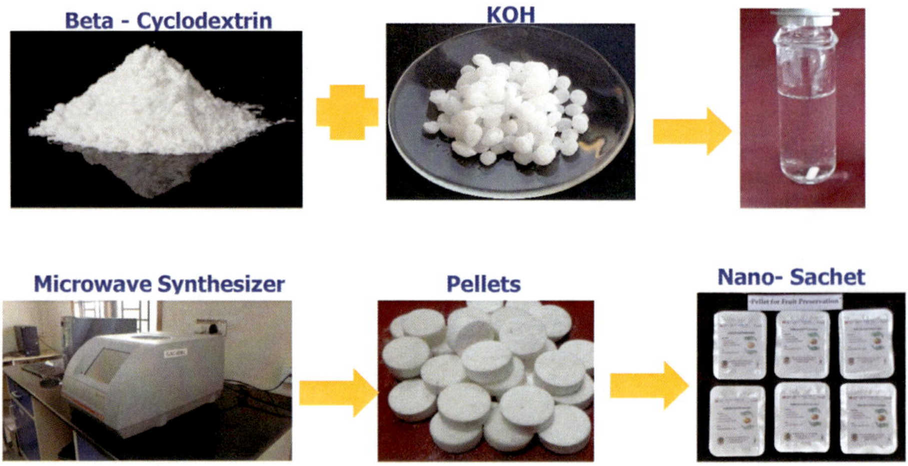

Fig. 10: Steps involved in the production of nano-sachet/nano-pellets to preserve fruits.

Nano-stickers and nano-pellets have been commercialized since 2019 and it is being sold out at the Center for Agricultural Nanotechnology, Tamil Nadu Agricultural University, Coimbatore, under Venture Capital Scheme @ Rs 7 per piece to preserve 2-3 kgs of mangoes or bananas.

Biosafety of nanotechnologies used in fruit preservation: Hexanal is chemically hexanaldehyde, a six carbon compound, naturally produced by plants in trace quantities. When the plants are wounded or cut, there is a grassy odour which is chemically referred as hexanal. Apart from a wound response, hexanal is widely produced by fruits such as apple, apricot, banana, sweet and sour cherries, citrus peel oils and juices, berries, guava, tomato and much more, as a component of flavor volatiles produced during ripening. This naturally producing volatile compound has an effect on fruit ripening processes when it is externally sprayed.

Hexanal inhibits the phospholipase – D enzyme in the skin of the fruits besides slowing down the evolution of ethylene during the ripening processes. These physiological mechanisms facilitate preservation of perishables under storage conditions. The scientists have found that hexanal based effective fresh formulations as a pre-harvest sprays or dip treatment before storage can extend the shelf-life of 2-3 weeks without loss fruit quality, flavor and colour. Even, the vapour of hexanal has a potential to alter the ripening of fruits that resulted in the extension of shelf-life. Such shelf-life extension will help the Asian countries where fruits and vegetables are produced aplenty but the per capita availability is just 50 per cent of their daily requirement due to the post-harvest losses.

The Tamil Nadu Agricultural University, Coimbatore, has undertaken assiduous efforts to determine the safety of the hexanal and its formulations. The safety tests were done as stipulated by the OECD (Organization for Economic Cooperation and Development). In order to test the safety of the chemical, a set of protocols were adopted to examine the ill-effects against beneficial micro-organisms, honey bees, natural enemies, earthworms, fish and human cell lines. In all the cases, a standard set of treatments were exercised (Gunasekaran *et al.*, 2018). The data have unequivocally demonstrated that hexanal and its formulations had no ill-effects against all trophic levels even after three times recommended doses of hexanal formulations were tested. The orchards in which hexanal spray was taken up had least impact on the occurrence of honeybees or natural enemies. Interestingly, while beneficial organisms had no deterrence against hexanal or its formulations, the incidence of pests such as fruit fly and post-harvest disease causing pathogens. More research is requited to precisely predict the mechanism involved deterrence against pests while allowing natural enemies and honey bees freely move around in the sprayed orchards. The exposure to various types of human cell lines were tested at various concentrations at various types of cells such as cancer cells, liver cells and epithelial cells in order to ensure that the new chemical formulations is absolutely safer. This tests help to understand the safety to humans in the event of direct exposure to hexanal formulations.

The biosafety manual on hexanal is designed to deliver the readers about the biochemistry, mode of action, its interaction with insect ecology, soil fauna and flora besides humans (Gunasekaran *et al.*, 2018). This is base document with a strong scientific support, we are well positioned to indicate the safety of hexanl against vast spectrum of organisms at all trophic levels. This will provide a data base for commercialization of the product.

Safety to environment: The safety tests were done as stipulated by the OECD (Organization for Economic Cooperation and Development). In order to test the safety of the chemical, a set of protocols were adopted to examine the ill-effects against beneficial micro-organisms, honey bees, natural enemies, earthworms, fish and human cell lines (Rajkishore *et al.*, 2013; Gunasekaran *et al.*, 2015). In all the cases, a standard set of treatments were exercised. The data have unequivocally demonstrated that hexanal and its formulations had no ill-effects against all trophic levels even after three times recommended doses of hexanal formulations were tested. The orchards in which hexanal spray was taken up had least impact on the occurrence of honeybees or natural enemies. Interestingly, the beneficial organisms had exhibited no deterrence against hexanal or its formulations, while the incidence of pests such as fruit fly and post-harvest diseases causing pathogens were kept at the bay. More research is requited to precisely predict the mechanisms involved in the deterrence against harmful insect pests while allowing natural enemies and honey bees freely move around in the sprayed orchards. The exposure to various types of human cell lines were tested at various concentrations in different types of cells such as cancer cells, liver cells and epithelial cells in order to ensure that the new chemical formulations is absolutely safer. This tests help to understand the safety to humans in the event of direct exposure to enhanced freshness formulations.

***Safer to soil beneficial microbes*:** In soil, two groups of living creatures namely microbes and earthworms are the biological indicators that are being impacted by the EFF. Microbes are the most sensitive organisms that can truly reflect any disturbance in the ecosystem. The nano-toxicity is often measured by the microbial activities EFF carries a bio-molecule (hexanal) which is encapsulated with surfactant (Tween 20) and co-surfactant (ethanol). Seventeen micro-organisms have been identified by the OECD to be used tested against to test for safety. Our studies have clearly shown that the EFF had absolutely no detrimental effect on these microbes as indicated by the total microbial activities using an enzyme assay. Indeed, micro-organisms are able to use EFF as a food source and were able to grow much faster compound to unsprayed orchards.

***Safer to earthworms*:** Earthworms are "farmers friend" and a good indicator of soil health. Whenever people buy farm lands, they look for the number of earthworms per unit weight of soil. Any pesticides, fertilizers or foreign materials added in excess to the soil, will cause a decrease in earthworm populations and health. Our researchers have introduced the earthworms in soils treated or untreated with EFF. Continuous monitoring of earthworm activities has shown that EFF had no negative impact on physical activities or the life cycle itself. In fact, earthworms grew much faster and robust in EFF sprayed mango orchard soils than their unsprayed counterpart. These biological measurements indicate that the EFF is safe macro-faunas in soil

***Nurtures natural enemies*:** Natural enemies are important constituents in insect ecology. We have looked at the insect ecology in mango orchards. Any new molecules that are introduced into the orchard should be safe beneficial insects that act as predators and parasites. Predators and parasites feed on detrimental insect pests and keep the crop free from the harmful predators are usually larger insects than the smaller ones that are prey. For instance, lace wing bug (*Chrysoperla* spp) eats the smaller aphids and Jassids. On the other hand, parasites are smaller than their prey and feed on eggs of the insect pests.

***Attracts honey bees, distracts fruit flies*:** Honey bees are unique creatures capable of discrimination contaminated and healthy environmental conditions. It is widely believed that honey bees exploit nanotechnology for its own survival. Honeybees have to visit and siphon off nectar from at least 1000 flowers to synthesize a drop of honey. Consequently, honey has the longest shelf-life of more than 1000 years naturally without any loss of quality. Indeed, this was a reason why traditional medicines are honey-based. Honey bees are often used as a model system to study the effect of toxicity of metal oxide nano-particles. In our study, we tested 4 species of honey bees that have exhibited no deterrence against EFF. Despite the honey bees are attracted by EFF, one of the potential pests impeding exports of mangoes to the other countries is fruit fly which is found to be deterred by EFF (Karthika *et al.*, 2015). After the application of EFF in the orchards, there is a trace this contradicts the statement that hexanal dissipates quickly around the orchards that repel the potential pest at the bay. In addition, some of mealy bugs in mango orchards get washed away by the formulation.

Accustomed to aquatic animals: Many agricultural inputs either sprayed on trees or applied to the soil gets into water bodies through runoff water that eventually affect aquatic flora and fauna including fishes. In order to test the aquatic toxicity, the OECD suggests testing against zebra fish (*Danio rerio* Hamilton). Zebra fish is a very sensitive organism to any foreign materials and is used as a model system to study the environmental toxicity. Our study have shown that EFF had no ill-effects on zebra fish even if the concentration was 10 times greater than the normal levels.

Safer to humans: Safety of any agricultural inputs to humans is to be tested using cultured cells of target sites. Agriculture involves human interference and it is important to understand the side-effects associated with the new formulations introduced in the crop production systems. The OECD has evolved protocols to test the intensity of toxicity using cultured cells. Target sites such as skin, lungs, kidneys and liver are likely to expose to the formulation directly or indirectly. Consequent, cultured cells of these organs are used to assess the likely toxicity at the standard assays using cytotoxicity and genotoxicity tests. In fact, we had examined the likely impacts and interactions of EFF on human cell lines and found no adverse effects.

Overall, the data suggest that the hexanal formulation as a fruit preservative as a pre-harvest spray or post-harvest dip or as any other mode of delivery had absolutely no ill-effects either in the mango orchard ecosystem. Beneficial microbes and earthworms in the soil, natural enemies (predators and parasites) and honey bees on the EFF sprayed trees besides safety to humans unequivocally demonstrated that the formulation safer to all spectrum of constituents in the ecosystem. Further, the quality of fruits in sprayed orchards had improved and no residues detected that fulfills the requirement of domestic and international markets.

Farmers feedbacks: Survey was conducted in four major mango producing domains of Tamil Nadu *viz,* Krishnagiri, Dharmapuri, Theni, Dindigul districts of Tamil Nadu and Kolar district of Karnataka where EFF technology was practiced by large number of farmers and acceptance behavior. The EFF was distributed at free of cost to get the feedback of the technology from the users About 3640 mango farmers had undertaken the EFF trial during the phase II (Dec 2014 - March 2018) of the project. The distribution of EFF had taken place in Periyakulam Mango Workshop (52), Tamil Nadu Innovative Initiative (TANII) meet at RRS, Paiyur (40), Banana Festival (347), Department of Nano Science and Technology, TNAU (68), Krishnagiri (752), Dharmapuri (519), Theni (500), Dindigul (441), Kolar (300), Farmers Day in TNAU (28) and Krishnagiri Packhouse Meet (281). The feedback from the users were collected by personal interview method. Collected data were classified, tabulated and statistically analyzed. The findings of the study revealed that post-harvest losses had been reduced by 10-12 per cent in the EFF sprayed fields in comparison to control. Due to the reduction in losses at the field level farmers got about 700 kg of additional yield of fruits. After the adoption of EFF technology which indicates the positive impact of EFF technology in mango. Our feedback survey suggested that 55 per cent of the respondents got the premium price for the EFF spray, 31 per cent expressed that the ripening of mango fruits were delayed, and 6 per cent indicated that there was no difference between treated and control. The data clearly suggests that more than 80 per cent of the mango of fruit growers benefitted from the EFF technology.

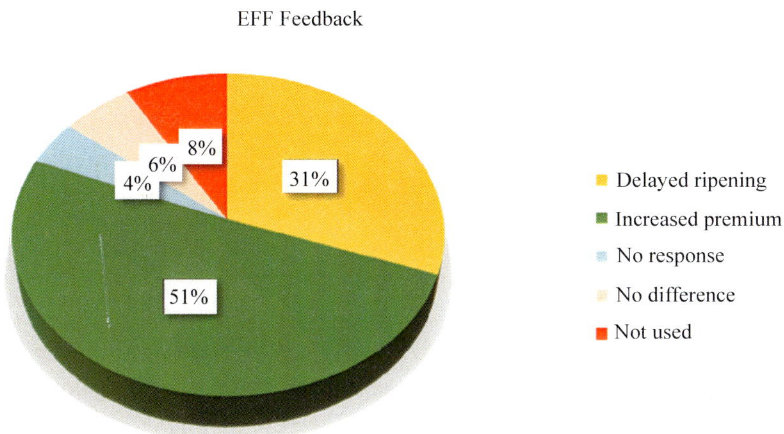

Nano packaging

Biodegradable films for packaging: Plastics causes extensive damage to the environment and human health as it stays in the ecosystem for more than 1000 years and the plastics are commonly used materials for packaging of fruits and vegetables. In order to overcome the problems, biodegradable films have been developed using the nano-fibrillated cellulose (NFC) extracted from banana pseudostem using a novel approach of one step, bleaching agent and alkali free acid hydrolysis method coupled with ultrasonication (Priya *et al.*, 2017). The treatment was effective in the depolymerization and defibrillation of banana pseudostem fiber and formation of nanofibrilated cellulose which was confirmed by several physio-chemical techniques. The XRD analysis revealed an increased crystallinity of cellulose by almost 20 per cent in nanofibrillated cellulose than their respective raw banana fibers. The FT-IR and TGA analysis confirmed the absence of lignin, hemicelluloses and pectin in the nanofibrillated samples. Thermal analysis showed the increased thermal stability of nanofibrillated cellulose. The NFC with a compatibilizer biodegradable films were developed that are known to be UV protectant and can be used for developing nano-packaging materials NFC also derived from waste land weeds like *Agave americana* and developed as a biodegradable film that can extended shelf-life of vegetables particularly tomato for a period of two weeks (Ponni *et al.*, 2020; Krishna Dev and Subramanian, 2020, 2021). The NFC was also derived from wood wastes and made as a biodegradable film that can decompose within 40-60 days (Vishnu *et al.*, 2023). These literatures clearly demonstrate that plant derived nano-fibrillated cellulose can be exploited to develop biodegradable films as a packaging material to preserve fruits and vegetables while ensuring environmental safety.

Improved gas barrier and controlled micro-environment: Compared to traditional packaging materials, nanocomposites incorporating particulate nanomaterials exhibit superior gas barrier properties due to enhanced tortuosity and interfacial interactions (Sanchezgarcia *et al.*, 2010). This advantage stems from the nanoparticles' high surface-to-volume ratio, thus enabling strong interfacial interactions with the polymer matrix despite requiring only a small loading (~5 % w/w) (Khan *et al.*, 2012). Consequently, these nanocomposites effectively maintain a stable micro-environment within the packaging by controlling gas composition, humidity, and temperature (Mihindukulasuriya and Lim, 2014). This controlled environment reduces the respiration rate of fresh fruits and vegetables, leading to a significant extension of their shelf-life.

Mechanical properties: The inclusion of nanoparticles in packaging materials for F and V offers a promising avenue to improve mechanical properties, bolstering the overall stability and protective capacity of the packaging. This enhancement arises from two primary mechanisms: (i) Favorable interfacial interactions: Nanoparticles readily form strong bonds with the polymer matrix and with other nanoparticles through interactions like hydrogen bonding and ionic interactions (Sanchezgarcia, Lagaron, and Hoa, 2010). These bonds effectively integrate the nanoparticles into the composite structure, distributing stress more uniformly and enhancing the material's overall mechanical strength and (ii) Stress transfer at the interface: The high surface area-to-volume ratio of nanoparticles creates a large interfacial zone between the nanoparticles and the matrix (Abdollahi *et al.*, 2013). This interface acts as a stress transfer zone, efficiently transferring applied forces from the matrix to the nanoparticles. This ability to distribute stress more effectively leads to improved mechanical properties like tensile strength, hardness, and elasticity. Studies have demonstrated the effectiveness of this approach, particularly with cellulose nanoparticles. For instance, Khan *et al.* (2012) reported a 26 per cent increase in tensile strength of chitosan films reinforced with just 5 per cent nanocrystalline cellulose. Similarly, Cheng *et al.* (2016) observed enhanced hardness and elasticity in guar gum films containing nanocrystalline cellulose. While increasing nanoparticle content generally improves mechanical properties, it's crucial to avoid exceeding an optimal threshold. Excessive filler can lead to aggregation, negating the reinforcing effect and potentially compromising other properties. Therefore, finding the optimal balance between nanoparticle content and desired mechanical properties is essential for maximizing the benefits.

Thermal properties: The ordering effect of nanoparticles on polymer molecules within the matrix enhances its thermal stability, as evidenced by changes in glass transition temperature and heat distortion temperature. Interactions between nanoparticles and the polymer matrix reorganizes polymer molecules into a more ordered state, increasing the packing efficiency and modifying the local energy landscape, which consequently enhances the thermal properties of the matrix. This reorganization translates into a clearer crystallization temperature, a higher heat distortion temperature, and a higher melting temperature for the composite material. These improved thermal properties are crucial in post-processing treatments like pasteurization, often used to extend the shelf-life of fresh fruits and vegetables, as they allow for higher processing temperatures without compromising the integrity of the packaging material or the produce itself.

Antimicrobial properties: Inhibiting microbial growth and viability, antibacterial nanomaterials hold promise for extending the shelf-life of fruits and vegetables (Kvitek *et al.*, 2008). These nanomaterials fall into two main categories based on their active components (Liu and Gu, 2009; Rossi *et al.*, 2014): metal ionic nano-materials in which potent ions like Ag, Cu, or Zn is incorporated into mineral nano-carriers like silicates for sustained release and interaction with bacterial functional groups (Brody *et al.*, 2011) and oxide photocatalytics which rely on light-activated radicals that reacts with biological objects for antibacterial action. While the precise molecular pathway behind the antimicrobial action remains under investigation, three distinct mechanisms have emerged as potential contributors: oxide photocatalysis as mentioned above, metal ion dissolution, and contact reaction. Metal ions disrupt bacterial energy metabolism and cell membranes, while contact reaction utilizes charge transfer for direct cell disruption (Liu and Gu, 2009; Montazer *et al.*, 2016; Barani *et al.*, 2011).

Conclusion and way forward

Nanotechnology is a powerful tool to enable atomic manipulation to develop processes and products impossible to achieve through conventional systems. The post-harvest losses in the fruits and vegetables are a global challenge and almost all countries across the globe facing the problem. This chapter brought basket of nanotechnologies such as pre-harvest sprays, post-harvest dip, nano-stickers, nano-pellets, hexanal vapour besides nano packaging materials with anti-microbial, UV protection and biodegradable nano-films. These technologies can be adopted singly or in combinations to minimize the loss of perishables from farm to fork.

The Department of Biotechnology has set a Regulatory Guidelines for the use of nano-products in agriculture and food systems in 2020 (Aholeya *et al.*,

2020). This is the base document used for the registration and notification of nano-fertilizers in the country. The registration of nano-products is yet to take off for paradoxical reasons. One key concern is the potential migration of nanocarriers, containing hexanal for example, into fruits and vegetables through processes like absorption, dissolution, and diffusion. The safety implications of these tiny carriers within food products remain unclear. Therefore, future research must prioritize legal considerations and safety evaluations. Regulatory frameworks and consumer concerns regarding nanomaterials in food also require careful consideration. Computational modeling can be a powerful tool for predicting the impact of nanotechnology on shelf-life and product quality. Increased focus on such research is crucial for a comprehensive understanding of this technology. Finally, scaling up production of nano-packaging materials presents a significant challenge due to its complexity and high technological demands. Optimizing existing processing techniques and developing cost-effective methods for mass production are essential for successful commercialization. Farmers throughout the world focus on using new innovations and technologies for enhancing the production of crops through intensive and extensive agriculture. Indeed, whenever there is a price hike in fruits and vegetables the policy makers in India are looking for innovative technologies to combat post-harvest losses to preserve them naturally or minimal cost to benefit. Many feel that nanotechnology will offer solution to unresolved challenging issues. The Government should initiate large scale use of innovative technologies to help the farmers to derive benefit from the cutting edge technologies.

References

Ajith T, 2016. Enhanced shelf-life of mango through hexanal dipping MSc (Horti) Thesis submitted to the Tamil Nadu Agricultural University, Coimbatore

Amanatidou A, Slump R A, Gorris, L G M and Smid E J, 2000. High Oxygen and High Carbon Dioxide Modified Atmospheres for Shelf-life Extension of Minimally Processed Carrots *Journal of Food Science*, 65(1): 61–66

Anusha B, 2014. Biological management of post-harvest pathogens of mango PhD (Pathology) Thesis submitted to the Tamil Nadu Agricultural University, Coimbatore

Anusuya P, Nagaraj R, Janavi, G J, Subramanian K S, Paliyath G and Subramanian J, 2016. Pre-harvest sprays of hexanal formulation for extending retention and shelf-life of mango (*Mangifera indica* L) fruits *Scientia Horticulturae*, 211: 231–240 https://doiorg/101016/jscienta201608020

Ashwini T, Ganapathy S, Subramanian K S, Rani C I and Meenakshi G G, 2018. Effect of Hexanal Vapour on Postharvest Quality and Shelf Life of Banana var Grand Naine https://doiorg/1020546/ijcmas2018702297

Baggio J S, Lourenço S D A and Amorim L, 2014. Eradicant and curative treatments of hexanal against peach brown rot *Scientia Agricola*, 71(1):72–76 https://doiorg/101590/S0103-90162014000100010

Bisht B, Bhatnagar P, Gururani P, Kumar V, Tomar, M S, Sinhmar R, Rathi N and Kumar S, 2021. Food irradiation: Effect of ionizing and non-ionizing radiations on preservation of fruits and vegetables– a review *Trends in Food Science* and *Technology*, 114, 372–385 https://doiorg/101016/jtifs202106002

Biswal B and Subramanian K, 2019. Slow Release of Hexanal By Biodegradable Electrospun Nanofibres for Increasing Shelf-Life of Harvested Mango Fruits *Madras Agricultural Journal*, 106: 633–642

Boonthanakorn J W, Daud A, Aontee and C Wongs-Aree, 2020. Quality preservation of fresh-cut durian cv 'Monthong' using micro-perforated PET/PE films *Food Packaging and Shelf Life* 23:100452 doi: 101016/jfpsl2019100452

Cefola M B, Pace L, Sergio F, Baruzzi M A, Gatto A, Carito V, Linsalata N A, Cascarano, and D Di Venere, 2014. Postharvest performance of fresh-cut 'Big Top' nectarine as affected by dipping in chemical preservatives and packaging in modified atmosphere *International Journal of Food Science and Technology* 49 (4):1184–95 doi: 101111/ijfs12415

Cheema A, Padmanabhan P, Subramanian J, Blom T and Paliyath G, 2014. Improving quality of greenhouse tomato (*Solanum lycopersicum* L) by pre- and postharvest applications of hexanal-containing formulations *Postharvest Biology and Technology*, 95: 13–19 https://doiorg/101016/jpostharvbio201403012

DeBrouwer E J, Sriskantharajah K, El Kayal W, Sullivan J A, Paliyath G and Subramanian J, 2020. Pre-harvest hexanal spray reduces bitter pit and enhances post-harvest quality in 'Honeycrisp' apples (Malus domestica Borkh) *Scientia Horticulturae*, 273, 109610 https://doiorg/101016/jscienta2020109610

Del-Valle V, P Hernandez-Munoz R, Catal ~ a, and Gavara R, 2009. Optimization of an equilibrium modified atmosphere packaging (EMAP) for minimally processed mandarin segments Journal of Food Engineering 91 (3):474–81 doi: 101016/jjfoodeng200809027

Dhakshinamoorthy D, Sundaresan S, Iyadurai A, Subramanian K S, Janavi G J, Paliyath G and Subramanian J, 2020. Hexanal Vapor Induced Resistance against Major Postharvest Pathogens of Banana (*Musa acuminata* L) *The Plant Pathology Journal*, 36(2):133–147 https://doiorg/105423/PPJOA0320190072

Dhanalakshmi V, 2018. Effect of hexanal nanoemulsion to extend the Shelf- life of acid lime MSc (Horti) Thesis submitted to the Tamil Nadu Agricultural University, Coimbatore

Ebrahimi A, Zabihzadeh Khajavi M, Ahmadi S, Mortazavian A M, Abdolshahi A, Rafiee S and Farhoodi M, 2022 Novel strategies to control ethylene in fruit and vegetables for extending their shelf life: A review *International Journal of Environmental Science and Technology*, 19(5):4599–4610 https://doiorg/101007/s13762-021-03485-x

Gill K S, Dhaliwal, H S, Mahajan, B V C, Paliyath G and Boora R S, 2016. Enhancing postharvest shelf life and quality of guava (*Psidium guajava* L) cv Allahabad Safeda by pre-harvest application of hexanal containing aqueous formulation. *Postharvest Biology and Technology*, 112, 224–232 https://doiorg/101016/jpostharvbio201509010

Gunasekaran, K, Karthika, S, Nandakumar, N B, Subaramanian K S, Gopinathan Paliyath and Jayasankar Subaramanian, 2015. Biosafety of Hexanal pp:89.

Henig Y S and Gilbert S G, 1975. Computer analysis of the variables affecting respiration and quality of produce packaged in polymeric films *Journal of Food Science*, 40:1033-1035.

Hutchinson M J, Ouko J R, Ambuko J, Owino W O and Subramanian J, 2018. Effects of hexanal dip on the post-harvest shelf life and quality of papaya (*Carica papaya* L) fruit *Tropical Agriculture* https://journalsstauwiedu/ojs/indexphp/ta/article/view/6689

Jan N A, Subramanian K S, Ganapathy S, Mohanraj J and Govindaraju K, 2022. Nano-fiber enabled regulated release of hexanal vapor and its impact on shelf life of mango fruits *Polymer Bulletin*, 1-17.

Janik M, Hanula M, Khachatryan K and Khachatryan G, 2023. Nano-/Microcapsules, Liposomes, and Micelles in Polysaccharide Carriers: Applications in Food Technology *Applied Sciences*, 13(21): Article 21 https://doiorg/103390/app132111610

Jayathunge K G, Gunawardhana D K, Illeperuma D C, Chandrajith, U G, Thilakarathne B M, Fernando, M D and Palipane K B, 2014. Physico-chemical and sensory quality of fresh cut papaya (Carica papaya) packaged in micro-perforated polyvinyl chloride containers *Journal of Food Science and Technology* 51 (12): 3918–25 doi: 101007/s13197-013-0952-8

Jincy M, Djanaguiraman M, Jeyakumar P, Subramanian K S, Jayasankar S and Paliyath G, 2017. Inhibition of phospholipase D enzyme activity through hexanal leads to delayed mango (*Mangifera indica* L) fruit ripening through changes in oxidants and antioxidant enzymes activity *Scientia Horticulturae*, *218*, 316–325 https://doiorg/101016/jscienta201702026

Kanmani, 2016. Effect of hexanal on extending shelf-life of banana MSc (Horti) Thesis submitted to the Tamil Nadu Agricultural University, Coimbatore.

Kaur K, Kaur G and Brar J S, 2020. Pre-harvest application of hexanal formulations for improving post-harvest life and quality of mango (*Mangifera indica* L) cv Dashehari *Journal of Food Science and Technology*, *57*(11), 4257–4264 https://doiorg/101007/s13197-020-04464-9.

Lu H, W Wu, J Limwachiranon D Yang, G Xiao, Z Luo and Li L, 2018. Effect of micro-perforated film packing on fatty acid-derived volatile metabolism of "red globe" table grapes *Food and Bioprocess Technology* 11 (10):1807–17 doi: 101007/s11947-018-2142-1

Kanithi Prasanna Kumar, 2018, Studies on Assessing Release Kinetics and upscaling of β-Cyclodextrin Metal Organic Frame Work - Hexanal complex Technology MSc (Nano Science and Technology) Thesis submitted to the Tamil Nadu Agricultural University, Coimbatore.

Koutsimanis G, Harte J and Almenar E, 2015. Freshness maintenance of cherries ready for consumption using convenient, microperforated, bio-based packaging *Journal of the Science of Food and Agriculture* 95: (5):972–82 doi: 101002/jsfa6771

Krishna Dev P, 2016. Hexanal Formulation to extend shelf-life of tomato fruits MSc (Nano Science and Technology) Thesis submitted to the Tamil Nadu Agricultural University, Coimbatore.

Krishnadev P, Subramanian K S, Janavi G J, Ganapathy S and Lakshmanan A, 2020. Synthesis and Characterization of Nano-fibrillated Cellulose Derived from Green *Agave americana* L Fiber *BioResources*, 15(2): 2442-2458

Krishnadev P, Subramanian K S, Lakshmanan A, Ganapathy S, Raja K and Rajkishore S K, 2021. Hydroxypropyl Methylcellulose Nanocomposites Containing Nano Fibrillated Cellulose (NFC) from Agave americana L for Food Packaging Applications BioResources, 16(4): 8125-8151.

McKeen L, 2012. Introduction to Food Irradiation and Medical Sterilization *The Effect of Sterilization on Plastics and Elastomers*, 1–40 https://doiorg/101016/B978-1-4557-2598-400001-0

Murr D P, Handa A K and Paliyath G, 2008. *Postharvest biology and technology of fruits, vegetables, and flowers* Iowa, US: Blackwell Publishing

Mutharasu P, 2018. Hexanal formulation on Shelf-life of Banana (*Musa spp*) cultivars MSc (Horti) Thesis submitted to the Tamil Nadu Agricultural University, Coimbatore

Nandhini P, 2017. EFF to extend Shelf-life of Guava (*Psidium guajava* L) MSc (Horti) Thesis submitted to the Tamil Nadu Agricultural University, Coimbatore.

Paliyath G and Droillard M J, 1992. The mechanisms of membrane deterioration and disassembly during senescence Plant Physiology and Biochemistry Paris https://eurekamagcom/research/002/523/002523989php

Paliyath G, Murr D P, Handa A K and Lurie S, 2009. *Postharvest Biology and Technology of Fruits, Vegetables, and Flowers* John Wiley and Sons

Paliyath G, Subramanian J, Lim L-T, Subramanian K S, Handa A K and Mattoo A K, 2018. *Postharvest Biology and Nanotechnology* John Wiley and Sons.

Phillips C A, 1996. Review: Modified Atmosphere Packaging and its effects on the microbiological quality and safety of produce *International Journal of Food Science* and *Technology*, 31(6):463–479 https://doiorg/101046/j1365-2621199600369x

Pinela J, Barreira J C M, Barros L, Antonio A L, Carvalho A M, Oliveira M B P P and Ferreira I C F R, 2016. Postharvest quality changes in fresh-cut watercress stored under conventional and inert gas-enriched modified atmosphere packaging *Postharvest Biology and Technology*, 112: 55–63 https://doiorg/101016/jpostharvbio201510004

Ponni P, Subramanian K S, Janavi G J and Subramanian J, 2020. Synthesis of nano-film from nanofibrillated cellulose of banana pseudostem (*Musa* spp) to extend the shelf life of tomato, *BioResources* 15(2): 2882-2905

Pradeep D, 2018. Development of Cellulosic Nano - Matrix to Extend shelf- life of fruits MSc (Nano Science and Technology) Thesis submitted to the Tamil Nadu Agricultural University, Coimbatore

Preetha S, 2017. Nano-matrix to Extend the Shelf-life of Fruits MSc (Nano Science and Technology) Thesis submitted to the Tamil Nadu Agricultural University, Coimbatore.

Preethi P, Soorianathasundaram K, Sadasakthi A, Subramanian K S, Vijay Rakesh Reddy S, Paliyath G and Subramanian J, 2021. Preharvest Application of Hexanal as a Surface Treatment Improved the Storage Life and Quality of Mango Fruits Coatings, 11(10): 1267.

Priya G, Subramanian K, Paliyath G and Subramanian J, 2017. «Genotypic variations in characteristics of nano-fibrillated cellulose derived from banana pseudostem,» *BioRes* 12(4);6984-7001.

Qu P, Zhang M, Fan K and Guo Z, 2022. Microporous modified atmosphere packaging to extend shelf life of fresh foods: A review *Critical Reviews in Food Science and Nutrition*, 62(1): 51–65 https://doiorg/101080/1040839820201811635

Ranjan S, Chandrasekaran R, Paliyath G, Lim L T and Subramanian J, 2020. Effect of hexanal loaded electrospun fiber in fruit packaging to enhance the post-harvest quality of peach *Food Packaging and Shelf Life*, *23*: 100447 https://doiorg/101016/jfpsl2019100447

Rocculi P, Romani S and Rosa M D, 2005. Effect of MAP with argon and nitrous oxide on quality maintenance of minimally processed kiwifruit *Postharvest Biology and Technology*, 35(3): 319–328 https://doiorg/101016/jpostharvbio200409003

Samwel J, Msogoya T, Kudra A, Mtui H D, Baltazari A, Sullivan J A, Subramanian J and Mwatawala M W, 2020. Pre-harvest field application of enhanced freshness formulation reduces yield loss in orange *Chemical and Biological Technologies in Agriculture*, 7(1): 25 https://doiorg/101186/s40538-020-00191-7

Shanumugapriya R, 2016. Development of Electrospun fibre matrix to extend shelf-life of fruits MSc (Nano Science and Technology) Thesis submitted to the Tamil Nadu Agricultural University, Coimbatore.

Sharma M, Jacob J K, Subramanian J and Paliyath G, 2010. Hexanal and 1-MCP treatments for enhancing the shelf life and quality of sweet cherry (*Prunus avium* L) *Scientia Horticulturae*, 125(3):239–247 https://doiorg/101016/jscienta201003020

Sholberg P L and Randall P, 2007. Fumigation of Stored Pome Fruit with Hexanal Reduces Blue and Gray Mold Decay *HortScience*, 42(3): 611–616 https://doiorg/1021273/HORTSCI423611

Tamileezham N, 2017. Hexanal Formulation to extend Shelf-life of Grapes (*Vitis vinifera* L) MSc (Horti) Thesis submitted to the Tamil Nadu Agricultural University, Coimbatore.

Villalobos M d C, Serradilla M J, Martın A, Ruiz-Moyano S, Pereira C and Cordoba M d G, 2016a. Synergism of defatted soybean meal extract and modified atmosphere packaging to preserve the quality of figs (*Ficus carica* L) Postharvest Biology and Technology 111:264–73 doi: 101016/jpostharvbio201509025

Vishnu R, Revathi Revathiswaran, K S Subramanian, Kalappan Thangamuth Parthiban, Kalichamy Chandrakumar, Elaveetil Vasu Anoop and Cintil Jose Chirayil, 2023. Isolation and characterization of nanocellulose from selected hardwoods, viz, *Eucalyptus tereticornis* Sm and *Casuarina equisetifolia* L, by steam explosion method *Scientific Reports* 13(1): 1-15

Vivekumar R, 2015. Hexanal and beta cyclodextrin complex for slow release of hexanal to enhance shelf-life of mango MSc (Nano Science and Technology) Thesis submitted to the Tamil Nadu Agricultural University, Coimbatore.

Yumbya P M, Hutchinson M J, Ambuko J, Owino W O, Sullivan A, Paliyath G and Subramanian J, 2018. Efficacy of hexanal application on the post-harvest shelf life and quality of banana fruits (*Musa acuminata*) in Kenya *The Journal of the Faculty of Food and Agriculture*, 14

Zhang B-Y, Samapundo S, Pothakos V, Sürengil G and Devlieghere F, 2013. Effect of high oxygen and high carbon dioxide atmosphere packaging on the microbial spoilage and shelf-life of fresh-cut honeydew melon *International Journal of Food Microbiology*, 166(3):378–390 https://doiorg/101016/jijfoodmicro201308002

6

Cold Chain Management, Specifications, and Quarantine Requirements for Fruits in Domestic and Export Markets

Suresh Kumar P, Pramod Shelake, Naveen Jose*, Mousumi Sabat** and Shankar Ganesh K***

ICAR National Research Centre for Banana, Tiruchirappalli - 620 102 Tamil Nadu
**ICAR National Institute of Natural Fibre Engineering and Technology Kolkatta-700040, West Bengal*
***ICAR- Central Institute of Agricultural Engineering, Bhopal-462 038 Madhya Pradesh*
****ICAR- Central Institute of Cotton Research, Regional Centre-641 003 Coimbatore, Tamil Nadu*

Introduction

The cold chain is a meticulously managed series of actions using specialized equipment to ensure products remain within a specific low-temperature range from production to consumption (ASHRAE, 2024; IIR, 2024). This unbroken sequence encompasses refrigerated production, storage, and distribution activities, safeguarding the safety and quality of perishable and thermo-sensitive goods like agricultural commodities (such as fresh produce, meat, dairy, seafood, chemicals, pharmaceuticals, flowers, wine, *etc.*), processed foods and pharmaceuticals (Aung and Chang, 2014; Awan, 2023). For climacteric fruits like bananas, custard apples, apples, pears, tomatoes, mangoes, and avocados, cold storage facilitates controlled ripening, extending the time between harvest and consumption. Cold chain management, integral to the overall supply chain, requires expertise in maintaining temperature, air quality, and adherence to regulatory standards. Any disruption in this process can lead to product spoilage and financial implications (FAO, 2019). The management of the cold chain system is crucial for preserving the freshness, taste, and nutritional value of fruits from harvest to consumption (Kader, 2003). Lower temperatures preserve nutritional integrity by slowing the degradation of vitamins, antioxidants, and other nutritional components, reducing microbial growth, enzymatic reactions and ensuring optimal health benefits (Makule *et al.*, 2022; Erkmen and Bozoglu, 2016). This reduction in pathogen activity extends shelf-life, allowing for longer storage, transportation, and display without compromising quality.

As per the International Plant Protection Convention (IPPC), Quarantine treatments of fresh fruits and vegetables involve killing, removing, or preventing reproduction by an undesirable living organism on living host material (Mangan and Hallman, 2019). These measures play a pivotal role in averting the introduction and spread of pests and diseases in newer areas where it doesn't exist and act as a crucial safeguard for the agricultural industry, curtailing the potential havoc wreaked by exotic pests on local crops and ecosystems, thereby mitigating economic losses (Chand *et al.*, 2017). By controlling the influx of potential threats, quarantine measures ensure the vitality and productivity of local crops, minimizing the necessity for expensive pest control measures and averting crop losses. Additionally, these measures contribute to biodiversity conservation by thwarting the introduction of non-native species that might disrupt native flora and fauna. Thus, they are instrumental in maintaining a stable and secure food supply, preventing widespread crop failures and food shortages resulting from the introduction of harmful pathogens (Karuppuchamy and Venugopal, 2016).

Several international organizations and agreements set standards and guidelines for the movement of plant products, including fruits. The primary international agreement addressing plant quarantine and the movement of plants and plant products is the International Plant Protection Convention (IPPC), governed by the Food and Agriculture Organization (FAO) of the United Nations (IPPC, 2017). Exporters and importers of fruits must be familiar with the specific phytosanitary requirements of the importing country. Compliance with international regulations, including those established by the IPPC and International Standards for Phytosanitary Measures (ISPMs), facilitates smooth international trade while protecting global plant health (ISPMs, 2023). National Plant Protection Organizations (NPPOs) play a key role in implementing and enforcing these regulations at the country level. Similarly, in India, the regulations for the quarantine of fruits are primarily governed by the Plant Quarantine (Regulation of Import into India) Order, 2003, issued under the Destructive Insects and Pests Act, 1914 (Act 2 of 1914) (PQO, 2003). It was enforced by the Plant Protection Advisor to the Government of India, Ministry of Agriculture and Farmers Welfare, and administered by the

Department of Agriculture, Cooperation, and Farmers Welfare. The order aims to prevent the introduction and spread of plant pests and diseases that could harm agriculture and plant health in the country. Considering the importance, this chapter has been divided into two sections. The first section details the cold chain management system, its components, and factors influencing the supply chain management (SCM) in fruits. While the other part covers the quarantine requirements for fruits in the domestic as well as export markets.

Cold chain management

Components of cold chain management for fruits: Cold chain management of fruits involves a series of interconnected components, each crucial in preserving the quality, safety, and freshness of fruits from pre-harvest to retail display. The vital stages that need to be managed properly to maintain fruit quality during the cold chain (Anonymous, 2024a) are described in Table 1. Proper refrigeration facilities, including storage warehouses and transportation vehicles, equipped with advanced temperature control systems, and monitoring devices, such as data loggers and sensors, play a vital role at different stages of the supply chain. Strict standard operating procedures and quality assurance protocols help ensure compliance with temperature requirements and industry regulations. Additionally, a well-trained workforce is indispensable for handling, loading, and unloading temperature-sensitive goods.

Factors influencing cold chain management for fruits: Effective cold chain management for fruits is influenced by many factors that affect the seamless preservation and transportation of temperature-sensitive produce. The success of this management lies in a careful consideration of variables such as the type of cargo, required temperature and humidity levels, size of the shipment, duration of transit, and external environmental conditions. Additionally, the intricacies of cold transport, employing various modes such as road, rail, air, and sea, necessitate the use of specialized vehicles and equipment.

Temperature control: Temperature directly impacts the rate of fruit ripening, enzymatic reactions, and microbial growth. Maintaining the correct temperature throughout the supply chain is essential to slow down these processes and extend the shelf life of fruits (Paltrinieri *et al.*, 2014; FAO, 2004). Variations in temperature during transportation or storage can compromise fruit quality especially the texture and taste of fruits, resulting in a shorter shelf-life and reduced consumer satisfaction.

Humidity levels: Humidity levels influence the moisture content of fruits during storage. Maintaining optimal humidity helps prevent dehydration or excessive moisture, both of which can affect the texture, appearance, and overall quality of fruits. Inadequate humidity control can lead to issues such as shrivelling, wilting, or mould growth (Paltrinieri *et al.*, 2014).

Table 1. Ideal storage environment for fresh fruits.

Fruit type	Ideal temperature range	Relative humidity	Storage duration
Apples	0°C to 4°C (32°F to 39°F)	90per cent to 95per cent RH	4 to 6 months
Oranges	4°C to 7°C (39°F to 45°F)	85per cent to 90per cent RH	2 to 3 weeks
Bananas	13°C to 16°C (55°F to 61°F)	85per cent to 95per cent RH	1 to 2 weeks
Grapes	0°C to 1°C (32°F to 34°F)	90per cent to 95per cent RH	1 to 2 months
Strawberries	0°C to 2°C (32°F to 36°F)	90per cent to 95per cent RH	5 to 7 days
Pineapples	7°C to 10°C (45°F to 50°F)	85per cent to 90per cent RH	1 to 2 weeks
Watermelons	7°C to 10°C (45°F to 50°F)	70per cent to 75per cent RH	2 to 3 weeks
Kiwifruit	-0.5°C to 0.0°C (31°F to 32°F)	90per cent to 95per cent RH	3 to 4 weeks
Mangoes	10°C to 13°C (50°F to 55°F)	85per cent to 90per cent RH	1 to 2 weeks
Avocados	7°C to 10°C (45°F to 50°F)	85per cent to 90per cent RH	3 to 5 days

Packaging: Packaging plays a crucial role in insulating fruits, protecting them from physical damage, and maintaining the desired temperature and humidity levels. Well-designed packaging prevents bruising, crushing, and exposure to external factors (Paltrinieri *et al.*, 2014; Schuur, 1988). Inadequate or improper packaging can lead to temperature fluctuations, increased exposure to external elements, and compromised quality.

Initial fruit quality: The shelf-life of fruits in the cold chain process is heavily influenced by their initial quality at the time of harvest. Critical factors include the maturity at harvest, careful handling, and adherence to stringent quality standards. Fruits of sub-optimal quality or those harvested prematurely are more prone to issues such as decay and bruising during the cold chain journey. Initiating the process with high-quality fruits not only enhances their preservation but also boosts their marketability (Nissen, 2018). It is imperative to incorporate thorough sorting and quality control measures to promptly identify and eliminate any damaged or subpar fruits from the supply chain, ensuring that only top-quality produce progresses through the cold chain with the highest standards of freshness and integrity.

Table 2: Components of cold chain management system for fruits.

Component	Role	Remark	Reference
Pre-harvest practices	They set the foundation for quality by implementing practices before fruit harvesting.	• Determine optimal harvest time. • Planning of harvesting time. • Use of proper handing techniques to minimize physical damage. • Maintaining field hygiene to reduce cross-contamination risk.	Anonymous, 2024a
Harvesting and post-harvest handling	It ensures careful and efficient harvesting, followed by proper handling to maintain fruit quality.	• Use gentle harvesting techniques to minimize bruising. • Immediate sorting of the harvest. • Proper cleaning and sanitation of harvesting tools, containers, *etc.* • Rapid cooling of fruits to slow down metabolic processes. • Record-keeping for traceability.	Palipane and Rolle, 2008
Packaging	Careful selection of appropriate packaging, containers, and refrigeration methods helps in attaining optimal temperature for shipments.	• Type of cargo being transported. • Specific temperature and humidity levels are required. • Size of the shipment. • Anticipated duration of transit. • External temperature conditions.	Anonymous, 2024b
Transportation	Cold transport encompasses the utilization of diverse transportation solutions to convey temperature-sensitive products from storage facilities to the consumer market. This involves the use of various modes of transportation, including road, rail, air, and sea.	• Prevent physical damage during transportation. • Monitoring temperature conditions during transit. • Minimizing transit time to reduce exposure to unfavorable conditions. • Securing a stable loading to prevent shifting and damage.	Anantharaman and Balasubramanian, 2015, Hundy *et al.*, 2008

Component	Role	Remark	Reference
Storage	Warehousing within the cold chain is a multifaceted system encompassing various processes, from storage and consolidation/deconsolidation to loading and unloading.	• Strategic design of facilities with precise climate control for both storage and loading areas. • Implement specific practices in the handling of perishable goods, ensuring the integrity of the cold chain throughout. • The facility should have a good temperature and control. system to slow down ripening and prevent dehydration or excess moisture, respectively. • Adequate ventilation in the storage system ensures proper air circulation. • To ensure freshness, the FIFO (First In, First Out) inventory management system should be adapted.	Anonymous, 2024a
Distribution	An effective cold supply chain for fruits involves the seamless movement of products from storage facilities to distribution centres or retail outlets with a strong emphasis on coordination.	• Use of refrigerated vehicles for distribution to ensure optimal temperature conditions and minimize handling to prevent bruising. • Real-time monitoring of transport conditions.	Hundy *et al.*, 2008
Retail display	Adequate refrigeration at retail establishments is essential to preserve the freshness of the arranged fruits. Careful handling during display is crucial to prevent any damage to the products.	• Regular stock rotation practices to ensure that older stock is sold before newer arrivals. • Educating store staff on proper handling techniques. • Continuous monitoring and adjustment of display conditions to align with the specific requirements of different fruits that meet both aesthetic and quality standards.	Hundy *et al.*, 2008

Several cold storage-related disorders can affect fruits if proper storage conditions are not maintained. These disorders highlight the importance of carefully managing temperature, humidity, and handling practices throughout the cold chain to minimize post-harvest losses and maintain fruit quality.

Table 3: Common storge related injuries associated with fruits.

Disorder	Description	Examples
Chilling injury	Occurs when fruits are exposed to temperatures below their optimum chilling threshold for an extended period, leading to physiological changes.	Tomatoes, cucumbers, citrus fruits.
Freezing injury	Caused by exposure to freezing temperatures, resulting in ice crystal formation within the fruit's cells and subsequent cell damage.	Strawberries, peaches, leafy greens.
Browning and decay	Improper humidity levels or prolonged storage can promote the growth of fungi and bacteria, leading to mold growth, rotting, and decay.	Berries, stone fruits, leafy greens.
Ethylene damage	Accumulation of ethylene in enclosed storage environments can cause premature ripening, color changes, and softening of fruits.	Bananas, avocados, tomatoes.
Moisture loss	Inadequate humidity levels can lead to moisture loss in fruits, resulting in shriveling, wilting, and loss of firmness.	Grapes, apples, pears.
Carbon dioxide injury	Excessive buildup of carbon dioxide in sealed storage environments can cause off-flavors, uneven ripening, and tissue damage in fruits.	Apples, berries, stone fruits.
Bruising and damage	Rough handling during harvesting, packing, or storage can cause bruising, punctures, and mechanical injuries in fruits.	Apples, bananas, citrus fruit.

Logistics and infrastructure requirements: A robust cold chain relies on efficient transportation and storage infrastructure to seamlessly move fruits from the harvest point to consumers, all while maintaining the necessary temperature conditions (Hundy *et al.*, 2008). The integrity of this process is contingent on proper logistics, and any inadequacies in infrastructure or logistics can give rise to delays, temperature fluctuations, and mishandling. These complications pose a significant risk to the overall quality of fruits, potentially resulting in spoilage and substantial financial losses. To mitigate such risks, the presence of well-maintained cold storage facilities, refrigerated transportation, and meticulous handling practices becomes indispensable for the successful management of the cold chain, ensuring that fruits reach consumers with their quality and freshness intact.

Monitoring and technology: The use of technology, such as real-time monitoring systems, helps track temperature variations, humidity levels, and other conditions during transportation and storage (Cil *et al.*, 2022). Lack of monitoring or outdated technology can result in a lack of visibility into the conditions of the cold chain. This increases the risk of unnoticed temperature deviations, potentially leading to quality issues and financial losses.

Agencies governing the quality requirements of fruits: The marketability of fruits depends on their adherence to some specifications that determine their quality, appearance, and overall consumer appeal. For fruits to command a strong presence in the market, various factors such as size, colour, ripeness, and flavour must align with consumer preferences and industry standards (Rahman *et al.*, 2021). Meeting these specifications not only ensures a visually appealing and flavourful product but also contributes to the fruit's longevity and resistance to deterioration during the storage and transportation phases of the supply chain. As such, understanding and adhering to these specifications play a pivotal role in enhancing the marketability of fruits, meeting consumer expectations, and facilitating successful integration into the competitive marketplace.

In India, the adherence to stringent quality standards for fruits is overseen by two prominent regulatory bodies: The Agricultural and Processed Food Products Export Development Authority (APEDA) and the Food Safety and Standards Authority of India (FSSAI) (Anonymous, 2024c, Anonymous, 2024d). For exporters, adhering to the quality standards is imperative to gain access to global markets. Although the particulars of these standards may differ based on the country and the type of fruit, compliance is universally essential for exporters seeking successful engagement in the international marketplace (Carter, 1997). In the international trade, the Codex Alimentarius Commission, a collaborative effort between the Food and Agriculture Organization (FAO) and the World Health Organization (WHO), is instrumental in developing international food standards, encompassing aspects like general specifications, labelling, and pesticide residues, with relevance to fruits. Private sector-led initiatives such as GlobalGAP and EUROPGAP establish voluntary standards for certifying agricultural products, including fruits, focusing on areas like food safety, traceability, environmental sustainability, and social welfare. The United States Food and Drug Administration (USFDA) sets stringent standards for the quality and safety of imported fruits into the United States, covering issues such as pesticide residues, food additives, and general food safety requirements. Japan's standards, established by the Ministry of Agriculture, Forestry, and Fisheries (MAFF), define criteria for quality, safety, and labelling of various agricultural products, including fruits. Australia and New Zealand have their standards and regulations governing the quality and safety of imported fruits, which may encompass considerations like maximum residue limits (MRLs) for pesticides and microbiological safety. The International Organization for Standardization (ISO) contributes to the global framework with standards related to the quality and safety of food products, exemplified by ISO 22000 focusing on food safety management systems. Importing countries may additionally have specific regulations covering pest control, quarantine measures, and unique quality parameters for imported fruits. Overall, compliance with these diverse international standards is paramount for exporters seeking access to global markets, ensuring the safety and quality of fruits for consumers worldwide.

These standards serve as comprehensive guidelines to ensure that fruits meet specific criteria concerning quality, safety and hygiene, catering not only to domestic consumption but also influencing the international market for exported fruits (Kenny, 2002). Major considerations in these standards include the maturity and ripeness of fruits, with a focus on harvesting at the appropriate stage to suit the intended market or distribution channel. Size and shape conformity are emphasized, with uniformity preferred for enhanced market appeal. Colour consistency is deemed crucial for visual appeal and consumer acceptance, necessitating fruits to exhibit characteristic colours associated with their variety and maturity stage. Another critical aspect is the firmness and texture of fruits, requiring them to possess the appropriate attributes based on variety and intended use, as overly soft or mushy fruits may be rejected due to quality concerns (Prasad *et al.*, 2018).

Ensuring freedom from defects is paramount, and quality standards specify tolerance levels for allowable defects, including bruises, cuts, scars, insect damage, and diseases. The packaging of fruits is also under scrutiny, with an emphasis on the cleanliness and suitability of materials to protect against physical damage and contamination (Prasad *et al.*, 2018). Chemical residues, including pesticides, are regulated through compliance with maximum residue limits (MRLs) to ensure the safety of fruits for human consumption. Microbiological standards are in place to guarantee that fruits are free from harmful microorganisms, contributing to overall food safety (Prasad *et al.*, 2018). Labelling requirements are stringent, with accurate information on variety, origin, packing date, and other relevant details mandated to comply with domestic and international regulations. Storage, transportation, and handling conditions are specified to preserve fruit quality, and certain fruits, such as bananas and tomatoes, may have unique requirements regarding exposure to ethylene gas during storage and transportation (Paltrinieri *et al.*, 2014).

Specifications for fruits in export markets

International standards:

Harmonization of standards: Various inter-governmental regulatory agencies play essential roles in ensuring the global and regional standards for plant protection and food safety. The International Plant Protection Convention (IPPC) takes the lead in setting the International Standards for Phytosanitary Measures (ISPMs), playing a crucial role in guaranteeing that exported horticultural produce complies with these standards for entry into international markets. The International Plant Protection Convention (IPPC) outlines crucial phytosanitary requirements aimed at safeguarding plant health. These requirements are detailed in the International Standards for Phytosanitary Measure 01, commonly known as ISPM 01. The primary objective of ISPM 01 is to prevent the introduction and spread of agricultural pests through plants and plant products during international trade activities (Fulano *et al.*, 2021). Specifically, when dealing with fresh fruits and vegetables, adherence to plant health inspections and the acquisition of phytosanitary certificates before shipment are mandatory. The European and Mediterranean Plant Protection Organization (EPPO) functions as a regional body, fostering cooperation in plant protection across Europe and the Mediterranean. It contributes by developing regional standards for phytosanitary measures.

The Codex Alimentarius Commission (CODEX) focuses on food safety, actively developing and promoting standards, codes of practice, guidelines, and recommendations in this domain. The European Commission takes charge of establishing legislation on plant health within the European Union (EU), aiming to prevent the introduction and spread of harmful organisms to plants and plant products. The fresh produce targeted for European niche markets must align with the European Plant Health Directive 2000/29/EC, particularly for fresh fruits and vegetables. To meet these standards, fresh fruits and vegetables must be devoid of specific contaminants such as *Salmonella*, *Listeria monocytogenes*, *Enterobacter sakazakii*, *Staphylococcal enterotoxins*, and histamine (CBI, 2024). The Organization of Economic Cooperation and Development (OECD) provides a comprehensive and globally harmonized export quality inspection system for its member countries, contributing to international consistency in agricultural trade. Together, these agencies form a critical network to ensure the integrity of plant protection measures, food safety standards, and the facilitation of international trade in agricultural products (Fulano *et al.*, 2021).

Governments worldwide implement phytosanitary measures to prevent the introduction and dissemination of pests affecting plants and plant products. Notably, countries like Japan and the United States impose restrictions on imports based on the country of origin. In the United States, a system involves individual country listings for Fresh Fruits and Vegetables (FFV) approved for entry, restricting the importation of most FFV. Japan's Plant Protection Law categorizes plants and plant products as either indefinitely banned, subject to plant health inspection (applicable to FFV not banned), or not requiring inspection, such as tea. Numerous FFVs, including avocados, papayas, mangoes, and various tropical fruits, are banned from several countries. Although certain products face restrictions due to disease-related concerns, some prohibited items can still enter Japan under specific quarantine conditions, such as undergoing fumigation or vapour heat treatment (UNCTAD, 2007). In the European Union (EU), plant health legislation is governed by Council Directive 2000/29/EC, which prohibits the introduction of harmful organisms, plants, and plant products. Certain fruits and vegetables, including cherimoyas, mangoes, passion fruit, and guavas, require a phytosanitary certificate and plant health inspections upon arrival in the EU. The EU implements restrictive measures only when specific problems are identified (Pay, 2005).

In the United States, regulated by the Plant Protection Act, the American Plant Health Inspection Service (APHIS) has the authority to restrict plant and plant product imports. While some FFVs can be imported from all countries, others must be on the approved list for the individual exporting country. Limited FFV, such as coconut, peeled garlic, cloves, and mushrooms, can be imported from all countries, while others depend on USDA regulations. Certain fruits and vegetables from China, the Philippines, and Thailand have approval, subject to specific conditions like inspection and treatment requirements. For example, mangoes from the Philippines must undergo vapour heat treatment at an APHIS-approved facility (FAO, 2006; ISPM, 2006). The microbiological hygiene criteria are formulated in alignment with the food safety standards of Codex Alimentarius, Good Hygiene and Manufacturing Practices, and the principles of Hazard Analysis Critical Control Point. HACCP is a methodical approach for identifying, evaluating, and managing food safety hazards. This system mandates the identification and control of potential food hazards at specific points within the commodity value chain. The attainment of food safety certification necessitates the implementation of a HACCP-based food safety management system, duly recognized by the Global Food Safety Initiative (GFSI) (CBI, 2024).

The harmonization of standards and compliance with global certifications are crucial for establishing a common and evidence-based framework for assessing and ensuring the quality and safety of exported fruits. Export documentation, particularly phytosanitary certificates, serves as a scientific assurance that the exported fruits conform to the plant health standards of both the exporting and importing countries. Overall, this scientific approach is essential for fostering international trade in fruits while safeguarding plant health and ensuring food safety (CBI, 2024).

Private sector requirements: The concept of the "Global Gap" entails the development of sound agricultural practices that can be universally adopted throughout the entire value chain of fruits and vegetables. It emphasizes the need for standardized practices that ensure the quality and safety of agricultural produce on a global scale (Global-GAP, 2024; USAID-KHCP, 2011). On the other hand, the term "EU organic" pertains to regulations stipulating that areas dedicated to the cultivation of organic vegetables within the European Union should be reasonably free from synthetic agrochemicals. This underlines the commitment to organic farming principles, promoting environmentally friendly and sustainable agricultural methods while ensuring the production of organic vegetables aligns with stringent standards (Fulano *et al.*, 2021).

Several standards significantly impact horticultural exports to the EU and the US. EurepGAP, a pre-farm-gate standard, encompasses the entire agricultural production process, from pre-planting to harvest. It operates as a supply-chain partnership involving retailers, produce suppliers/growers, and associate members in the agricultural input and service sectors. The British Retail Consortium (BRC) has developed post-farm-gate Food Technical Standards which are used to assess manufacturers of retailers' own-brand food products. The Safe Quality Food 2000 (SQF, 2000), administered by the Food Marketing Institute in the US, is recognized by the Global Food Safety Initiative. It provides a comprehensive Code specifying food safety and quality system requirements for all sectors of the food industry. The SQF standards, utilizing HACCP, good manufacturing practice (GMP), good hygiene practice (GHP), and GAP principles, are primarily employed by US and Australian retailers, with SQF 1000 for pre-farm gate and SQF 3000 for retail outlet levels (Humphrey, 2008; UNCTAD, 2007).

Compliance with global certifications: To comply with the global certifications and requirements, a four-tier approach is followed which includes establishing organized grower systems for exporting, enforcement of standards in a strict manner, more focus on research and adoption of innovative technologies for the preservation and implementation of training and generation of awareness

(Lengai *et al.*, 2022). Structured grower systems should incorporate integrated pest and crop management, along with good agricultural and handling practices (Jaffe and Masakure, 2005). The emphasis on sanitary and phytosanitary standards is geared towards sustainably producing safe, organically grown produce without chemical residues for the global population. Achieving this goal requires a combination of strategies, including integrated pest management and integrated crop management, which focus on preventing pest populations, economically employing pest management strategies, ensuring human and environmental safety, and producing safe and healthy food. The adoption of biopesticides, derived from plant parts with bioactive compounds, is highlighted for their effectiveness, biodegradability, varied modes of action, and economic affordability. Biocontrol agents, including micro-organisms like *Trichoderma* and *Bacillus*, are also advocated for effective pest and disease management, with a reduced likelihood of pest resistance due to the biochemical nature of the interaction (Kunbhar *et al.*, 2018; Dehariya *et al.*, 2018).

Export documentation:

Phytosanitary certificates: Phytosanitary certificates are official documents issued by plant health authorities. Scientifically, these certificates affirm that the exported fruits have undergone inspections and testing to confirm their freedom from pests, diseases, or contaminants that might pose a risk to the agricultural ecosystems of the importing country.

Under Article V of the International Plant Protection Convention (IPPC), a phytosanitary certificate is a document through which the authorities of the country of origin affirm that the specified products are in a healthy condition and free from plant diseases. This certificate is a requisite for both the export and import of these products. The IPPC outlines specific provisions for phytosanitary certification, with the primary goal of ensuring that plants, plant products, and other regulated articles intended for export adhere to the certifying statement specified in Article V(2)(b). The procedures for issuing phytosanitary certificates include the stipulation that inspections and related activities must be conducted exclusively by or under the authority of the official national plant protection organization. Any uncertified alterations or deletions render the certificates invalid. The certification process is entrusted to technically qualified public officers duly authorized by the official national plant protection organization. The export phytosanitary inspection is typically carried out upon specific requests from the importing country and is commonly conducted at the Plant Protection Station of the seaport or airport from where the shipment originates (Kumar and Puri, 2011).

Compliance with destination country requirements: Compliance with destination country requirements involves meeting specific criteria related to the quality, safety, and characteristics of the exported fruits as dictated by the importing country. This could include adherence to certain chemical residue limits, packaging standards, and other scientifically determined factors that ensure the fruits meet the standards of the destination country (FAO, 2013). Compliance with destination country requirements for the export of fruits, particularly from developing countries, presents significant challenges. In the context of the United States, the permit application process is deemed simple, but the initial admissibility for new products can be time-consuming and costly. The intricate process of risk assessment, rule change, and publication can extend to at least 18 months, resulting in a substantial backlog of permit requests. Similar challenges are observed in Japan, where stringent phytosanitary measures necessitate expensive protocols and onsite inspections for farms in regions with known diseases, adding to the overall cost of imported fruit.

The European Union, Japan, and the United States have implemented regulations on pesticides, which can unintentionally discriminate against developing countries. The high costs associated with complying with regulations, particularly for minor crops exported by developing nations, may limit the availability of certain active substances, resulting in higher residual pesticide limits. Initiatives, such as the COLECAP Pesticides Initiative Programme, aim to help developing countries comply with European food safety requirements. Additionally, recent amendments to regulations, like Regulation (EC) No 882/2004, may place a burden on exporting countries to control food safety, potentially affecting smaller nations entering global markets.

Quarantine requirements:

Definition and purpose of quarantine in fruit trade:

Preventing the spread of pests and diseases: Quarantine measures in the context of fruit trade refer to a set of regulations and practices implemented to prevent the introduction and spread of pests and diseases that may affect crops. The primary objective is to control and limit the movement of plants, plant products, and other potentially contaminated materials to contain the risk of transmitting harmful pests or diseases. By imposing quarantine restrictions on certain fruits or regions, authorities aim to safeguard the agricultural sector from the introduction of new and destructive pests or diseases that could have severe consequences on crop yields and overall ecosystem health (van Klinken *et al.*, 2023).

In the realm of plant health, as delineated by the International Plant Protection Convention (IPPC) and the Food and Agriculture Organization of the United Nations (FAO), the crucial role of pest risk management is to mitigate the risk associated with trade or other pathways to a level deemed acceptable by the importing country. While it is acknowledged that the risk level can never be reduced to zero, it must be carefully weighed against the benefits of trade, coupled with the availability of measures to diminish either the likelihood or impact of introducing a new plant pest that could potentially survive and proliferate in the importing country. This preventive objective, distinct from other pest management goals, aims to avert the introduction of new pests to a specific area or region. Unlike strategies primarily focused on reducing economic losses for producers and ensuring societal food security, the pivotal decision-maker in determining an acceptable risk level is the recipient of the consignment of agricultural goods or other materials susceptible to plant pest infestation or infection, including dunnage and wood packaging (Quinlan *et al.*, 2020).

Protecting local crop husbandry: The purpose of quarantine in the fruit trade extends to the protection of local agriculture. By establishing quarantine protocols, authorities seek to shield domestic crops from potential threats posed by pests and diseases that may be introduced through imported fruits. Local agriculture is vulnerable to the invasion of new pests or diseases, which can spread rapidly and cause widespread damage to crops. Quarantine measures, therefore, act as a preventive barrier, ensuring that fruits entering the country or specific regions comply with strict standards and do not carry harmful agents that could jeopardize the health of local plants, ecosystems, and agricultural productivity. This proactive approach helps maintain the integrity and sustainability of local agriculture (Martin *et al.*, 2016).

International quarantine regulations and standards:

Pest risk analysis: The goals of a Pest Risk Analysis (PRA) are, within a defined area, to recognize pests and/or pathways posing quarantine concerns and assess their associated risks. Risk analysis is a methodical approach to collect, assess, document, and share information, ultimately resulting in actionable recommendations. It also aims to pinpoint vulnerable regions and, when applicable, propose risk management strategies. In the context of Sanitary and Phytosanitary (SPS) measures and the International Plant Protection Convention (IPPC), pest risk analysis comprises two key components. Firstly, it evaluates the likelihood that the pest may be transported on the exported product, survive the journey to the importing country, and establish itself there. Secondly, it assesses the potential impact of the identified pest in the importing country, taking into account the presence of suitable hosts and the prevailing climatic conditions (Khetarpal and Kavita, 2002).

Conducting effective risk analysis relies heavily on having comprehensive information about the pest status in both the exporting and importing countries. Understanding the pest status of the exporting country is crucial to identifying and assessing listed pests that may potentially accompany the exported commodities.

The PRA process for quarantine pests adheres to a structured three-stage framework:

- Stage 1 (initiating the process), the focus is on identifying pests and pathways that raise quarantine concerns, warranting consideration for risk analysis within the specified PRA area.
- Stage 2 (risk assessment) commences with categorizing individual pests to determine if they meet the criteria for being a quarantine pest. The risk assessment process further entails evaluating the likelihood of pest entry, establishment, and spread, along with assessing their potential economic consequences, including environmental impacts.
- Stage 3 (risk management) involves the identification of potential strategies for mitigating the risks identified in Stage 2. These management options undergo evaluation based on their effectiveness, feasibility, and overall impact, facilitating the selection of appropriate measures (ISPM, 2006).

The International Standards for Phytosanitary Measures (ISPM) No. 8, titled 'Determination of Pest Status in an Area,' outlines the procedures for recording information about a pest and categorizing its status. This standard permits claims of pest absence based on the absence of records, although all such claims must be substantiated by published documents. Additional guidance on general and specific surveillance practices is provided in ISPM-6 (Guidelines for Surveillance), offering detailed insights into best surveillance practices, technical requirements for diagnostic services, and the importance of comprehensive record-keeping.

Inspection and certification: National Plant Protection Organizations (NPPOs) bear the crucial responsibility of equipping plant quarantine inspectors with the necessary skills and knowledge. In fulfilling this role, NPPOs can leverage various educational resources, including educational institutions, specialized training institutes, modular programs, and in-service systems. The training curriculum encompasses diverse subject areas, covering human resource management to address personal skills and occupational health and safety concerns. Additionally, inspectors are trained in plant health and its impact on trade and the environment, encompassing fundamental aspects such as international agreements, standards, phytosanitary principles, legislation, regulations, national policies, pest risk analysis, regulated pests, and product identification. Moreover, inspectors are equipped with technical procedures such as pest diagnosis, sampling regimes, inspection procedures for various commodities, electronic detection systems usage, phytosanitary actions and treatments, and post-entry quarantine and passenger systems. Through comprehensive training, NPPOs empower inspectors to safeguard plant health, facilitate trade, and uphold phytosanitary standards effectively (ISPM, 2006).

Pest-free areas and pest-free production sites: Pest-free place of production refers to a location in which a specific pest is scientifically proven not to occur, and this condition is officially maintained for a defined period. It offers a mechanism for an exporting country to guarantee if required by an importing country, that consignments of plants, plant products, or regulated articles originating from this place are free from the specified pest. Pest freedom is determined through surveys, growing season inspections, and supported by documentation.

Depending on the pest, local circumstances, and acceptable risk levels, various measures, from growing-season inspections to comprehensive survey systems, may be employed to ensure an adequate level of security. This concept applies to any premises or collection of fields managed as a single production unit. It distinguishes itself from a pest-free area and involves individual management by the producer under the National Plant Protection Organization's supervision. The choice between a pest-free place of production and a pest-free area depends on the pest's distribution, characteristics, and administrative considerations. Key factors for establishing and maintaining pest-free places of production include pest characteristics, the site's features and the producer's operational capabilities. Buffer zones may be required based on the pest's biology, with specific measures and periodic reviews documented. Additional declarations on phytosanitary certificates may be necessary, and information on pest-free places of production should be available to importing countries upon request (ISPM, 2006).

National quarantine measures: Ensuring the effectiveness of quarantine treatment in preventing the spread of pests to new territories is imperative, requiring a 100 per cent success rate against the targeted insects or pathogens. However, the susceptibility of insects to such treatments varies among species, making it challenging to achieve complete mortality for all pest development stages (Moore, 2021). To determine the quarantine significance of a pest, pest risk analysis is essential, and treatment efficacy should be tailored to each species. The development of quarantine treatments involves diverse methods and testing approaches, influenced by experimentation goals, available data, and the scale of the study. When defining pests and plant products as targets, choosing suitable treatment types becomes relatively straightforward. However, for export products requiring disinfestation schedules, the treatment must meet the importing country's specified efficacy levels. Complying with phytosanitary conditions is crucial, particularly for banned products seeking export. To achieve successful disinfestation without compromising product quality, a thorough understanding of the importing country's quarantine requirements is essential. The entire process involves studying the importing country's quarantine regulations, selecting appropriate experimental designs and tools, and implementing methods that ensure complete disinfestation while preserving product integrity. The required efficacy varies based on the specific quarantine requirements of each country (Theyse, 2009).

Pest and disease control: This encompasses strict regulations on the import and export of plants and agricultural goods, along with risk assessments, surveillance, and diagnostic facilities to identify and respond to potential threats. Inspection services play a vital role in monitoring shipments and enforcing compliance, while quarantine facilities may isolate and examine imported materials. Public awareness campaigns, collaboration with international bodies, and rapid response mechanisms contribute to an effective defence against outbreaks (Waage and Mumford, 2008).

Certification and inspection processes: Export inspection and certification processes in India are governed by the National Plant Protection Organization (NPPO), headquartered in Faridabad. The NPPO issues standard formats and guidelines, providing exporters or their agents with a comprehensive framework for submitting applications. The process of export inspection and certification involves a systematic sequence of actions to ensure the quality and adherence to phytosanitary standards of plants and plant products destined for export. The initial step is the submission of a detailed application by the exporter, including essential documents such as invoices, packing lists, and relevant permits. Upon scrutiny of the application, the inspection fee is determined, considering specific requirements and conditions (DPPQS, 2023).

Table 4. A list of common quarantine insects and pests for fruits

Name of the crop	Exporting countries	Export potential	Major pest/pathogen affecting the export quality of the fruits	Symptoms
Pomegranate	United Arab Emirates (UAE), Bangladesh, Nepal, Netherlands, Saudi Arabia, Sri Lanka, Thailand, Bahrain, and Oman.	India exported approximately 62,280 tons of pomegranates, valued at USD 58.36 million, to various countries during 2022-23.	Pomegranate fruit borer (*Deudorix isocrates*).	Odious smell and excreta of caterpillars coming out of the entry holes of fruits which ultimately leads to fruit rot.
			Thrips (*Scirtothrips dorsalis* and *Rhipiphorothrips cruentatus*).	Rasping the surface and sucking the oozing cell sap leading to drying and shedding of flowers. Scrapping on fruits leads to scab formation, reducing market value.
			Fruit sucking moth *(Othreis fullonia, O. materna* and *O. homaena).*	The moth pierces through the rind and sucks the juice with their long proboscis leading to entry of microorganisms from the punctured regions and ultimately the fruits drop occurs.
			Colletotrichum rot (*Colletotrichum gloeosporioides*).	The rot extends beyond rind into the arils which disintegrate and are dark grey/brown black colour but are not watery causing maximum damage among all fruit rot.
			Scab (*Sphaceloma punicae*).	The fruits are initially small brown raised and rough to touch, giving russet scab appearance to the skin and later enlarge to form larger spots with light centre and dark edge. Flower buds or small fruits when attacked become deformed.

Name of the crop	Exporting countries	Export potential	Major pest/pathogen affecting the export quality of the fruits	Symptoms
Mango	United Arab Emts, U.K., USA, Qatar, and Kuwait.	The country has exported 22963.76 MT of fresh Mangoes to the world for the worth of Rs. 378.49 crores/ 48.53 USD Millions during the year 2022-23.	Fruit fly (*Bactrocera dorsalis*).	Injury to fruit occurs through oviposition punctures by females and subsequent larval tunneling.
			Mango mealybug (*Drosicha mangiferae*).	The nymphs of this pest suck sap from leaves and inflorescence causing dryness leading to flower drop and negligible fruit set. Development of sooty mould on the fruit.
			Mango nut weevil (*Sternochaetus mangiferae*).	Grub makes zigzag tunnels in pulp and eats unripe tissue and bore into cotyledons. oviposition injuries on marble sized fruits and fruit dropping occurs.
			Mango Thrips (*S. dorsalis*).	It mainly feed on inflorescence and fruits which show discoloured tissues that subsequently turn brown.
			Anthracnose (*Colletotrichum gloeosporioides*).	Numerous oval or irregular brown to deep brown spots of variable size appear on the tender peanut as well as mature fruits. Fruits are blacken and heavy fruit drop occurs during severe infection.
Banana	Iran, Turkey, Qatar, Pakistan, Singapore, United Arab Emirates, Afghanisthan.	The country has exported 232648.78 MT of fresh or dried Bananas to the world for the worth of Rs. 74089.99 lakh during the year 2020-2021.	Thrips (*Chaetanaphothrips signipennis*).	Distinctive reddish brown oval stains on the finger, which can extend the entire length. In severe cases peel splits and the exposed flesh quickly discolors.
			Anthracnose (*Colletotrichum musae*)	Small, circular, black spots develop on the affected fruits. Severe infestation leads to shriveled and black coloured rotten fruits.
			Cigar end rot (*Verticillium theobromae*).	Black necrosis from the perianth to the tip of immature fruits causing shrinkage and folding of the tissues. Fingers that appear rotten and look like the ash of cigar.

Name of the crop	Exporting countries	Export potential	Major pest/pathogen affecting the export quality of the fruits	Symptoms
Apple	Kuwait, UK, Saudi Arbia, Turkey, United Arab Emirates, Bangladesh, Bhutan.	The country has exported 30680.52 MT of fresh apples to the world for the worth of Rs. 10665.16 lakh during the year 2020-2021.	Codling moth (Cydia pomonella).	Caterpillar bore into the fruits and feed on the pulp. Infested leaves and fruits show tunnels with frass.
			Scab (Venturia inaequalis).	Fruits show small, rough, black circular lesions. The centre of the spots become corky and on mature fruits and yellow halo is seen around the lesions.
			Soft rot (Penicillium expansum).	Young spots start from stem end of the fruit as light brown watery rot. As the fruit ripens area of the rotting increases and skin becomes wrinkled.
			Bitter rot (Glomerella cingulate).	Faint, light brown discolouration beneath the skin develops. The discolouration expands in a cone shape. The lesions increased and covers entire areas of fruits.
Papaya	Germany, Oman, Italy, France, UK, Nepal, Italy, Taiwan.	The country has exported 7667.47 MT of fresh or dried papaya to the world for the worth of Rs. 1888.10 lakh during the year 2020-2021.	Fruit fly (Bactrocera dorsalis).	Maggots puncture into semi-ripe fruits with decayed spots. Oozing of fluid and brownish rotten patches on fruits and dropping of fruits occurs.
			Papaya ring spot Papaya ring spot virus.	On fruits circular concentric rings are produced.
			Anthracnose (Colletotrichum gloeosporioides).	On fruit initially brown superficial discoloration of the skin develops which are circular and slightly sunken. Fruits mummified and deformed.

Name of the crop	Exporting countries	Export potential	Major pest/pathogen affecting the export quality of the fruits	Symptoms
Guava	Germany, Oman, Italy, France, Bhutan, UK, Nepal, Italy.	The country has exported 2886.37 MT of fresh guavas to the world for the worth of Rs. 939.69 lakh during the year 2020-2021.	Fruit Borer (Congethes punctiferalis).	Caterpillar bores into young fruits. Feeds on internal contents (pulp and seeds). Dry up and fall off in without ripening.
			Fruit fly (Bactrocera diversus).	Adults and maggots attack semi – ripe fruits. Oviposition punctures on fruits.Maggots destroy and convert pulp into a bad smelling.
			Tea mosquito bug (Helopeltis antonii).	Nymphs and adults make punctures on petiole, tender shoots and fruits. Brownish – black necrotic patches develop on foliage. Elongate streaks and patches develop on shoots. Corky scab formation on fruits.
			Anthracnose (Colletotrichum gloeosporioides).	The characteristic symptoms consist of sunken, dark colored, necrotic lesions on mature fruits on the tree
Grapes	Netherland, Banglades, United Arab Emts, UK and Russia.	The country has exported 267,950.39 MT of fresh or dried grapes to the world, worth Rs. 2,543.42 crores/ 313.70 USD millions during the year 2022-23.	Mealy bug (Maconellicoccus hirsutus).	Nymphs and adults suck the sap from leaves, shoots and fruits. Honey dew – development of sooty mould on leaves, shoots and branches and rotting of berries occurs.
			Berry plume moth (Oxyptilus regulus).	Caterpillar bore into green berry fruit and tunnelling and feeds on internal content. Affected berries – known as "stung berries.
			Downy mildew (Plasmopara viticola).	White growth of fungus on berries which subsequently becomes leathery and shrivels. Later infection of berries result in soft rot symptoms.
			Bird's eye spot/anthracnos (Gloeosporium ampelophagum).	The disease appears first as dark red spots on the berry. Later, these spots are circular, sunken, ashy-gray and in late stages these spots are surrounded by a dark margin which gives it the "bird's-eye rot" appearance.

Name of the crop	Exporting countries	Export potential	Major pest/pathogen affecting the export quality of the fruits	Symptoms
Pineapple	Nepal. Maldives, Afghanisthan, Jordan, Russia, Pakistan, turkey.	The country has exported 4545.32 MT of fresh or dried pineapple to the world for the worth of Rs. 1985.01 lakh during the year 2020-2021.	Mealybug (Pseudococcus brevipes).	Nymphs and adults congregate on fruits and suck the sap affecting the market value.It is vector of pine apple wilt disease.
			Thielaviopsis rot (Thielaviopsis paradoxa).	Slight skin darkening due to water soaking of the skin over rotted portions of the flesh. As the flesh softens, the skin above readily breaks under slight pressure.
			Heart rot (Phytophthora parasitica).	The disease causes complete rotting of the central portion of the stem and shows sign of rotting with foul odour.
Oranges	Bangladesh, Apple, Bhutan, Singapore, UK, USA, Malaysia.	The country has exported 164585.79 MT of fresh or dried oranges to the world for the worth of Rs. 45784.99 lakh during the year 2020-2021.	Fruit sucking moth (*Eudocima maternal*)	The punctures produced during feeding by the adult moths leads to fruit rot.
			Rust mite (*Phyllocoptruta oleivora*).	Rust mite infestation produces a multitude of brown–reddish black spots on the fruit surface. brown irregular patches on them that fetches low market price to the fruits.
			Thrips (Thrips nilgiriensis).	Nymphs and adults suck the sap from fruits. Ring like appearance on the fruit. Irregular mottled patches on rind.
			Citrus canker (*Xanthomonas citri subsp. citri*).	Young lesions are raised or pustular, particularly on the lower leaf surface. The pustules later became corky and crateriform, with a raised margin and a sunken centre.
			Greening disease (*Candidatus liberibacter*).	Infected fruits are small and mis-shapen, many fall prematurely, while those that remain on the tree do not colour properly.

Name of the crop	Exporting countries	Export potential	Major pest/pathogen affecting the export quality of the fruits	Symptoms
Sapota	United arab Emirates, Canada, Netherland, USA, France, Oman, Norway.	The country has exported 657.50 MT of fresh sapota to the world for the worth of Rs. 491.48 lakh during the year 2020-2021.	Fruit fly (*Bactrocera dorsalis*).	Maggot bore into semi-ripen fruits with decayed spots and dropping of fruits. Oozing of fluid. Brownish rotten patches on fruits.
			Sooty mould (*Capnodium* sp.).	It is a fungal disease developed on honeydew-like excretion secreted by aphids and scale insects results in reduced translocation of food to the fruits, which leads to reduction in their size.

(FAO, 2013)

Following the issuance of a quarantine order, the consignment undergoes a thorough inspection, including sampling and laboratory testing. The sampling process is conducted under established international standards, such as ISTA Rules for seed propagation and Bureau of Indian Standards for various consumable products. Visual inspections, X-Ray examinations, and other relevant procedures ensure the consignment's quality and compliance with phytosanitary regulations. In cases where live insect infestation is identified, the exporter or their agent is responsible for arranging fumigation. This may take place at the exporter's premises or another approved location, supervised by a qualified pest control operator under the scrutiny of the Plant Quarantine (PQ) officer. The fumigation process incurs associated charges, including supervision fees and, if applicable, storage charges. Upon successful completion of the inspection and any necessary treatments, the Phytosanitary Certificate (PSC) is issued in duplicate, with one copy provided to the exporter and the other retained for official records. It signifies that the consignment meets the required standards and is free from quarantine pests. If needed, re-issuance of the PSC can be requested for corrections or amendments within a specific timeframe.

However, if the inspection reveals issues such as prohibited commodities, quarantine pest presence, or challenges in fumigation, the issuance of the PSC is rejected. The exporter or agent is promptly informed in writing of the reasons for rejection, and this information is communicated to customs and port authorities to prevent the export of non-compliant products (Kumar and Puri, 2011).

Impact of non-compliance with quarantine requirements: Non-compliance with quarantine requirements in the export of fruits can have significant negative impacts on both the exporting country and the global trade network. It may lead to the imposition of trade barriers by importing countries. Fruits that do not meet the required quarantine standards may be rejected, causing financial losses for exporters. Repeated rejections can harm the reputation of the exporting country's agricultural products. Exporters may incur financial losses due to rejected shipments and the costs associated with meeting quarantine standards retroactively. This includes expenses for retesting, fumigation, or disposal of non-compliant produce. It can also result in lost revenue and market share. Persistent non-compliance can damage the reputation of the exporting country's agricultural sector. Importing countries may lose trust in the quality and safety of the products, impacting future trade relationships. Non-compliance with quarantine requirements can lead to the introduction of pests, diseases, or contaminants into the importing country. This poses a threat to the local agriculture and ecosystem, potentially causing widespread damage to crops, increased pesticide use, and environmental degradation. This can further impede the export of fruits and vegetables and increase the complexity and cost of meeting regulatory requirements. Non-compliance with quarantine regulations may result in legal actions or penalties. Exporting countries and individual exporters may face fines, bans, or legal proceedings, depending on the severity of the violation and the importing country's regulations (Sun *et al.*, 2021). Importing countries may seek alternative sources for their agricultural products, leading to a loss of market share for the exporting country and potentially affecting the livelihoods of local farmers and businesses.

Regulatory compliance for export:

Phytosanitary certificates: These certificates are issued by an official of an exporting country or the country of re-export, certifying the absence of pests and suitability for entry into the destination country for designated plants or plant products. Operating at the port of entry, these documents serve several key purposes, including:

- Identifying the types of plants and plant products involved.
- Specifying the geographical area where the plants or plant products were cultivated.
- Verifying whether the plants or plant products underwent treatment in the country of origin and, if so, detailing the specific treatment applied.
- Assessing compliance with quarantine requirements.
- Confirming that the plant products meet the precise certification criteria mandated for entry into the destination country.

Phytosanitary certificates serve the purpose of affirming that shipments containing plants, plant products, or other regulated articles comply with specified phytosanitary import standards and align with the certifying statement outlined in the relevant model certificate. The issuance of phytosanitary certificates should strictly align with this intended purpose.

Inspections and associated procedures, culminating in the issuance of phytosanitary certificates, are exclusively entrusted to the official national plant protection organization or those acting under its explicit authority. The issuance of phytosanitary certificates is a responsibility designated to public officers possessing the requisite technical qualifications. These officers must be duly authorized by the official of the National Plant Protection Organization (NPPO) to act on its behalf, operating under its supervision and control. The appointed officers must possess sufficient knowledge and information, instilling confidence in the importing contracting parties regarding the reliability of the phytosanitary certificates as credible and trustworthy documents (Brunel *et al.*, 2013).

Compliance with importing country regulations: Importing nations often outline specific requirements related to the preparation and issuance of phytosanitary certificates. These commonly encompass:

Language: Importing countries may mandate certificates to be completed in a designated language or one from a predefined list, with encouragement to include one of the official languages of FAO.

Period of validity: Specifications may include the allowed timeframe for certificate issuance post-inspection or treatment, the duration between dispatch and the consignment's departure from the origin country, and the overall validity of the certificate.

Completion: Countries may insist on certificates being completed through typing or legible handwritten capital letters.

Units: Requirements may dictate that descriptions of the consignment and declared quantities be expressed in specified units.

In India, to import agricultural commodities, an importer must obtain an import permit in advance for items listed in Schedule V and VI of PQ Order, 2003. The application process is detailed in Annexure-II. Imported agricultural commodities undergo inspection at the port of entry to prevent the introduction of exotic pests and diseases. This process adheres to the DIP Act, of 1914, and the Plant Quarantine (Regulation of Import into India) Order, of 2003. Import clearance involves multiple steps, including sampling, detailed testing, and post-entry quarantine (PEQ) testing at the importer's premises (DPPQS, 2007).

Export quarantine protocols: These protocols encompass a range of regulatory actions to exclude animal and plant pests, as well as pathogens, from entering specific sites, areas, countries, or groups of countries. Quarantine protocols also involve the detection and control of exotic pests already introduced, utilizing domestic regulations. Post-entry quarantine inspections are performed for identified planting materials, and Pest Risk Analysis (PRA) is conducted to determine and finalize phytosanitary requirements for importing plant and plant materials. Overall, these protocols represent a comprehensive strategy to safeguard agricultural interests, promote international trade, and manage the associated risks effectively (FAO, 2013).

Phytosanitary standards: This may involve inspections, testing, and certification processes to verify that the fruits comply with internationally recognized phytosanitary norms. The international standards include the World Trade Organization (WTO), Maximum Residual Limit (MRL) established by CODEX, Sanitary and Phytosanitary measures (SPS), Phytosanitary certification (PC), Pest Risk Analysis (PRA), North American Plant Protection Organization (NAPPO), International Plant Protection Convention (IPPC).

Within national standards, a key component is the Post Entry Quarantine (PEQ) process, where plants are cultivated in isolation for a specified duration in a designated facility or nursery area approved by the Plant Protection Adviser. The primary objectives of this practice include:

- Facilitating the adoption of standardized operating procedures among all inspection authorities designated by the Ministry of Agriculture.
- Preventing the introduction and dissemination of harmful pests that have the potential to impact plants and other plant materials.
- Conducting thorough examinations of containers and fruits to detect any pesticide residue (FAO, 2013).

Fruits designated for export markets must adhere to a comprehensive set of both legal and non-legal technical quality requirements (CBI, 2024). Legal requisites encompass Maximum Residue Limits (MRLs) for pesticides and contaminants, traceability, product conformity, integrity, size specifications, labelling, and packaging. Non-legal requirements encompass aspects of social and environmental compliance. Rigorous official control procedures are implemented to ensure food safety and conformity with legal requirements, incorporating documentary checks, identity verification, traceability, and adherence to market standards. Essential documentation includes the bill of landing, phytosanitary certificate, Codex Alimentarius food safety standards certification, packing list, customs clearance, and a unique traceability code, commonly the lot number or Global-GAP Number (GGN). MRLs define the legally permissible levels of pesticide residues or contaminants in food or feed, with Regulation (EC) 1881/2006 specifying the limits (limit of 0.10 mg/kg for most fresh fruits and vegetables), emphasizing safety for consumers (European Commission, 2009).

Compliance with MRLs ensures that the produce does not contain prohibited or elevated levels of agrochemical pesticide residues. Fruits destined for European Union markets must meet specific labelling and packaging standards following food labelling legislation (CBI, 2024). Packaging materials should be new, clean, transparent, and preserve pod integrity without causing damage. Each package must carry a sticker label with the packer's name and address, produce name and variety, country of origin, class and size, lot number, or Global-GAP number for traceability, and an official control mark (Fulano *et al.*, 2021). Premium quality requirements dictate that the fruits should be fresh, intact, free of foreign matter and pests, aesthetically pleasing, devoid of unusual smell and taste, and without abnormal external moisture or parchment. Non-legal requirements extend to social and environmental compliance, with specific codes of conduct varying based on the export niche market. Examples include the Sustainable Agriculture Code, Ethical Trading Initiative (ETI), and Business Social Compliance Initiative (BSCI). Social and environmental compliance further entails meticulous record-keeping of all farm operations, adherence to environmental standards, and the well-being of farm workers (FAO, 2013; CUTS, 2009).

Pre-shipment treatment: Pre-shipment treatments are procedures conducted on fruits before they are exported to eliminate or minimize the risk of carrying pests or diseases. The pre-shipment treatment of fresh fruits involves various methods to ensure compliance with phytosanitary regulations, control pests, and maintain the quality of the produce during international trade.

The radiation treatment utilizes doses of 50–400 Gy to impede the hatching and development of pests like fruit flies, codling moths, and weevils. In the irradiation process, electromagnetic radiation such as gamma radiation, microwaves, X-rays, infrared rays, and electron beam treatment, is used for pest control. It disrupts the reproductive capabilities of pests, reducing the risk of infestation during transport. While this approach is cost-effective in the long run, it requires a substantial upfront investment in irradiation facilities (Hallman, 2017; Hennessey *et al.*, 2014). Vapour Heat treatment (TPT-VH), involves subjecting fruits like papaya, melon, and mango to elevated temperatures (40-50 °C) in a hot chamber for durations of up to one hour, effectively targeting pests such as melon flies and fruit flies. However, this method may cause damage to sensitive produce (Hennessey *et al.*, 2014).

Fumigation involves exposing fruits to gaseous substances such as phosphine, sulfuryl fluoride, and carbonyl sulphide to eliminate pests. Fumigation methods, including ethylene dibromide (EDB), methyl bromide, and fungicides are also very common to fruits. Even though Methyl bromide is effective, is regulated due to its ozone-depleting properties, and chemical fumigation is generally avoided due to strict requirements on pesticide residues. The fumigation process is found to be effective against a wide range of pests and has fewer environmental concerns (Suwapanich *et al.*, 2019; Papoutsis *et al.*, 2019). Cold treatment, maintains temperatures between 0 to 3 °C for an extended period, proving to be environmentally friendly and enhancing the shelf-life of citrus fruits, specifically lemon, sweet orange, tangerine, pomelo and clementine (Dohino *et al.*, 2019; Hennessey *et al.*, 2014).

Cold Storage Phytosanitary Irradiation (CPI) involves exposing fruits to irradiation at low temperatures to control pests. This method is effective for disinfestation while minimizing the impact on fruit quality. Under CPI, processes such as the High-pressure/vacuum process, Double vacuum process, Hot and cold open tank process, and Sap displacement method are commonly employed (Hallman *et al.*, 2016). Storage environments are modified to control temperature, humidity, and gas composition to slow down the ripening and ageing processes. A controlled atmosphere helps in extending the shelf-life of fruits and inhibiting the development of certain pests. The utilization of cold treatment is increasingly becoming a favoured phytosanitary intervention for maintaining the physical quality of fresh fruits and vegetables. Additionally, it serves as an effective deterrent against insect pest infestation, thereby ensuring prolonged marketability of the produce. By implementing pre-shipment treatments, exporting countries demonstrate their commitment to providing safe and pest-free produce to the importing nations, ensuring that the fruits meet the required phytosanitary standards.

Documentation: Documentation is a crucial aspect of export quarantine protocols. It involves the preparation and submission of various certificates and paperwork to verify compliance with phytosanitary standards. This documentation may include phytosanitary certificates, detailing the results of inspections and treatments, as well as additional declarations ensuring that the exported fruits meet the regulatory requirements of both the exporting and importing countries.

The State/Central Government phytosanitary certification agencies will adhere to standardized formats for export inspection and certification, following established operational procedures by the Directorate of Plant Protection, Quarantine and Storage (NPPO) for consistency. These agencies will maintain a technical folder for receiving and filing information from the Dte of PPQS, including export ban lists, regulated pests, prohibited import commodities, specific phytosanitary requirements from bilateral agreements, and accredited fumigation agencies/treatment facilities. Any proposed changes to procedures or document revisions must be communicated to the Dte of PPQS with technical justifications for approval. The Dte of PPQS will promptly communicate modifications to authorized officers, who will ensure the replacement of obsolete documents.

The accessible document will facilitate compliance with export inspection procedures. Each agency will maintain detailed records for each consignment, organized by registration number, containing original applications, checklists, inspection reports, treatment certificates, phytosanitary certificates, and relevant records. The records, including pest interceptions and identification, will be retained for at least one year and should be easily retrievable when needed, ensuring comprehensive documentation and compliance with export standards. Proper documentation facilitates the smooth flow of trade by providing transparency and assurance regarding the safety and quality of the exported fruits (DPPQS, 2007).

Inspection procedures: Inspection procedures are conducted at various stages of the export process to assess the condition of the fruits and ensure compliance with phytosanitary standards. This involves thorough examinations of the fruits for any signs of pests, diseases, or other contaminants. Inspectors may also verify that pre-shipment treatments have been applied appropriately. These inspection procedures are essential for identifying and addressing potential risks before the fruits are shipped, contributing to the overall effectiveness of export quarantine protocols in safeguarding the health of plants, ecosystems, and agricultural industries in both exporting and importing regions (FAO, 2013).

Exporters seeking certification under the CWI System must submit an "Application for Inspection" to the Export Inspection Agency (EIA) along with necessary documents and fees. The application timeline depends on the inspection location, with a minimum of three working days for inspections at the same station as the EIA office and seven working days for other locations. Inspections can occur at the port of shipment or exporter's premises, requiring facilities provided by the exporter. The inspection process involves physical verification, sampling following Indian standards, and preparation of a Field Inspection Report (FIR). Samples are sent to laboratories for analysis based on importing country standards, and the FIR is used to issue a Certificate of Inspection if the consignment meets specifications. The certificate's validity is 45 days, and health certificates are issued based on laboratory tests, conforming to parameters described in standards and importer requirements, as well as the Field Inspection Report. The EIA communicates refusals in writing within ten days, and the entire process ensures compliance with international standards for export (EIC, 2011).

Conclusion

The significance of cold chain management in preserving fruit quality lies in its ability to minimize spoilage, retain nutritional value, control ripening, reduce pathogen growth, extend shelf life, preserve aesthetic appeal, enhance market accessibility, provide consistent quality, and comply with regulatory standards. Securing the seamless progression of the cold chain management process is crucial to avoid spoilage, losses, and financial setbacks at various stages such as transit, processing, storage, and display. Precise control of temperature and product conditions is pivotal for the supply chain's success, affecting all involved parties, including manufacturers, shippers, and end-customers. Utilizing advanced thermal packaging technology and real-time temperature monitoring facilitates ongoing maintenance, monitoring, adjustment, and tracking of the cold chain, ensuring its integrity from production through final delivery and display.

Effective cold chain management, adherence to specifications, and stringent quarantine requirements are indispensable pillars in ensuring the quality and safety of fruits for both domestic consumption and export markets. Cold chain management plays a vital role in preserving the freshness and extending the shelf life of fruits by maintaining optimal temperature, humidity, and handling conditions throughout the supply chain. Compliance with specifications ensures that fruits meet the desired quality standards regarding size, color, ripeness, and packaging, meeting consumer expectations and regulatory requirements. Moreover, strict adherence to quarantine protocols safeguards against the spread of pests and diseases, safeguarding agricultural ecosystems and trade relationships. Together, these components form a comprehensive framework that not only enhances the marketability and competitiveness of fruits but also promotes food safety, sustainability, and economic viability in the global fruit industry.

References

Agricultural and Processed Food Products Export Development Authority, http://apeda.gov.in

Anantharaman S and Balasubramanian N, 2015. Reefer vehicles and cold chain transport. Retrieved from https://www.motorindiaonline.in/reefer-vehicles-and-cold-chain-transport/ on 21st January, 2024.

Anonymous, 2024a. Cold Storage of Agricultural Products. Retrieved from https://energypedia.info/wiki/Cold_Storage_of_Agricultural_Products#cite_note-Thompson-5 on 21st January, 2024.

Anonymous, 2024b. Cold Chain Logistics Management: How to Handle Temperature-Sensitive Goods. Retrieved from https://www.altexsoft.com/blog/cold-chain-logistics/ on 21st January, 2024.

Anonymous, 2024c. About APEDA. Retrieved from https://apeda.gov.in/apedawebsite/about_apeda/about_apeda.htm on 24th January, 2024.

Anonymous, 2024d. About FSSAI. Retrieved from https://www.fssai.gov.in/cms/about-fssai.php on 24th January, 2024.

ASHRAE, 2024. ASHRAE Terminology. Retrieved from https://www.ashrae.org/technical-resources/free-resources/ashrae-terminology on 31st January, 2024.

Aung M M and Chang Y S, 2014. Temperature management for the quality assurance of a perishable food supply chain. *Food Control*, *40*, 198-207. https://doi.org/10.1016/j.foodcont.2013.11.016.

Awan N, 2023. Cold Chain Management: What it is and it's important. Retrieved from https://www.linkedin.com/pulse/cold-chain-management-what-its-important-nabil-awan/ on 31st January, 2024.

Brunel S, Horn N M, Unger J G and Arnitis R, 2013. Implementation of international standards for phytosanitary measures no. 7 phytosanitary certification system and no. 12 phytosanitary certificates. *EPPO Bulletin*, 43(2): 309-315.

Carter S, 1997. Global agricultural marketing management. Food and Agriculture Organisation.

CBI (Centre for the Promotion of Imports from Developing Countries), 2024. What Requirements Should Fresh Fruits or Vegetables Comply with to be Allowed on the European Market. Retrieved from https://www.cbi.eu/market-information/fresh-fruitvegetables/buyer-requirements/ on 16th January, 2024.

Chand P, Singh A, Vishwakarma R and Singh C K, 2017. Plant Quarantine: An Effective approach for prevention of alien pest and disease. *Bulletin of Environment, Pharmacology and Life Sciences*, 6(11):07-12.

Cil A Y, Abdurahman D, and Cil I, 2022. Internet of Things enabled real time cold chain monitoring in a container port. *Journal of Shipping and Trade*, *7*(9), 1-26. https://doi.org/10.1186/s41072-022-00110-z.

CUTS. Standards and Market Access: Opportunities and Challenges for East African Exports into European Union, 2009. Retrieved from http://www.cuts-international.org/ARC/Nairobi/BIEAC/PDF/BIEACStandards_and_Market_Access.pdf on 20th January, 2024.

Dehariya S K, Shukla A, Barde S K and Ahirwar K, 2018. Efficacy of botanical pesticides against sucking insects' pests in Brinjal. *International Journal of Current Microbiology and Applied Sciences*, 7(01): 1930-1935.

do Nascimento Nunes M C, 2008. Impact of environmental conditions on fruit and vegetable quality. *Stewart Post-harvest Review*, *4*(4), 1-14.

Dohino T, Suzuki N, Miyazaki I, Tanno M, Yoshinaga M and Shukuya T, 2019. Comparison of cold tolerance of eggs and larvae of Bactrocera dorsalis (Diptera: Tephritidae) among citrus fruits. *Research Bulletin of the Plant Protection Service (Res. Bull. Pl. Prot. Japan)*, 55, 1-10.

DPPQS (Directorate of Plant Protection, Quarantine and Storage), 2007. Standard Operating Procedures for Export Inspection and Phytosanitary Certification of Plants/ Plant Products and other Regulated Articles. Final Document. Department of Agriculture and Cooperation. Ministry of Agriculture. GOI.

DPPQS (Directorate of Plant Protection, Quarantine and Storage), 2023. Import and Export Procedure. Retrieved from https://ppqs.gov.in/divisions/plant-quarantine/import-export-procedure. Department of Agriculture and Cooperation. Ministry of Agriculture. GOI on 24th December, 2023.

EIC, (Export Inspection Council), 2011. Certification Scheme for Fruit Products (CSFP). *Document No. EIC/CSFP-Ex. Instructions/2011-12-Issue 1.* Ministry of Commerce and Industry, Govt. of India.

Erkmen O and Bozoglu T F, 2016. Food preservation by low temperatures. *Food Microbiology: Principles into Practice. John Wiley and Sons, Ltd, Chichester, UK.* https://doi.org/10.1002/9781119237860.ch29.

European Commission. The European Commission Directive 2009/128/EC on Maximum Residue Levels, 2009. Retrieved from https://ec.europa.eu/food/plant/pesticides /max_residue_levels_en on 23rd January, 2024.

FAO, 2004. Manual for the preparation and sale of fruits and vegetables – From field to markets, Rome: FAO. Retrieved from http://www.fao.org/docrep/008/y4893e/y4893e00.HTM on 24th January, 2024.

FAO, 2006. International Standards for Phytosanitary Measures. 1 to 24. (2005 edition). Produced by the Secretariat of the International Plant Protection Convention. Food and agriculture organization of the United Nations, Rome. Retrieved from https://www.fao.org/3/a0450e/a0450e00.pdf on 26th January, 2024.

FAO, 2013. Manual of Phytosanitary Inspection Procedures. Sub-Regional Office for the Caribbean, Bridgetown, Barbados Food and Agriculture Organization of the United Nations, Rome. Retrieved from https://www.fao.org/fileadmin/templates/agphome/documents/caribbeanspci/ManPSInspProc.pdf on 26th December, 2023.

FAO, 2017. Integrated pest management of major pests and diseases in eastern Europe and the Caucasus. Food and Agriculture Organization of the United Nations. Budapest. Retrieved from https://www.fao.org/3/i5475e/i5475e.pdf on 13th January, 2024.

Fulano A M, Lengai G M and Muthomi J W, 2021. Phytosanitary and technical quality challenges in export fresh vegetables and strategies to compliance with market requirements: case of smallholder snap beans in Kenya. *Sustainability*, 13(3):1546.

Global-GAP, 2024. What We Do. Retrieved from https://www.globalgap.org/uk_en/what-we-do/ on 19th January, 2024.

GlobalGAP, 2024. About Global G.A.P. Retrieved from https://www.globalgap.org/about/ on 21st January, 2024.

Hallman G J, 2017. Process control in phytosanitary irradiation of fresh fruits and vegetables as a model for other phytosanitary treatment processes. *Food Control*, *72*, 372-377.

Hallman G J, Hénon Y M, Parker A G and Blackburn C M, 2016. Phytosanitary irradiation: An overview. *Florida Entomologist*, 1-13.

Hennessey M K, Jeffers L, Nendick D, Ken G, Floyd L, Hansen J. D, Bailey WD, Winborne, I, Bartels D and Ramsey C, 2014. Phytosanitary Treatments. *In* The Handbook of Plant Biosecurity; Gordh G, McKirdy S, (Eds.); Springer: Dordrecht, The Netherlands, pp 102-115.

Henson S, 2008. The role of public and private standards in regulating international food markets. *Journal of International Agricultural Trade and Development*, 4(1): 63-81.

Horticultural statistics at a glance, 2021

https://www.citrusfreight.com/resource/blog/pomegranate-export-from-india

Humphrey J, 2008. Private standards, small farmers and donor policy: EUREPGAP in Kenya. Working paper series, 308. Brighton.

Hundy G F, Trott and A R, Welch T C, 2008. Refrigerated Transport, Handling and Distribution. In Refrigeration and Air-Conditioning- Fourth Edition, pp. 214-225 Elsevier. https://doi.org/10.1016/B978-0-7506-8519-1.00017-7.

IIR, 2024. IIR International Dictionary of Refrigeration. Retrieved from http://dictionary.iifiir.org/search.php on 31st January, 2024.

IPPC, 2017. International Plant Protection Convention. Retrieved on https://www.ippc.int/en/structures/ on 31st January, 2024.

ISPM, 2006. ISPM No. 11 Pest Risk Analysis for Quarantine Pests including Analysis of Environmental Risks and Living Modified Organisms. *International Standards for Phytosanitary Measures*, 114.

ISPM, 2001. ISPM No. 12: Guidelines for phytosanitary certificates. Retrieved from https://www.ippc.int/largefiles/adopted_ISPMs_previousversions/en/ISPM_12_2001_En_2006-05-03.pdf on 31st January, 2024.

ISPMs, 2023. Adopted Standards (ISPMs). Retrieved on https://www.ippc.int/en/core-activities/standards-setting/ispms/ on 31st January, 2024.

Jaffee S and Masakure O, 2005. Strategic use of private standards to enhance international competitiveness: Vegetable exports from Kenya and elsewhere. *Food Policy*, 30(3): 316-333.

Kader AA, Perspective on Post-harvest horticulture, 2003. *Hortscience* -2003, 38:1004–1008. Retrieved from https://journals.ashs.org/hortsci/downloadpdf/journals/hortsci/38/5/article-p1004.xml on 31st Jan 2024.

Karuppuchamy P and Venugopal S, 2016. Integrated pest management. In *Ecofriendly pest management for food security* (pp. 651-684). Academic Press. https://doi.org/10.1016/B978-0-12-803265-7.00021-X.

Kenny M, 2002. Quality and safety of fresh fruits and vegetables along the production chain. *Food Nutrition and Agriculture*, (31):78-86.

Khetarpal R K and Kavita G, 2002. Implications of sanitary and phytosanitary agreement of WTO in Plant Protection in India. *Annual Review of Phytopathology*, 1: 1-26.

Kumar J and Puri S, 2011. Standard Operating Procedures for Export Inspection and Phytosanitary Certification of Plants/Plant Products and other Regulated Articles. *Centre of Advanced Faculty Training in Plant Pathology*.

Kunbhar S, Rajput L B, Gilal A A, Channa G A and Sahito J G M, 2018. Impact of botanical pesticides against sucking insect pests and their insect predators in brinjal crop. *Journal of Entomology and Zoology Studies*, 6(2): 83-87.

Lee G C H, 2006. Private food standards and their impacts on developing countries. European Commission DG Trade Unit G, 2.

Lengai G M, Fulano A M and Muthomi J W, 2022. Improving access to export market for fresh vegetables through reduction of phytosanitary and pesticide residue constraints. *Sustainability*, 14(13), 8183.

Makule E, Dimoso N and Tassou S A, 2022. Precooling and cold storage methods for fruits and vegetables in Sub-Saharan Africa—A review. *Horticulturae, 8*(9):776. https://doi.org/10.3390/horticulturae8090776.

Mangan R L, and Hallman G J, 2019. Temperature treatments for quarantine security: new approaches for fresh commodities. In *Temperature Sensitivity in Insects and Application in Integrated Pest Management* (pp. 201-234). CRC Press.

Martin R R, Constable F and Tzanetakis I E, 2016. Quarantine regulations and the impact of modern detection methods. *Annual Review of Phytopathology*, 54:189-205.

Moore S D, 2021. Biological control of a phytosanitary pest (*Thaumatotibia leucotreta*): A case study. *International Journal of Environmental Research and Public Health,* 18(3): 1198.

National Horticulture Board, http://nhb.gov.in

Nissen R, Bound S, Adhikari R and Cover I, 2018. Factors affecting Post-harvest management of apples: a guide to optimising quality. National Library of Australia Cataloguing -in-Publication entry. Retrieved from https://www.utas.edu.au/__data/assets/pdf_file/0003/1662186/FGT-Post-Harvest-Manual-2018-Web.pdf on 2nd February, 2024.

Palipane K B and Rolle R, 2008. Good practice for assuring the post-harvest quality of exotic tree fruit crops produced in Jamaica. Retrieved from https://www.fao.org/fileadmin/user_upload/ags/publications/exotic_fruit_book_web.pdf on 31st January, 2024.

Paltrinieri G and Staff F A O, 2014. Handling of fresh fruits, vegetables and root crops: A training manual for grenada. *Rome, Italy: Food and Agriculture Organization of the United Nations*. https://www.fao.org/3/au186e/au186e.pdf.

Panghal A, Chhikara N, Sindhu N and Jaglan S, 2018. Role of Food Safety Management Systems in safe food production: A review. *Journal of Food Safety*, 38(4): e12464.

Papoutsis K, Mathioudakis M M, Hasperué J H and Ziogas V, 2019. Non-chemical treatments for preventing the Post-harvest fungal rotting of citrus caused by *Penicillium digitatum* (green mold) and Penicillium italicum (blue mold). *Trends in Food Science and Technology*, 86:479-491.

Pay E, 2005. Overview of the sanitary and phytosanitary measures in QUAD countries on tropical fruits and vegetables imported from developing countries (No. 1). Research Paper.

PQO, 2003. Plant Quarantine (Regulation of Import into India) Order. Retrieved on https://ibkp.dbtindia.gov.in/DBT_Content_Test/CMS/Guidelines/20181115121925276_PlantQuarantine_order_2003.pdf on 31st January, 2024.

Prasad K, Jacob S and Siddiqui M W, 2018. Fruit maturity, harvesting, and quality standards. In *Preharvest Modulation of Post-harvest Fruit and Vegetable Quality* (pp. 41-69). Academic Press. https://doi.org/10.1016/B978-0-12-809807-3.00002-0

Quinlan M M, Leach A, Jeger M and Mumford J, 2020. Pest risk management in trade: The opportunity from using integrated combined measures in a systems approach (ISPM 14). *Outlooks on Pest Management*, 31(3): 106-112.

Rahman S M E, Mele M A, Lee Y T and Islam M Z, 2021. Consumer preference, quality, and safety of organic and conventional fresh fruits, vegetables, and cereals. *Foods*, *10*(1), 105. https://doi.org/10.3390/foods10010105

Schuur C C, 1988. Packaging for fruits, vegetables and root crops. *Food and Agriculture Organization of the United Nations, Barbados: Bridgetown.*

Sun D, Liu Y, Grant J, Long Y, Wang X and Xie C, 2021. Impact of food safety regulations on agricultural trade: Evidence from China's import refusal data. *Food Policy*, 105: 102185.

Suwapanich R, Krusong W and Thompson A K, 2019. Post-harvest control of anthracnose in mangoes by fumigation with vinegar and ethanol vapours. *International Journal of Post-harvest Technology and Innovation*, 6(3):179-191.

UNCTAD- United Nations Conference on Trade and Development, 2007. Food Safety and Environmental Requirements in Export Markets - Friend or Foe for Producers of Fruit and Vegetables in Asian Developing Countries. Retrieved from https://unctad.org/system/files/official-document/ditcted20068_en.pdf.UNCTAD/DITC/TED/2006/8. United Nations Publication, New York and Geneva on 05th January 2024.

USAID (United States Agency for International Development)-Kenya Horticulture Competitiveness Project (USAID-KHCP). The EU Market for Green Beans, 2011. Retrieved from https://pdf.usaid.gov/pdf_docs/PA00KC9M.pdf on 18th January 2024.

van Klinken R D, Kingham L, Hill M P and Collins K, 2023. A menu of measures to manage trade-related plant pest risks, and a review of methods for demonstrating measure efficacy. *Biological Invasions*, 25(4): 1227-1248.

Waage J K and Mumford J D, 2008. Agricultural biosecurity. *Philosophical Transactions of the Royal Society B: Biological Sciences*, 363(1492): 863-876.

7

Controlled Atmosphere Storage of Fruits

Sandeep Mann and _Gurjeet Kaur_*

Principal Scientist, Agriculture Structures and Environmental Control ICAR-Central Institute of Post Harvest Engineering and Technology Ludhiana, Punjab-141004, India

[*]Senior Research Fellow ICAR-Central Institute of Post Harvest Engineering and Technology, Ludhiana, Punjab-141 004, India

Introduction

Perishable nature of horticultural produce is a major factor contributing in high post-harvest losses. The metabolic processes such as respiration, transpiration, and ethylene production remain active after the harvesting of produce, which drive ripening and senescence Fig. 1. Thus, an effective post-harvest handling is crucial to delay these changes. The rate of these processes depends on the produce's physiological characteristics, composition, and storage conditions like temperature and humidity. Low-temperature storage is the simplest and most efficient method to slow down deterioration Escobedo-Avellaneda and Welti-Chanes, 2016. Combining low temperatures with methods that limit oxygen exposure, which fuels many enzymatic reactions, further extends shelf-life and maintains food quality. Reduced oxygen levels decrease respiration and the activity of enzymes like polyphenol oxidase and peroxidase.

Controlled atmosphere storage

Applying controlled atmosphere (CA) at low oxygen levels combined with reduced temperatures helps slow food senescence. CA technology alters the gas composition around foods to reduce metabolic reactions, maintain freshness, and extend shelf-life. This involves adjusting the concentrations of oxygen, carbon dioxide, ethylene, and/or nitrogen, with constant monitoring and adjustment of CO_2 and O_2 levels. Lowering oxygen levels reduces ethylene production, while increasing carbon dioxide levels diminishes ethylene's effects. Although CA is more expensive due to gas control, it is ideal for long-term storage. The other advantages and disadvantages associated with CA technology are discussed in Table 1. CA can be implemented after packaging or during bulk storage by modifying the gas composition in storage rooms to meet the specific needs of the stored foods Fragoso and Mújica-Paz, 2016.

Respiration

- Oxidizes carbohydrates using O_2.
- Produces ATP, CO_2, and water.
- ATP promotes ripening reactions.

Transpiration

- Water loss.
- Resulting in softening, shrivelling, shrinkage, wilting.
- Loss of juiciness, nutritional value.

Ethylene

- Regulates growth, development, and senescence.
- Physiologically active at very low concentrations.

Fig. 1: Metabolic processes and their associated effects on fruit quality.

Table 1: Advantages and disadvantages of CA storage.

Advantages	**Disadvantages**
Prolongs shelf-life of foods.	Costlier than other preservation methods.
Avoids microbiological reactions.	Off-flavors are produced by fermentation at very low O_2 levels.
Controls pests including insects, rats, and mice.	Reduce the synthesis of substances that are volatile and impact sensory quality.
Reduce ethylene biosynthesis retarding ripening.	To prevent gas leaks, the storage room needs special seals.
Improves color stability due to a low effect on chlorophylls.	Long-time of exposure needed 8–12 days.
Low effect on acidity, sugars, and vitamins.	Requirement of product specific concentration of gases .

In CA storage, the desired atmosphere is initially achieved by purging specific gases and then maintained by controlling the rate of gas input. This rate is influenced by the room's capacity, gas tightness, humidity, temperature, and the metabolism of the stored foods. The optimal gas concentrations depend on various factors, including the type of produce, species, maturity stage, and growing conditions. Differences in biochemistry and anatomy, such as intracellular space size and gas diffusion rates, also affect the required gas concentrations. Ideally, fruits and vegetables should be stored at low oxygen levels, near the anaerobic compensation point ACP. An increase in oxygen levels above the ACP can increase the respiration rate, while levels below it can cause fermentation, which negatively impact fruit metabolism.

Gases used in CA storage : Various gases are used in CA storage to control the metabolic processes in foods, outbreak of micro-organisms, insects and pests. The choice of low or high CO_2 levels depends on treatment time, room design, gas costs, and control systems. Controlling O_2 and levels can synergistically extend the shelf-life of foods. The role of these gases in avoiding deterioration and extension of fruits is discussed below:

Carbon dioxide CO_2 : CO_2 inhibits the growth of aerobic bacteria and fungi and prevents oxidation. Increased CO_2 levels can decrease synthetic reactions in climacteric fruit, delay ripening, inhibit enzymatic reactions, reduce organic volatile production, modify organic acid metabolism, reduce pectin breakdown, inhibit chlorophyll breakdown, retard fungal growth, inhibit ethylene effects, and preserve tenderness and color.

Oxygen O_2 : Oxygen supports aerobic bacteria growth and enzymatic reactions, affecting product quality. While it shouldn't be eliminated to avoid fermentation by anaerobic bacteria, reducing O_2 levels can lower respiration rates, decrease oxidation, delay climacteric fruit ripening, extend shelf-life, slow chlorophyll breakdown, reduce ethylene synthesis, alter fatty acid synthesis, slow soluble pectin degradation, decrease off-flavors, and change texture.

Nitrogen N_2: Nitrogen is used as a filler gas to prevent package collapse, as it is inert and nonreactive. It can replace oxygen to avoid lipophilic compound oxidation.

Ethylene C_2H_4: This hormone accelerates ripening and senescence, reducing quality and promoting decay when present in storage environments.

Types of Controlled Atmosphere: CA technology has become a key post-harvest technique since its inception in the 20th century. Jacques Etienne Berard noted in the early 1800s that fruit maintain their quality in low oxygen atmospheres. In 1927, Kidd and West linked reduced respiration rates in apples to extended storage life Yahia *et al.*, 2019. Over time, researchers have optimized gas concentrations for various horticultural products. CA storage has evolved from a static system to a dynamic one. The evolution of CA over time has been discussed as below Bodbodak and Moshfeghifar, 2016; Yahia *et al.*, 2019.

Static CA:

Conventional CA: Traditionally used for storing apples and pears, this system relies on the fruit's respiration to modify the atmosphere in a sealed room, reducing O_2 and increasing CO_2 levels. These gas levels are monitored and controlled daily, with air introduced to maintain O_2 levels and CO_2 removed via scrubbing. It typically takes about 10 days to establish the desired gas levels 3 kPa O_2 and 3 kPa CO_2 in a CA room.

Rapid CA: This system quickly establishes the desired atmosphere immediately after sealing the CA room, extending the product's shelf-life and maintaining quality. Developed about four decades ago, rapid CA reduces oxygen levels from 21 per cent to approximately 4 per cent within 12–36 hours. It is widely used for storing apples, pears, and cabbages.

Low oxygen CA: The CO_2 is allowed to rise as a result of respiration up to 2 kPa in this type of CA, while the oxygen is decreased and kept at about 2 kPa. Following that, analyzers are used to keep a track on the two gases. Adsorbers are used to remove CO_2, and atmospheric air is added to regulate the O_2 levels. The rooms must be constructed with features for gas tightness and equipped with the right equipment in order to maintain the necessary gas concentrations.

Ultra-low oxygen CA: In the 1990s, it was shown that fruits like apples can be stored at oxygen levels as low as 0.5 per cent. Traditionally, apple storage recommended oxygen levels around 2 kPa, but storage below 2.5 kPa is now considered ultralow oxygen ULO storage. Despite higher costs, ULO storage has been proven to retain better firmness and quality. By maintaining oxygen levels below 2 kPa, storage life can be extended, and fruit quality can be preserved. This system reduces oxygen to 1–1.3 kPa while keeping CO_2 levels no higher than 2–2.5 kPa, requiring efficient monitoring to prevent harmful oxygen levels.

Hyper low oxygen CA: Hyper Low Oxygen HLO CA is an advanced form of ULO, maintaining O_2 and CO_2 levels below 1 per cent, near anaerobic conditions. This requires gas-tight rooms, specialized technology, and a sophisticated automatic gas analysis system with frequent calibrations and remote control. HLO aims to extend produce life and manage physiological disorders like scald in apples and pears by pushing respiratory conditions to their limits.

Low oxygen stress: Low oxygen stress LOS extends ULO by keeping produce in induced anaerobiosis $O_2 < 0.7$ kPa for 2–15 days, with CO_2 levels below 0.1 kPa. This process, facilitated by N_2 generators and rapid oxygen reduction 24–36 hours, must maintain optimal gas-tightness. After LOS, the storage continues under ULO conditions.

Initial low oxygen stress: Initial low oxygen stress ILOS involves reducing oxygen to 0.5 kPa for 15–20 days at the start of storage, conditioning fruit to resist further abiotic stresses. Effective against superficial scald, ILOS avoids chemical treatments and is mainly used in Great Britain for certain apple varieties. Following ILOS, typical ULO conditions are restored.

Both CA and ULO storage are static systems where the atmosphere is set to optimal levels and doesn't adjust based on product response. This requires precise adjustment for each produce type and condition, complicates access to the fruit without disturbing the atmosphere, and lacks real-time information.

Dynamic controlled atmosphere: Dynamic Controlled Atmosphere DCA <0.6kPa of O_2 and <0.9kPa of CO_2 : Controlled atmosphere storage has advanced with the development of dynamic controlled atmosphere systems. DCA storage aims to maintain the lowest possible oxygen levels, similar to ultralow oxygen ULO storage, but dynamically adjusts gas concentrations based on the physiological responses of the produce. If the system detects low-oxygen stress, it increases the oxygen level until the produce's response returns to the optimal threshold. This method leverages existing CA technology, enhancing it with near real-time control, which extends produce storage life beyond traditional CA methods. DCA also mitigates storage disorders like superficial scald in apples and pears. Historically, superficial scald was managed using post-harvest antioxidants such as diphenylamine DPA or ethoxyquin for pears, but their use is now prohibited in the European Union.

Controlling gas levels in CA storage: CA storage rooms are designed to maintain specific gas compositions to extend the shelf-life of stored foods. These rooms are constructed with insulated panels and gastight locking mechanisms to ensure airtight conditions. The walls and doors, often made from metalpanels, are sealed with rubber gaskets to prevent gas leaks. Foods are placed in these insulated, airtight rooms, where the atmosphere is controlled to include low oxygen levels, enriched carbon dioxide, and balanced nitrogen Yahia *et al.*, 2019. Various systems and equipment are utilized to establish and maintain the desired atmospheric conditions for the required duration as discussed in Table 2.

Monitoring gas levels in CA storage: To achieve precise gas control in CA storage rooms, continuous monitoring is essential to detect stress signals. In dynamic controlled atmospheres, gas concentrations are regularly monitored by sensors. Key parameters measured include ethanol production, chlorophyll fluorescence, and the respiratory quotient Mditshwa *et al.*, 2018. There are three primary types of sensors used in DCA technology: chlorophyll fluorescence CF, respiration quotient RQ, and ethanol ET.

Chlorophyll fluorescence sensor: This technology determines the lowest oxygen level LOL that the fruit can tolerate during cold storage. O_2 levels are adjusted based on the fruit's physiological response, monitored non-destructively by CF sensors, which are highly sensitive to O_2 stress. These sensors automatically increase O_2 levels to prevent irreversible O_2 injury. The commonly used sensors in apples are HarvestWatch™. Research has shown that DCA-CF effectively maintains apple quality post-harvest. Recent studies on 'Gala', 'Red Delicious', 'Golden Delicious', 'Braeburn', 'Fuji' and 'Cripps Pink Zanella' and 'Rossi', 2015 'Elstar, PCP' Schlie *et al.*, 2020, 'Royal Gala' Both *et al.*, 2017 and 'Red Delicious', 'Granny Smith' Stürz *et al.*, 2024 apples indicate that DCA-CF is more effective in preserving quality than other techniques.

Respiration quotient sensor: DCA-RQ, a recent advancement in LOS technology, monitors and identifies when anaerobic respiration is attained by stored fruit. It records the respiration quotient RQ = CO_2 produced/O_2 consumed, which rises as O_2 concentration falls below the acceptable limit. Sensors like SafePod are primarily used to monitor the RQ of apples during DCA storage. The metabolic activity varies within the fruit canopy, with the sun-exposed side showing higher activity. Consequently.

Chlorophyll fluorescence can vary within a fruit, potentially causing DCA-CF to misjudge the low oxygen limit and affect fruit quality and shelf-life. Thus, this technology provides better monitoring and quality maintenance. A novel dynamic controlled atmosphere RQ-DCA system was developed by Bessemans *et al.*, 2016. 'Granny Smith' apples stored under RQ-DCA were compared to those in regular CA storage and CA storage with ethylene blocker 1-methylcyclopropene 1-MCP. RQ-DCA effectively controlled superficial scald and maintained higher firmness and skin color compared to regular CA storage, comparable to 1-MCP-treated fruit after 7 days at 18°C. Another similar study has been reported by Thewes *et al.*, 2018 on 'Fuji' apples indicating apples stored under DCA-RQ 1.5 are lesser responsive to 1-MCP application as compared to the other apple cultivars studied Fig. 2. This fact may be related to the extremely low ethylene production rate by 'Fuji' apple. Similarly, Weber *et al.*, 2019 examined dynamic controlled atmosphere monitored by respiratory quotient DCA-RQ and chlorophyll fluorescence DCA-CF on 'Nicoter' 'Kanzi®' apples at elevated 3.0°C and low 1.0°C storage temperatures. They found that fruit stored under DCA-RQ had higher anaerobic metabolites but similar electrolyte leakage, while storage at 3°C reduced physiological disorders and maintained better firmness after storage plus shelf-life. The study suggests that 'Nicoter' apples can be stored at 3°C with a DCA system based on CF or RQ to conserve energy.

Table 2: Control of gases in controlled atmosphere storage Escobedo-Avellaneda and Welti-Chanes, 2016

O_2 control	CO_2 control	Ethylene control
External burners: Burning propane/natural gas generates CO_2 and water vapour; require CO_2 scrubbers and cooling systems to control CO_2 levels and air temperature before and after combustion. Types: Open flame burners, catalytic burners.	**Water CO_2 scrubbers:** Brine solutions pumped/sprayed over evaporator coils absorb CO_2 due to pressure differences; can cause corrosion, better to use dry cooling units.	**Chemical adsorption:** Uses potassium permanganate and is typically applied in smaller stores or for modest ethylene reduction; non-reversible process requires frequent replacement of the adsorber.
Ammonia burners: NH_3 reacts with O_2 to produce N_2 and H_2; H_2 reacts to form H_2O; system avoids producing CO/ethylene but poses a risk of NH3 leakage.	**Calcium hydroxide:** Lime bags placed in CA room or through pipes; air is forced through these pipes and lime to reduce CO_2 levels.	**Catalytic destruction:** Most efficient method, maintaining low ethylene levels <0.02 ppm ; machine installed outside the room, draws in air to filter over a high-temperature catalytic bed, pre-cools it, and reintroduces it into the room.
Flushing with liquid or Gaseous N_2**:** by flushing N_2 directly into the storage room until the desired O_2 concentration is reached; method is simple but relatively expensive per use.	**Molecular sieve scrubbing:** efficient but costly system uses crystalline metal aluminosilicates with tiny pores to trap CO_2 molecules; operates in two circuits, one absorbing CO_2 while the other regenerates.	**UV light:** Ethylene is eliminated by the formation of ozone from the reaction of UV light and O_2.
Gas separator systems: Systems lower O_2 levels by increasing N_2 concentration 97-99 per cent **Pressure swing absorption PSA :** compress and filter dry air through a carbon bed, absorbing O_2 and enhancing N_2.	**NaOH solution:** Dissolved in water removes CO_2 by circulating air through open tubes, with removal proportional to NaOH exposure time.	
Hollow fiber membrane separators HFMS : Relies on different gas diffusion rates through membranes. **High-temperature ammonia cracking systems HTAC :** Uses anhydrous NH3 to react with air, yielding H_2 and N_2, with H_2 reacting with O_2 to form water vapor.	**Activated carbon:** With high porosity, captures CO_2 by adsorption; based on activated carbon releases CO_2 when flushed with CO_2-free air.	

Fig. 2: Apple at maturity and different stages during storage Adapted from Thewes *et al.*, 2018.

Ethanol sensor: DCA-ET is another recent development in LOS technology, measuring the lowest oxygen limit by detecting ethanol accumulation. Ethanol, a known fermentation product, is monitored until it reaches an unacceptable level around 1 ppm for fruit quality. The DCS™ sensor is widely used in The Netherlands for long-term storage of 'Elstar' and 'Jonagold' apples (Prange *et al.*, 2013). Although documented research on the efficient utilization of DCA-ET in maintaining post-harvest quality of fruits is limited, it is gaining popularity in the industry.

Effects of CA on quality of fruits: The concentration of CO_2 and O_2, along with the duration of exposure, plays a significant role in determining the occurrence and severity of storage disorders. Additionally, various other factors, such as biochemical and anatomical variations among cultivars and individual fruits, contribute to the varying susceptibility to such disorders. Oxidation of phenolic compounds is a primary cause of external skin injuries or internal disorders, often manifesting as visible brown spots or tissue cavities. Researchers have extensively studied the effects of CA storage on fruits like apples, pears, guava, and others as summarized in Table 3. Below, we delve into the latest studies to analyse how CA impacts the quality and storage stability of different fruits.

Apple: Controlled atmosphere CA storage is a vital technique in the post-harvest management of apples, aimed at preserving their quality and extending their shelf-life. By controlling oxygen, carbon dioxide, and other gases, CA storage creates optimal conditions that slow down physiological processes, reducing decay and maintaining fruit firmness and flavor. In this context, Kweon *et al.*, 2013 studied the impact of controlled atmosphere CA storage on 'Fuji' apples Table 3. They found that flesh browning and ethylene production were lowest at 1.5 kPa CO_2, while watercore disappeared more rapidly at <1 kPa CO_2. The study recommends storing 'Fuji' apples at 0°C in 1-2 kPa O_2/<1 kPa CO_2 if harvested before 190 days after full bloom. Similarly, 'Jonagold' apples were examined for metabolic changes during initial periods of controlled atmosphere CA storage in 1 kPa O_2/3 kPa CO_2, 3 kPa O_2/3 kPa CO_2, 1 kPa O_2/10 kPa CO_2, and regular air 20.8 kPa O_2/0.03 kPa CO_2 as a control (Bekele *et al.*, 2016). The stress response patterns were similar for both immediate and delayed CA storage. Specifically, aspartate and 1-aminocyclopropane-1-carboxylic acid were positively correlated with O_2 levels early in storage, while glucose-6-phosphate and certain amino acids like proline, alanine, and threonine were negatively correlated. These metabolic changes highlight the apples' adaptation mechanisms to CA storage. In a subsequent study on 'Jonagold' apples stored under regular air RA and various controlled atmosphere CA conditions 10.5 kPa O_2 + 0.5 kPa CO_2, 22.0 kPa O_2 + 1.0 kPa CO_2, 31.0 kPa O_2 + 3.0 kPa CO_2, and 4 0.5 kPa O_2 + 6.0 kPa CO_2 for six months at 0 ± 0.3°C, Saquet (2016) found that lower O_2 or higher CO_2 in CA treatments suppressed ethylene production and respiration rate, with the 0.5 kPa O_2 + 6.0 kPa CO_2 condition being most effective. 'Jonagold' apples were tolerant to all CA conditions, showing improved flesh firmness, greener skin color, and higher titratable acidity, with no significant effect on total soluble solids.

Along with CA, DeEll *et al.*, 2016 studied the impact of multiple 1-Methylcyclopropene 1-MCP which is a ethylene blocker on apple quality and disorder development. After harvesting the apples, they applied 1-MCP to McIntosh, Empire, and Northern Spy firstly after harvest and then after 4 month of CA storage. CA conditions for "McIntosh" apples were 2.5 kPa O_2 + 2.5 kPa CO_2 in the first month, 3.5 kPa CO_2 in the second month, and 4.5 kPa CO_2 afterwards. 2.5 kPa O_2 + 2 kPa CO_2 and 2.5 kPa O_2 + 2.5 kPa CO_2 were the CA conditions for the apples "Empire" and "Northern Spy," respectively. In CA, apples were stored at 3°C for McIntosh, Empire, and, Northern Spy at 0°C for the period of 9 months. The results showed that 1-MCP reduced ethylene production, improved firmness and acidity, and decreased external CO_2 injury in 'Northern Spy' apples. A second 1-MCP treatment after 4 months of CA storage improved firmness in some cases but could lead to variable responses in disorder susceptibility, particularly during long-term CA storage. To analyse injury development in stored fruits, Lumpkin *et al.*, 2015 analyzed the volatile compounds of 'Fuji' apples during cold storage in air and low pO_2 CA with up to 5 kPa CO_2. Apples stored in CA produced fewer butyl, hexyl, and propyl esters but had increased methyl and ethyl esters compared to those stored in air. The highest CO_2 injury incidence occurred at 5 kPa CO_2, where increased methyl ester accumulation could indicate early CO_2 injury development.

Dynamic controlled atmospheres DCA was also explored to control superficial scald in 'Granny Smith' apples over two growing seasons (Mditshwa *et al.*, 2017). Apples were stored in DCA in which O_2 levels ranged between 0.3–0.5 per cent whilst CO_2 was maintained at 1 per cent and 95 per cent RH at 0°C for 5 days to 20 weeks indicated that DCA-treated apples could only withstand a 6-week shipping period without increased scald risk. DCA-stored apples had better firmness, higher ground color, and lower levels of ethylene, α-farnesene, and MHO compared to air-stored apples. Strong correlations were found between scald-associated metabolites and scald incidence. Anoxia 0 kPa O_2 can occur during DCA and CA storage if conditions are not accurately controlled, leading to increased fermentation end products like acetaldehyde, ethanol, and ethyl acetate. These compounds, in controlled concentrations, enhance fruit firmness and flavor. However, excessive fermentation products can result in off-flavor and other disorders, emphasizing the importance of precise storage conditions. Thus, Maree Wood *et al.*, 2022 studied anaerobic metabolism in various apple cultivars during 84 days of regular air RA and controlled atmosphere CA storage after 0-, 3-, 7-, and 14-days of anoxia Table 3. Prolonged anoxia increased anaerobic metabolite concentrations AMC in all cultivars. 'Golden Delicious' and 'Jonagold' had faster AMC reduction under CA, while 'Kanzi' showed the opposite trend. CA storage post-anoxia maintained better fruit quality and reduced AMC more effectively over 84 days compared to RA storage.

Avocado: To examine internal owxidative/fermentative behavior of 'Hass' avocado fruits at different maturity stages during post-harvest in context to changing external O_2 concentration, Valle-Guadarrama *et al.*, 2013 stored fruits for 16 days in natural air NA conditions at 20°C, compared to those in controlled atmosphere CA with 5.3 per cent O_2 for 24 hours. Internal O_2 levels decreased with tissue depth, staying above tolerance in NA fruits but dropping below in CA-exposed fruits. Fermentative activity was noted in both anaerobic and aerobic conditions due to low O_2 and pH levels, highlighting the influence of external O_2 concentration on fruit fermentative behavior.

whats Controlled atmosphere storage for blueberries involves optimizing gas composition to extend shelf-life and maintain fruit quality. Concha-Meyer *et al.*, 2015 studied the shelf-life of fresh highbush blueberries under different storage conditions Table 3. These included air control, CA sorage with 5 per cent O_2 : 15 per cent CO_2 : 80 per cent N_2, and ozone gas O_3 at 4 ppm 4°C or 2.5 ppm 12°C. After ten days at both temperatures, they discovered that CAS and O_3 had no effect on the growth of yeast and mold. Blueberries kept at 4°C, however, lost less weight than those kept at 12°C. When compared to other treatments, blueberries treated with O_3 at 12°C had less weight loss and firmness loss by day 10. This implies that proper refrigeration and low ozone gas concentrations can help preserve the freshness of blueberries while they are being stored. Similarly, Chiabrando and Giacalone 2011 investigated the use of 1-methylcyclopropene 1-MCP to control ripening in 'Lateblue' blueberry fruit. Fruits were treated with 1-MCP 0.3 and 0.6 µL L-1 post-harvest and stored in air at 0°C for 35 days or in a controlled atmosphere 3 kPa O_2 + 11 kPa CO_2 for 60 days. 1-MCP-treated blueberries exhibited reduced weight loss and lower soluble solids content compared to untreated fruit. However, 1-MCP had no significant impact on acidity, anthocyanins, phenolics, or antioxidant activities. The study suggests that while 1-MCP can preserve certain qualities in blueberries, it does not affect acidity or phytochemical content.

Blue honeysuckle berries *Lonicera caerulea* L. have rich taste profile and abundance of bioactive compounds. Dziedzic *et al.*, 2020 investigated methods to extend the post-harvest lifespan of *Lonicera caerulea* L., a new health-promoting fruit. Four cultivars were studied after short-term storage in CA 20 per cent CO_2 and 5 per cent O_2 and modified atmosphere using Xtend bags, compared to ambient conditions NA. Storage conditions and cultivars significantly influenced fruit quality, bioactive components, and antioxidant activity. 'Vostorg' maintained fruit firmness, while 'Indigo Gem' preserved soluble solids content despite substantial weight loss due to berry size and shape. CA reduced fruit respiration intensity, suggesting its potential to improve the post-harvest quality of honeysuckle fruit.

CA storage reduced firmness loss, ethylene production, and color changes compared to the control. The most effective CA treatment was 5 per cent CO_2 + 5 per cent O_2, extending shelf-life to 25 days compared to the control's 20 days.

Cherry: Like other fruits, sweet cherries are also highly perishable during post-harvest, posing challenges for packing houses. Therefore, low-temperature storage is essential to prevent fruit deterioration. Serradilla *et al.*, 2013 examined the effects of controlled atmospheres 3 per cent O_2 + 10 per cent CO_2, 5 per cent O_2 + 10 per cent CO_2 and 8 per cent O_2 + 10 per cent CO_2 on the microbial population of 'Ambrunés' sweet cherries over 30 days of storage and subsequent shelf-life.

Table 3. A summary of studies conducted on CA storage of Fruits.

Fruit	CA conditions	Storage conditions	Reference
Apple *Malus sylvestris* L. cv. Fuji.	• kPa O_2+3 kPa CO_2. • kPa O_2+3 kPa CO_2 • 1 kPa O_2+10 kPa CO_2	0°C for 20 days.	Kweon *et al.*, 2013
Apple fruit *Malus domestica* Borkh., cv. Jonagold.	• 1 kPa O_2+3 kPa CO_2 • 3 kPa O_2+3 kPa CO_2 • 1 kPa O_2+10 kPa CO_2	1°C for 36 days.	Bekele *et al.*, 2016
Apple *Malus domestica* cv. Galaxy.	• 1.2 kPa O_2+2.0 kPa CO_2 • DCA–RQ 1.5 DCA-RQ, 1.2 kPa CO_2	1.0°C; 1.5°C and 2.0°C for 9 months.	Both *et al.*, 2018
Apple *Malus domestica* Borkh. cv. BC2 Fuji.	• 1 kPa O_2 + 0.5, 1.5, or 5.0 kPa CO_2	0.5°C, 8 months.	Lumpkin *et al.*, 2015
Apple *Malus domestica* Borkh., cv. Jonagold.	• 0.5 kPa O_2 + 0.5 kPa CO_2 • 2.0 kPa O_2 +1.0 kPa CO_2 • 1.0 kPa O_2 + 3.0 kPa CO_2 • 0.5 kPa O_2 + 6.0 kPa CO_2	0 ± 0.3°C, 6 months.	Saquet 2016
Apple *Malus domestica* Borkh., cv. Jonagold, Delicious and Kanzi.	Delicious • 1 kPa O_2 +2.5 kPa Jonagold • 1 kPa O_2 +2.5 kPa Kanzi • 1 kPa O_2 +<0.7 kPa	Delicious: 1°C. Jonagold: 1°C. Kanzi: 3°C. 84 days.	Wood *et al.*, 2022
Avocado *Persea americana* cv. Hass.	• 5.3±0.3 per cent O_2+94.7 per cent N_2	20°C, 16 days.	Valle-Guadarrama *et al.*, 2013
Blueberries *Vaccinium corymbosum* L., cv Ozark Blue.	• 5 per cent O_2+15 per cent CO_2+ 80 per cent N_2	4°C or 12°C, 10 days.	Concha-Meyer *et al.*, 2015
Blueberry *Vaccinium corymbosum* L. cv Lateblue.	• 3 kPa O_2 + 11 kPa CO_2	0°C, RH 90–95 per cent for 35 days.	Chiabrando and Giacalone 2011

Fruit	CA conditions	Storage conditions	Reference
Blue honeysuckle. *Lonicera caerulea* L. cultivars Vostorg, Indigo Gem, Aurora and Honeybee .	• 5 per cent O_2+20 per cent CO_2	2°C, RH 90–92 per cent for 7 and 14 days.	Dziedzic *et al.*, 2020
Carambola *Averrhoa carambola* cultivar 'B17'.	• 5 per cent O_2+10 per cent CO_2 • 10 per cent O_2+10 per cent CO_2 • 10 per cent O_2+5 per cent CO_2 • 5 per cent O_2+5 per cent CO_2	10 °C for 25 days.	Boonyaritthongchai *et al.*, 2012
Cherry *Prunus avium* L.	• 3 per cent O_2 + 10 per cent CO_2 • 5 per cent O_2 + 10 per cent CO_2 • 8 per cent O_2 + 10 per cent CO_2	1°C, RH 95 per cent. for 15, 21, and 30 days.	Serradilla *et al.*, 2013
Dragon fruit *Selenicereus undatus.*	• 2 kPa O_2+0.5 kPa CO_2 • 5 kPa O_2+5 kPa CO_2	6 ± 0.5°C for 50 days.	Ho *et al.*, 2021
Feijoa fruit *Acca sellowiana*, cv. Unique.	• O_2 levels 2, 6, 11 and 21 kPa O_2 • CO_2 levels 0.2, 4 and 7 kPa CO_2	5 ∘C for 56 days.	East *et al.*, 2009
Fig *Ficus carica* L. cv. Sarılop.	• 1 per cent O_2 +12 per cent CO_2	10.3°C, RH 61.3 per cent for 4 months .	Sen *et al.*, 2010
Fig *Ficus. carica* L. cv. Ottomanit or Cyprus.	• 5 kPa O_2+5, 10 or 15 kPa CO_2	0°C, RH 95 per cent FOR 1 month.	Bahar and Lichter 2018
Guava *Psidium guajava* L. cv. Pedro Sato.	• 1, 5, 10, 15 21 kPa O_2	12.5 ± 0.8°C, RH 95 per cent for 28 days.	Teixeira and Durigan 2010
Guava *Psidium guajava* L. cv. Lucknow-49, Allahabad Safeda and Apple Colour.	• 2.5 kPa O_2 + 2.5 kPa CO_2 • 2.5 kPa O_2 +5kPa CO_2 • 5 kPaO_2 + 2.5 kPa CO_2 • 5kPaO_2 +5kPaCO_2 • 8kPaO_2 +5kPa CO_2 • 10 kPa O_2 +10 kPa CO_2	8 ∘C for 30 days.	Singh and Pal 2008

Fruit	CA conditions	Storage conditions	Reference
Kiwifruit *Actinidia deliciosa.*	• 1.5 per cent O_2 + 1.5 per cent CO_2	1°C, RH 85 per cent for 4 and 8 weeks.	Latocha *et al.*, 2014
Litchi fruit *Litchi chinensis* Sonn. cv. McLean's Red.	• 17 per cent O_2þ 6 per cent CO_2 • 7 per cent O_2 þ 3 per cent CO_2	2°C, RH 90 per cent for 21 days.	Sivakumar and Korsten 2010
Litchi *Litchi chinensis Sonn.* cv. Gola.	• 1 per cent O_2 + 3 per cent CO_2 • 2 per cent O_2 + 3 per cent CO_2 • 3 per cent O_2 + 3 per cent CO_2 • 1 per cent O_2 + 4 per cent CO_2 • 2 per cent O_2 + 4 per cent CO_2 • 3 per cent O_2 + 4 per cent CO_2 • 1 per cent O_2 + 5 per cent CO_2 • 2 per cent O_2 + 5 per cent CO_2 • 3 per cent O_2 + 5 per cent CO_2	5 ± 1°C, RH 90 ± 5 per cent for 35 days.	Ali *et al.*, 2016
Litchi *Litchi chinensis Sonn.* cv. Gola.	• 1 per cent O_2 + 5 per cent CO_2 • 0.25 per cent CN + CA • 0.25 per cent MN + CA • 0.25 per cent CN +0.25 per cent MN +CA.	5 ± 1°C, RH 90 ± 5 per cent for 28 days.	Ali *et al.*, 2019
Longan *D Dimocarpus longan* Lour. cv. Chuliang and Shixia.	• 5 per cent O_2 + 3 per cent CO_2 + 92 per cent N_2 • 5 per cent O_2 + 5 per cent CO_2 + 90 per cent N_2 • 5 per cent O_2 + 10 per cent CO_2 + 85 per cent N_2	4°C for 18 days.	Zhou *et al.*, 2016
Longan *Dimocarpus longan* L. cv. Daw.	• 5 per cent O_2 + 5 per cent CO_2 • 5 per cent O_2 + 10 per cent CO_2 • 5 per cent O_2 + 15 per cent CO_2	2°C, RH 90 per cent for 35 days.	Khan *et al.*, 2017
Mango, *Mangifera indica* L.	• 5 per cent O_2 + 5 per cent CO_2 • 3 per cent O_2 + 8 per cent CO_2	10°C, RH 90 per cent for18 days.	Sivakumar *et al.*, 2011

Fruit	CA conditions	Storage conditions	Reference
Medlar *Mespilus germanica* L.	• 2 per cent O_2+ 5 per cent CO_2 • 3 per cent O_2+ 10 per cent CO_2	0 ± 0.5°C, RH 90 ± 5 per cent for 60 days.	Selcuk and Erkan 2015
Pear fruits *P. bretschneideri* Rehd. cv. Zaosuli.	• 2.0 per cent O_2 + 7.0 per cent CO_2	1°C for 40 days.	Goliáš *et al.*, 2016
Pears *Pyrus communis* L. cv. Swiss Bartlett.	• 18 kPa or 2.5 kPa O_2+2 kPa CO_2	0°C for 167 days.	Lum *et al.*, 2017
European pear *Pyrus communis* L.	• 18 kPa O_2 +2 kPa CO_2 • 2.5 kPa O_2 + 2 kPa CO_2	0°C for170 days.	Flaherty *et al.*, 2018
Pears *Pyrus pyrifolia Nakai.*	• 11 per cent O_2,10 per cent CO_2, 79 per cent N_2	4°C, RH 85–90 per cent for 5 days	Wang *et al.*, 2021
Persimmon fruit *Diospyros kaki* cv. Fuyu.	• 0.15 kPa O_2 +2.0 kPa CO_2 • 0.15 kPa O_2 + 6.0 kPa CO_2	0, 5°C for 3 months.	Brackmann *et al.*, 2014
Plums *Prunus domestica* cv. Laetitia.	• O_2: 1 kPa O_2+3 kPa CO_2 • 1 kPa O_2 +5 kPa CO_2 • 2 kPa O_2 +5 kPa CO_2 • 2 kPa O_2 +10 kPa CO_2 • 11 kPa O_2 +10 kPa CO_2 • 1 kPa O_2+ 0 kPa CO_2 • 1 kPa O_2 +1 kPa CO_2 • 2 kPa O_2 +1 kPa CO_2 • 2 kPa O_2 +2 kPa CO_2 • 2 kPa O_2 +1 kPa CO_2	0.5°C±0.1°C, RH 96±2 per cent for 60 days.	Steffens *et al.*, 2014
Red raspberry *Rubus idaeus* L.	• 7.5 kPa O_2 +12.5 kPa CO_2	1 C, 95 per cent RH for 2 years.	Forney *et al.*, 2015
Strawberry *Fragaria x ananassa* cv. Minomusum.	• 20 per cent, 30 per cent, 40 per cent CO_2	5°C for 10 days.	Nakata 2020
Strawberry *Fragaria x ananassa* cv. Sonata and Avocado *Persea americana* cv. Hass.	• 5 kPa O_2+15 kPa CO_2 • 5 kPa O_2 +10 kPa CO_2	5°C for 14 days.	Alamar *et al.*, 2017

They identified various micro-organisms, including yeasts, molds, lactic acid bacteria, *Staphylococcus* spp., *Pseudomonas* spp., *Enterobacteriaceae* spp. and *coliforms*. Controlled atmospheres with 5 per cent and 8 per cent O_2 + 10 per cent CO_2 effectively controlled the growth of mesophilic aerobic bacteria, psychrotrophs, *Pseudomonas* spp., yeasts, and molds after 15 days, with yeast and mold counts between 1 and 1.75 log CFU/g. The study concludes that specific controlled atmospheres can effectively reduce microbial growth in sweet cherries, extending their shelf-life.

Carambola fruits: Another fruit which is commercially valuable in tropical and sub-tropical areas, Carambola, also known as starfruit, is sweet, juicy fruit rich in vitamins A and C. Boonyaritthongchai *et al*., 2012 studied the storage of carambola fruit in high CO_2 5 per cent and 10 per cent and/or low O_2 5 per cent and 10 per cent controlled CA at 10°C.

Dragon fruit: For dragon fruits, Ho *et al*., 2021 studied the optimal controlled atmosphere CA storage under four different CA conditions and ambient air up to 50 days (Table 3). Results indicated that CA storage significantly reduced scale yellowing and maintained acidity. The condition of 2 kPa O_2 + 5 kPa CO_2 at 6°C was found to be optimal. Normal senescence reduced many volatiles, while low O_2 and high CO_2 levels increased fermentation products. These changes did not significantly affect sensory quality, concluding that CA storage effectively preserves dragon fruit quality.

Feijoa fruit: The feijoa, also known as "pineapple guava," is an evergreen shrub from the Myrtaceae family, native to high-altitude areas in Uruguay and southern Brazil. East *et al*., 2009 monitored gas exchange rates and quality changes in feijoa fruit *Acca sellowiana* cv. Unique stored at 5°C under 16 different controlled atmosphere CA conditions to assess their commercial potential for extending storage life Table 3. The study found that a combination of low O_2 and low CO_2 was most effective, reducing weight loss from 1.7 per cent to 1.2 per cent, minimizing hue value changes from 4° to 2°, and decreasing the incidence of blemished fruit from 30 per cent to 20 per cent compared to regular air storage. This suggests that CA storage can significantly extend the shelf-life and maintain the quality of feijoa fruit.

Fig: Sun-drying is a longstanding practice in the Mediterranean Basin for preserving fruit, but storage pests can affect marketability. Methyl bromide MB was once a common fumigant for pest control, but its phaseout under the Montreal Protocol has created a need for alternatives in post-harvest treatment. As an alternative to methyl bromide MB for controlling storage pests in dried figs, Sen *et al*. (2010) tested controlled atmosphere CA with reduced O_2 10.5 per cent at 41°C against fig moth, Indian meal moth, and dried fruit beetle. The study also analyzed dried fig quality immediately post-treatment and after 4 months of storage, comparing it with MB-treated controls. CA treatment effectively controlled all tested pests, maintained or improved dried fruit quality, and required shorter treatment times about 26 hours compared to other MB alternatives. This research supports CA as a viable MB alternative for post-harvest pest control in dried figs, addressing the urgent need for alternatives to MB due to environmental concerns. Another study focusing on quality assessment of Ottomanit figs under controlled atmosphere CA was carried out by Bahara and Lichter, (2018). Figs were stored for 30 days at 0°C under four storage conditions O_2/CO_2 in kPa : 5/5, 5/10, 5/15, and 20/0 ambient atmosphere in the cold room. CA delayed ripening, as indicated by total soluble solids, acidity, and color. The 5/5 kPa CA maintained higher firmness. Higher CO_2 levels led to CO_2 injury, with least decay observed under 5 kPa CO_2. Fluorescence imaging suggested that 5/5 kPa O_2/CO_2 preserved fruit integrity compared to other conditions.

Guava: Guava *Psidium guajava* L. is an important fruit crop of the sub-tropical and tropical regions in the world. Its delicate nature, short post-harvest life, and susceptibility to chilling injury and diseases, limit the potential for commercialisation of guava fruit. The guava is a climacteric fruit with a short shelf-life and its shelf-life can be increased by using cold storage and controlled atmosphere CA. However, not much research has been done to assess how CA storage affects guavas, particularly when carbon dioxide CO_2 levels are high. Teixeira and Durigan (2010) studied the storage of "Pedro Sato" guava at 12°C for 28 days in CA with different oxygen levels air, 21, 15, 10, 5, and 1 kPa Table 3.

Reduced respiration rates and delayed ripening were the main effects of lower O_2 levels 1 and 5 kPa, which preserved the green color and increased sugar content. Fruits in 1, 5, 10, and 15 kPa O_2 atmospheres did not exhibit climacteric peaks, in contrast to control and high-O_2 21 kPa fruits Fig. 3. This study supports using low-O_2 atmospheres for extended storage of 'Pedro Sato' guava fruit, ensuring delayed ripening and maintaining quality during storage. In 2016, Teixeira and his team examined 'Pedro Sato' guava fruit quality during controlled atmosphere CA storage, varying O_2 5 kPa and CO_2 1, 5, 10, 15, or 20 kPa levels at 12.2°C for 28 days. Clear CO_2 injury occurred at 5 kPa O_2 + 15 or 20 kPa CO_2, showing increased pH and soluble pectin with reduced firmness. Optimal storage conditions for 'Pedro Sato' guavas involve 5 kPa O_2 and less than 5 kPa CO_2 to avoid CO_2 damage. Identifying metabolites triggering high CO_2 responses in guava fruit warrants further investigation.

Fig. 3: 'Pedro Sato' guava fruit storage under controlled atmosphere A Day 0, B Day 14, and C Day 28. Adapted from Teixeira and Durigan, 2010.

Kiwifruit: Latocha *et al*., 2014 compared kiwifruit *Actinidia arguta* and its hybrid fruit stored in air AS versus controlled atmosphere CA with 1.5 per cent O_2 + 1.5 per cent CO_2 at 1°C and 85 per cent RH for four and eight weeks (Table 3). They focused on 'Ananasnaya' and 'Bingo' at different maturity stages. Fruit firmness and acidity changed notably during long-term storage. Air storage sufficed for four weeks, while CA was more suitable for the hardy kiwifruit over eight weeks. Fruit stored in CA and ripened had sensory qualities akin to vine ripe fruit. However, long-term cold storage intensified bitter taste, marking a notable negative change. Optimal storage methods depend on the duration and desired sensory outcomes, with CA being beneficial for extended preservation.

Litchi: Litchi, a sub-tropical fruit native to south-east Asia, is prized for its juicy and fragrant flesh enclosed in a spiky red rind. Known for its sweet flavor and refreshing taste, litchi is often enjoyed fresh, in desserts, and as a flavoring in various culinary dishes. Sivakumar and Korsten (2010) studied the effect of 1-MCP pre-treatment and two 2 CA conditions CA-1: 17 per cent O_2 + 6 per cent CO_2; CA-2: 7 per cent O_2 + 3 per cent CO_2 on 'McLean's Red' litchi fruit quality and pericarp browning Table 3. Fruits treated with 1-MCP 500 nl/l were stored under CA-1 or CA-2 at 2°C and 90 per cent RH for 21 days. Compared to standalone CA-1, CA-2, the combination of 1-MCP and CA-1 effectively prevent browning, maintainted red color, ascorbic acid content, and taste. This treatment also reduced polyphenol oxidase PPO and peroxidase POD activity, retained membrane integrity, and preserved anthocyanin content without any off-flavors. CA-2 alone led to higher browning, increased PPO and POD activity, and loss of membrane integrity. Thus, 1-MCP pre-treatment and CA-1 best preserved fruit quality for 21 days. Ali *et al.* (2016) studied 'Gola' litchi fruits stored under 10 CA conditions at 5 ± 1°C, focusing on pericarp browning, biochemical quality, and antioxidative activities Table 3. Control fruits turned completely brown after 28 days and were excluded. The CA7 1 per cent O_2 + 5 per cent CO_2 significantly reduced weight loss, pericarp browning, and membrane leakage. These fruits also exhibited higher soluble solid content, titratable acidity, and ascorbic acid levels. The fruits stored under CA7 conditions maintained superior organoleptic quality. Therefore, 1 per cent O_2 + 5 per cent CO_2 conditions effectively delayed pericarp browning, preserved antioxidative activities, and maintained the biochemical and organoleptic quality of litchi fruits for 35 days.

Subsequently, Ali *et al.*, 2018 treated litchi fruit with 0.25 per cent methionine MN and 0.25 per cent cysteine CN alone or in combination and stored them under 1 per cent O_2 + 5 per cent CO_2 controlled atmosphere CA at 5 ± 1ºC for 28 days. In comparison to MN and CN + MN, the CN treatment was most successful in preventing browning under CA conditions. In addition to reducing disease incidence, weight loss, polyphenol oxidase PPO and peroxidase POD activities, the CN treatment significantly postponed the browning index. Additionally, it preserved increased concentrations of ascorbic acid, total phenolics TPC, superoxide dismutase SOD, ascorbate peroxidase APX, and catalase CAT enzymes, as well as DPPH radical scavenging capacity. Thus, pre-treating litchi fruit with 0.25 per cent CN is a viable way to control browning and maintain fruit quality while storing it in CA. In 2021, Ali and his team used chemical treatments such as ascorbic acid AA and oxalic acid OA on litchi fruit browning under 5 per cent CO_2 + 1 per cent O_2 controlled atmosphere CA storage at 5 ± 1°C for 28 days. The combined use of AA and OA significantly reduced browning, activities of polyphenol oxidase and peroxidase. This combination also decreased weight loss and maintained higher levels of anthocyanins, total phenolics. Additionally, the AA + OA + CA treatment resulted in lower levels of superoxide anion, and hydrogen peroxide, while enhancing ascorbic acid, titratable acidity, and sensory quality. In conclusion, the AA + OA combination effectively delayed browning and preserved the visual appearance of litchi fruits under CA storage.

Longan fruit: Longan *Dimocarpus longan* Lour. fruit usually only lasts 3–4 days at room temperature after being harvested. To prevent pericarp browning and preserve aril firmness, the longan industry uses CA storage, where it can withstand increased CO_2 6–8 per cent and decreased O_2 4–6 per cent. Zhou *et al.*, 2016 studied γ-aminobutyric acid GABA accumulation in 'Chuliang' and 'Shixia' Longan fruit during precooling and controlled-atmosphere CA storage. The fruits were stored at 4°C in 5 per cent O_2 with 3 per cent, 5 per cent, or 10 per cent CO_2. Results showed that 5 per cent CO_2 + 5 per cent O_2 delayed aril softening and pericarp browning. GABA levels and glutamate decarboxylase GAD activities decreased in higher CO_2 treatments, while GABA aminotransferase GABA-T activities fluctuated. GABA concentrations increased after pre-cooling and after 3 days in 10 per cent CO_2 + 5 per cent O_2, then declined over time. Differences in GAD and GABA-T activities were observed between cultivars. Overall, GABA accumulation was linked to increased GAD activity rather than decreased GABA-T activity. Sulphur dioxide has traditionally been used to control pericarp browning in longan fruit, but health concerns necessitate alternative treatments. Khan *et al.*, 2017 studied longan fruit tolerance to low O_2 2, 5, 10, 15 per cent and elevated CO_2 5, 10, 15, 20 per cent at 2°C Table 3. Results showed significantly higher pericarp browning and decay in the control group compared to CA treatments. CA storage reduced polyphenol oxidase PPO activity, maintained L^* value, and slowed the decrease in total phenolic contents TPC. Pericarp browning was strongly correlated with PPO activity, L^* value, and TPC.

Mango: Commercial techniques for managing post-harvest illnesses in mangoes include the post-harvest application of the fungicide prochloraz and hot-water dip. The need to find natural, environmentally friendly alternatives to conventional fruit production methods has grown, especially for the fruit industry, as consumer demand for fruit grown organically grows. Sivakumar *et al.*, 2011 evaluated the effect of 1-MCP and CA CA-1: 5 per cent O_2 + 5 per cent CO_2; CA-2: 3 per cent O_2 + 8 per cent CO_2 on hot-water treated mangoes cv. Kent. The 1-MCP+CA-2 treatment was more effective than 1-MCP+CA-1 in reducing anthracnose incidence, weight and firmness loss, and delaying skin and flesh color development. It also maintained ascorbic acid, carotenoid, total phenolic, and flavonoid contents, and antioxidant activity. Additionally, it prevented increases in ethanol, and acetaldehyde content. An untrained panel preferred 1-MCP+CA-2 treated fruit over other treatments. Thus, 1-MCP+CA-2 is a promising alternative to prochloraz for hot-water treated mangoes, suitable for organic export markets.

Medlar fruit: Medlar *Mespilus germanica* L. belongs to the Rosaceae family and is native to southeastern Europe, Anatolia, Crimea, Caucasia, and parts of Iraq and Iran. It grows as a small deciduous tree or shrub, reaching 2–3 m in the wild and 4–6 m when cultivated. In Turkey, especially in north-east Anatolia, Medlar fruit is widely consumed from both wild and cultivated varieties. The fruit is known for its astringency and is traditionally used as a remedy for enteritis, possessing various human healing properties. Selcuk and Erkan 2015 tested Medlar fruit in different storage conditions: 21 per cent O_2 + 0.03 per cent CO_2 PL-1, 2 per cent O_2 + 5 per cent CO_2 PL-2, 3 per cent O_2 + 10 per cent CO_2 PL-3, and modified atmosphere packaging MAP at 0°C for 60 days. They monitored weight loss, firmness, decay, browning, color, pH, acidity, soluble solids, phenolics, flavonoids, tannins, ascorbic acid, antioxidant activity, and sugars every 15 days. MAP maintained 7 per cent CO_2 and 14 per cent O_2, with the lowest weight loss <0.4 per cent at 60 days. PL-2 and PL-1 best preserved fruit quality, delaying ripening and browning. pH increased, acidity decreased, and nutrient levels fluctuated during storage. PL-2 showed higher acidity, soluble solids, sugars, and antioxidants, especially for ascorbic acid and tannins. Overall, 60 days at 0°C in palliflex storage-maintained Medlar fruit properties.

Pear: Goliáš *et al.*, 2016 stored mature pear fruits cv. Zaosuli at 1.0–1.5°C under CA with 2.0 per cent O_2 and 7 per cent CO_2, and regular atmosphere RA with 20.9 per cent O_2 and 0.1 per cent CO_2. They detected and quantified 64 volatile compounds including hydrocarbons, alcohols, aldehydes, ketones, esters, terpenes, lactones, and fatty acids. 13 compounds differentiated the pear treatments, with notable changes in n-hexanol, 2-furaldehyde, cis-geraniol, and α-damascenone after 40 days. CA reduced ethylene production and increased respiration due to elevated CO_2 levels, affecting volatile compound production during storage and post-warming. European pears *Pyrus communis* L. are stored cold to prolong shelf-life, yet they often develop senescent scald and internal breakdown over time. These issues can be mitigated using CA storage or the ripening inhibitor 1-MCP. Lum *et al.*, 2017 examined the impact of

1-MCP and CA storage on ‘Cold Snap’ and ‘Swiss Bartlett’ pears to mitigate post-harvest disorders like senescent scald and internal breakdown. They stored freshly harvested pears at 0°C in refrigerated air or CA 18 kPa or 2.5 kPa O_2, and 2 kPa CO_2 for over 167 days Table 3. Both 1-MCP and CA slowed ethylene production, delayed fruit softening and peel yellowing, and reduced disorders like senescent scald and internal breakdown. Ascorbate levels decreased rapidly in ‘Cold Snap’ but were better maintained with 1-MCP in ‘Swiss Bartlett’. In another similar study on ‘AC Harrow Crisp’ pears.

Flaherty *et al.*, 2018 evaluated amino acids, organic acids, and antioxidants during storage Table 3. Citrate, lactate, and fumarate increased during storage, influenced by 1-MCP differently. Aspartate and glutamate levels were higher with 1-MCP. γ-Aminobutyrate accumulated in damaged fruit. Principal component analysis linked citrate and fumarate changes to internal breakdown and senescent scald, while γ-aminobutyrate and alanine were tied to internal cavities. Antioxidant depletion correlated with all storage disorders. Wang *et al.*, 2021 explored the impact of 10 per cent CO_2 in controlled atmosphere CA packaging on fresh-cut pear respiration and energy metabolism over 5 days. The study found that 10 per cent CO_2 moderated the respiratory rate, maintained mitochondrial integrity, and increased energy charge by 1.6 times compared to the control. Enzymes related to energy metabolism showed varied activity and expression levels. High CO_2 CA packaging inhibited senescence, lowered energy levels, and enhanced stress resistance in fresh-cut pears. This research delved into the biochemical mechanisms of how pears respond to elevated CO_2 levels in CA packaging, focusing on respiration pathways and energy metabolism.

Persimmon fruit: Persimmon fruit, originating from East Asia, is renowned for its vibrant orange color and sweet, honey-like taste. Its distinct flavor, reminiscent of apricots and dates, makes it a popular choice for fresh consumption, baking, and inclusion in salads or desserts. Brackmann *et al.*, 2014 examined the impact of two CO_2 levels in ultra-low oxygen on ‘Fuyu’ persimmon characteristics during 12 weeks of controlled atmosphere CA storage at -0.5°C. They tested CA conditions: [1] 0.15 kPa O_2 + 6.0 kPa CO_2 and [2] 0.15 kPa O_2 + 2.0 kPa CO_2. Respiration CO_2 production accounted for 24 per cent of total mass loss, significantly influenced by the controlled atmosphere. Higher CO_2 2.0 kPa led to changes in respiratory quotient and increased mass loss due to respiration while reducing soluble solids after storage.

Plums: Steffens *et al.*, 2014 studied the impact of controlled atmosphere CA storage on ‘Laetitia’ plums, focusing on internal breakdown and quality preservation. In two experiments in 2010 and 2011, various CA conditions were tested alongside cold storage CS; 21.0 kPa O_2 + 0.03 kPa CO_2. Results showed that CA led to lower respiration and ethylene production, preserved firmness, texture, acidity, reduced skin redness and cracking compared to CS. In 2010, CA with 2+5, 1+5, and 1+3 delayed ripening but had higher internal breakdown. In 2011, 2+1 and 2+2 CA conditions delayed ripening with reduced internal breakdown. The optimal CA condition for 'Laetitia' plums at 0.5°C is 2 kPa O_2 + 2 kPa CO_2, balancing ripening delay and internal breakdown reduction.

Red raspberry: Forney *et al.*, 2015 evaluated nine red raspberry genotypes' market-life under air or controlled atmosphere CA; 12.5 kPa CO_2/7.5 kPa O_2 at 1°C with 95 per cent RH over two harvest years. They assessed firmness, soluble solids, acids, sugars, and volatile composition post-harvest, monitoring decay and physiological breakdown PB. CA significantly delayed decay lag period >45 days vs. 19–29 days in air and PB increased by 1.0 to 9.7 days. Firmness loss was similar in air and CA storage. Sugars were mainly fructose 51 per cent, glucose, 26 per cent and sucrose 23 per cent with acids primarily citric 75 per cent. Fruit volatiles were mostly C13 norisoprenoids 65–93 per cent. Firmness correlated with PB resistance, and PB related negatively to C13 norisoprenoid content. However, sugar and acid content showed no correlation with storage-life.

Strawberry: Fresh strawberries have a short post-harvest life because they are susceptible to mechanical injury, water loss, microbial decay, and physiological deterioration. Their quality attributes decline rapidly after harvesting, and they may only last a day at room temperature or 5–10 days at 0°C. Nakata and Izumi 2020 stored 'Minomusume' strawberries in varying CO_2 atmospheres 20, 30 and 40 per cent using controlled atmosphere CA and active modified atmosphere packaging MAP for 10 days at 5°C Table 3. CO_2 levels of 20 to 40 per cent in CA delayed fungal growth and mold mycelia formation but caused surface black discoloration above 30 per cent CO_2 due to CO_2 injury. MAP conditions stabilized at 20 per cent CO_2 and 2 per cent O_2, maintaining constant fungal counts and diversity. Overall, CO_2 atmospheres in MAP did not affect visual or physico-chemical quality except for firmer fruit with 20 and 30 per cent CO_2. However, upon transfer to ambient conditions, MAP with 30 and 40 per cent CO_2 led to mold growth and black discoloration. The study underscores the delicate balance needed in CO_2 levels to preserve strawberries without inducing surface defects. Alamar *et al.* (2017) studied British strawberries cv. Sonata and Chilean avocados cv. Hass in controlled atmospheres CA of 15 kPa CO_2 + 5 kPa O_2 5°C and 10 kPa CO_2 + 5 kPa O_2 5 or 20°C at various stages of post-harvest. They monitored real-time respiration rate RR, disease severity, color, ethylene production, and firmness. Continuous RR recordings in CA showed that strawberries treated midway with CA were firmer and had better color despite RR fluctuations. Just 2.5 days of CA extended strawberry shelf-life by 3 days. Avocado RR decreased under CA at 20°C, maintaining firmness, while early CA reduced internal discoloration at 5°C. The study suggests targeted CA use can extend shelf-life and reduce waste, with potential energy savings from shorter CA application times.

Conclusion and future trends

Controlled atmosphere storage is a critical technology in the preservation and post-harvest management of fruits. By meticulously regulating the composition of gases, temperature, and humidity within storage environments, this method effectively extends the shelf-life of various fruits, maintaining their quality and nutritional value. The reduction of oxygen levels and the careful management of carbon dioxide and ethylene concentrations slow down the metabolic processes of fruits, thereby delaying ripening and senescence. This technique not only minimizes post-harvest losses but also enhances the marketability and distribution range of fresh produce. The benefits of controlled atmosphere storage are particularly significant for fruits that are highly perishable or that are transported over long distances. For example, apples, pears, and berries show remarkable improvement in storage longevity and quality retention when stored under controlled conditions. Additionally, this method supports the global food supply chain by allowing fruits to reach consumers in distant markets while maintaining their freshness and appeal. However, the successful implementation of controlled atmosphere storage requires precise control systems and a thorough understanding of the specific respiratory characteristics and storage requirements of each fruit type. More research is needed to fully understand how different crops tolerate CA conditions, the effects on quarantine-relevant insect species, and the ideal combinations of temperature, gas composition, and treatment duration. The current knowledge gap regarding the post-CA behavior of fruits highlights the need for further investigation into the metabolic changes induced by CA storage and the development of appropriate handling methods. Additionally, comprehensive studies should examine the use of nonconventional gases such as nitrous oxide, nitric oxide, and carbon monoxide in combination with CA, especially during transit. There is also a need to investigate how various preharvest factors, such as growing environment and season of harvest, influence the post-harvest quality and storage potential of fruits and vegetables under CA conditions. This is particularly important for increasing the commercial use of CA technology for tropical fruits. This will improve the commercial viability and effectiveness of CA storage, ensuring better preservation and quality of produce.

References

Alamar M C, Collings E, Cools K and Terry L A, 2017. Impact of controlled atmosphere scheduling on strawberry and imported avocado fruit. *Post-harvest Biology and Technology*, 134:6-86.

Ali S, Khan AS, Malik A U and Shahid M, 2016. Effect of controlled atmosphere storage on pericarp browning, bioactive compounds and antioxidant enzymes of litchi fruits. *Food Chemistry*, 206:18-29.

Ali S, Khan A S, Malik A U, Anwar R, Anjum M A, Nawaz A and Naz S, 2021. Combined application of ascorbic and oxalic acids delays post-harvest browning of litchi fruits under controlled atmosphere conditions. *Food Chemistry*, 350: 129277.

Ali S, Khan A S, Malik A U, Nawaz A and Shahid M, 2019. Post-harvest application of antibrowning chemicals modulates oxidative stress and delays pericarp browning of controlled atmosphere stored litchi fruit. *Journal of Food Biochemistry*, 43 3 : e12746.

Baha, A and Lichter A, 2018. Effect of controlled atmosphere on the storage potential of Ottomanit fig fruit. *Scientia Horticulturae*, 227: 196-201.

Bekele E A, Ampofo-Asiama J, Alis R R, Hertog M L, Nicolai B M and Geeraerd A H, 2016. Dynamics of metabolic adaptation during initiation of controlled atmosphere storage of 'Jonagold'apple: Effects of storage gas concentrations and conditioning. *Post-harvest Biology and Technology*, 117: 9-20.

Bessemans N, Verboven P, Verlinden B E and Nicolaï B M, 2016. A novel type of dynamic controlled atmosphere storage based on the respiratory quotient RQ-DCA. *Post-harvest Biology and Technology*, 115:91-102.

Bodbodak S and Moshfeghifar M, 2016. Advances in controlled atmosphere storage of fruits and vegetables. In *Eco-friendly technology for post-harvest produce quality* pp. 39-76. Academic Press.

Boonyaritthongchai P, Wongs-Aree C and Kanlayanarat S, 2010. Effects of controlled atmosphere storage on the post-harvest quality of carambola fruits. In *Asia Pacific Symposium on Post-harvest Research, Education and Extension,* 943: 209-212.

Both V, Thewes F R, Brackmann A, de Oliveira Anese R, de Freitas Ferreira D and Wagner R, 2017. Effects of dynamic controlled atmosphere by respiratory quotient on some quality parameters and volatile profile of 'Royal Gala'apple after long-term storage. *Food Chemistry*, 215:483-492.

Brackmann A, Thewes F R, Anese R D O, Both V and Gasperin A R D, 2014. Respiration rate and its effect on mass loss and chemical qualities of'Fuyu'persimmon fruit stored in controlled atmosphere. *Ciência Rural*, 44: 612-615.

Chiabrando V and Giacalone G, 2011. Shelf-life extension of highbush blueberry using 1-methylcyclopropene stored under air and controlled atmosphere. *Food Chemistry*, 126 4 :1812-1816.

Concha-Meyer A, Eifert J D, Williams R C, Marcy J E and Welbaum G E, 2015. Shelf-life determination of fresh blueberries Vaccinium corymbosum stored under controlled atmosphere and ozone. *International Journal of Food Science*, 1-9.

DeEll J R, Lum G B and Ehsani-Moghaddam B, 2016. Effects of multiple 1-methylcyclopropene treatments on apple fruit quality and disorders in controlled atmosphere storage. *Post-harvest biology and technology*, *111*, 93-98.

Dziedzic E, Błaszczyk J, Bieniasz M, Dziadek K and Kopeć A, 2020. Effect of modified MAP and controlled atmosphere CA storage on the quality and bioactive compounds of blue honeysuckle fruits *Lonicera caerulea* L. *Scientia Horticulturae*, 265:109226.

East A R, Araya X T, Hertog M L A T M, Nicholson S E and Mawson A J, 2009. The effect of controlled atmospheres on respiration and rate of quality change in 'Unique'feijoa fruit. *Post-harvest Biology and Technology*, 53(1-2):66-71.

Escobedo-Avellaneda Z and Welti-Chanes J, 2016. Controlled atmosphere storage: applications for bulk storage of foodstuffs. *Encyclopedia of Food and Health*, 301-307.

Flaherty E J, Lum G B, DeEll J R, Subedi S, Shelp B J and Bozzo G G, 2018. Metabolic alterations in post-harvest pear fruit as influenced by 1-methylcyclopropene and controlled atmosphere storage. *Journal of Agricultural and Food Chemistry*, 66(49):12989-12999.

Forney C F, Jamieson A R, Pennell K D M, Jordan M A and Fillmore S A 2015. Relationships between fruit composition and storage life in air or controlled atmosphere of red raspberry. *Post-harvest Biology and Technology*, 110:121-130.

Fragoso A V and Mújica-Paz H, 2016. Controlled atmosphere storage: effect on fruit and vegetables. *Encyclopedia of Food and Health*, 308-311.

Goliáš J, Létal J, Balík J and Kožíšková J, 2016. Effect of controlled atmosphere storage on production of volatiles and ethylene from cv. Zaosuli pears. *Horticulture Science,* 43: 117-125.

Ho P L, Tran D T, Hertog M L and Nicolaï B M, 2021. Effect of controlled atmosphere storage on the quality attributes and volatile organic compounds profile of dragon fruit Hylocereus undatus. *Post-harvest Biology and Technology*, 173:111406.

Khan M R, Suwanamornlert P, Leelaphiwat P, Chinsirikul W and Chonhenchob V, 2017. Quality and biochemical changes of longan Dimocarpus longan Lour cv. Daw fruit under different controlled atmosphere conditions. *International Journal of Food Science* and *Technology*, 52(10): 2163-2170.

Kweon H J, Kang I K, Kim M J, Lee J, Moon Y S, Choi, C and Watkins C B, 2013. Fruit maturity, controlled atmosphere delays and storage temperature affect fruit quality and incidence of storage disorders of 'Fuji'apples. *Scientia Horticulturae*, 157: 60-64.

Latocha P, Krupa T, Jankowski P and Radzanowska J, 2014. Changes in post-harvest physicochemical and sensory characteristics of hardy kiwifruit Actinidia arguta and its hybrid after cold storage under normal versus controlled atmosphere. *Post-harvest Biology and Technology*, 88: 21-33.

Lum G B, DeEll J R, Hoover G J, Subedi S, Shelp B. J and Bozzo G G, 2017. 1-Methylcylopropene and controlled atmosphere modulate oxidative stress metabolism and reduce senescence-related disorders in stored pear fruit. *Post-harvest Biology and Technology*, 129: 52-63.

Lumpkin C, Fellman J K, Rudell D R and Mattheis J P, 2015. 'Fuji'apple Malus domestica Borkh. volatile production during high pCO2 controlled atmosphere storage. *Post-harvest Biology and Technology*, 100: 234-243.

Mditshwa A, Fawole O A and Opara U L, 2018. Recent developments on dynamic controlled atmosphere storage of apples—A review. *Food Packaging and Shelf-life*, 16:59-68.

Mditshwa A, Fawole O A, Vries F, van der Merwe K, Crouch E and Opara U L, 2017. Minimum exposure period for dynamic controlled atmospheres to control superficial scald in 'Granny Smith'apples for long distance supply chains. *Post-harvest Biology and Technology*, 127: 27-34.

Nakata Y and Izumi H, 2020. Microbiological and quality responses of strawberry fruit to high CO2 controlled atmosphere and modified atmosphere storage. *HortScience*, 55(3):386-391.

Prange R K, Wright A H, DeLong J M and Zanella A, 2013. June. A review on the successful adoption of dynamic controlled-atmosphere DCA storage as a replacement for diphenylamine DPA, the chemical used for control of superficial scald in apples and pears. In *XI International Controlled and Modified Atmosphere Research Conference,* 1071: 389-396.

Saquet A A, 2016. Storability of "Jonagold" apple fruit under extreme controlled atmosphere conditions. *Journal of Agricultural Science and Technology, B 6,* 262-268

Schlie T P, Köpcke D, Rath T and Dierend W, 2020. Importance of Individual Apple Fruit Characteristics for DCA-CF Storage on the Variety 'Elstar, PCP'. *Erwerbs-Obstbau,* 62:57-60.

Selcuk N and Erkan M, 2015. The effects of modified and palliflex controlled atmosphere storage on post-harvest quality and composition of 'Istanbul'medlar fruit. *Post-harvest Biology and Technology*, 99: 9-19.

Sen F, Meyvaci K B, Turanli F and Aksoy U, 2010. Effects of short-term controlled atmosphere treatment at elevated temperature on dried fig fruit. *Journal of Stored Products Research,* 46 (1): 28-33.

Serradilla M J, del Carmen Villalobos M, Hernández A, Martín A, Lozano M and de Guía Córdoba M, 2013. Study of microbiological quality of controlled atmosphere packaged 'Ambrunés' sweet cherries and subsequent shelf-life. *International Journal of Food Microbiology*, 166 (1): 85-92.

Singh S P and Pal R K, 2008. Controlled atmosphere storage of guava Psidium guajava L. fruit. *Post-harvest Biology and Technology*, 47(3):296-306.

Sivakumar D and Korsten L, 2010. Fruit quality and physiological responses of litchi cultivar McLean's Red to 1-methylcyclopropene pre-treatment and controlled atmosphere storage conditions. *LWT-Food Science and Technology*, 43(6) 942-948.

Sivakumar D, Van Deventer F, Terry L A, Polenta G A and Korsten L, 2012. Combination of 1-methylcyclopropene treatment and controlled atmosphere storage retains overall fruit quality and bioactive compounds in mango. *Journal of the Science of Food and Agriculture*, 92(4)821-830.

Steffens C A, Amarante C V D, Alves E O and Brackmann A, 2014. Fruit quality preservation of'Laetitia'plums under controlled atmosphere storage. *Anais da Academia Brasileira de Ciências*, 86(1): 485-494.

Stürz S, Sadar N, Folie I and Zanella A, 2021. August. The impact of DCA-CF, 1-MCP and waxing on apple quality under simulated extreme transport and shelf-life conditions. In *XIII International Controlled and Modified Atmosphere Research Conference-CAMA2021* 1386: 87-94.

Teixeira G H and Durigan J F, 2010. Effect of controlled atmospheres with low oxygen levels on extended storage of guava fruit Psidium guajava L.'Pedro Sato'. *HortScience, 45* (6):918-924.

Teixeira G H, Júnior L C C, Ferraudo A S and Durigan J F, 2016. Quality of guava Psidium guajava L. cv. Pedro Sato fruit stored in low-O2 controlled atmospheres is negatively affected by increasing levels of CO2. *Post-harvest Biology and Technology*, 111: 62-68.

Thewes F R, Brackmann A, de Oliveira Anese R, Ludwig V, Schultz E E and Berghetti M R P 2018. 1-methylcyclopropene suppresses anaerobic metabolism in apples stored under dynamic controlled atmosphere monitored by respiratory quotient. *Scientia Horticulturae*, 227: 288-295.

Valle-Guadarrama S, Morales-Cabrera M, Peña-Valdivia C B, Mora-Rodríguez B, Alia-Tejacal I, Corrales-García J and Gómez-Cruz A, 2013. Oxidative/fermentative behavior in the flesh of "Hass" avocado fruits under natural and controlled atmosphere conditions. *Food and Bioprocess Technology*, 6: 272-282.

Wang D, Ma Q, Li D, Li W, Li L, Aalim H and Luo Z, 2021. Moderation of respiratory cascades and energy metabolism of fresh-cut pear fruit in response to high CO2 controlled atmosphere. *Post-harvest Biology and Technology*, 172:111379.

Weber A, Thewes F R, Sellwig M, Brackmann, A, Wünsche, J N, Kittemann, D, and Neuwald, D A, 2019. Dynamic controlled atmosphere: Impact of elevated storage temperature on anaerobic metabolism and quality of 'Nicoter'apples *Food Chemistry*, 298, 125017

Wood R M, Thewes F R, Reynaud M, Kittemann D, Sautter C.K, Wünsche, J N and Neuwald D A, 2022. Apple fruit recovery from anoxia under controlled atmosphere storage *Food Chemistry*, 371, 131152.

Yahia E M, Fadanelli L, Mattè P and Brecht J K, 2019. Controlled atmosphere storage In *Post-harvest Technology of Perishable Horticultural Commodities*, 439-479. Woodhead Publishing.

Zanella A, and Rossi O, 2015. Post-harvest retention of apple fruit firmness by 1-methylcyclopropene 1-MCP treatment or dynamic CA storage with chlorophyll fluorescence DCA-CF *European Journal of Horticultural Science*, 80(1):11-17.

Zhou M, Ndeurumio K H, Zhao L, and Hu Z, 2016. Impact of precooling and controlled-atmosphere storage on γ-aminobutyric acid GABA accumulation in longan Dimocarpus longan Lour fruit. *Journal of Agricultural and Food Chemistry*, 64(33): 6443-6450.

8

Processed Fruit Products for Local and Export Markets

R K Goyal and Uma Prajapati

Maharana Pratap Horticultural University, Karnal-132 001, Haryana, India

Introduction

India is bestowed with varied agro-climatic conditions. Thus, horticultural crops have lot of potential in the country. India has witnessed voluminous increase in horticulture production over the last few years, which has undoubtedly doubled the income of farmers. As a result, the country has emerged as the 2nd largest producer of fruits (102.48 million tonnes) after China during 2020-21 (Anonymous, 2023). The abundance of fruits ranging from tropical delights like mango, banana, guava and papaya to temperate fruits such as apple, peach, plum, apricot and pear, have laid a robust foundation for the growth of the fruit processing industry. This chapter delves into the dynamics of fruit processed products in both local and international markets, exploring the challenges, opportunities and strategies for sustained success. Fruit processing plays a crucial role in adding value to horticultural produce, enhancing shelf-life and meeting the diverse demands of local and international markets. In India, a country blessed with abundant fruit varieties, the processed fruit industry has been witnessing significant growth, catering not only to the domestic market but also making strides in the global arena. The processing of these fruits into various products such as juices, pulps, concentrates, jams and dried fruits has emerged as a thriving industry. This chapter will provide insights into the current state of fruit processed products in India, their importance for both local consumption and export markets, challenges faced by the industry and potential avenues for the future growth.

Importance of fruit processed products

Processing industry, which is completely unorganized sector, is in fast developing phase in the country. These days, only 1.5 to 2 per cent of the total horticultural produce is being converted in value added durable products but value addition to food products is expected to increase from 2 to 25 per cent by the year 2025. It is a great need in today's time to increase the processing and value addition of fruits and vegetables to sort out the problem of post- harvest losses, glut in the market, nutritional shortage and unemployment Goyal, 2017). Food processing sector is considered as the fastest growing sector of the Indian economy. The Fruits and Vegetable Processing Industry (FVPI) in India is considered as one of the largest sectors in terms of production, consumption, economic growth and exports. India's total post-harvest loss ranges from 4.87 to 11.61 per cent in vegetables and 6.02 to 15.05 per cent in fruits (Anonymous, 2022). This loss is majorly contributed by weak cold chain infrastructure and improper post-harvest management practices. The export of processed goods including pickles, sauce, jelly, marmalade, jam and squash has lot of potential. In India, there is a large domestic market for dried and dehydrated fruits. Fruits are abundant, delicate, seasonal and perishable. Just 10 per cent of the India's horticultural output benefits from having a sufficient cold storage facility (Birthare, 2018). Thus, fruit processing offers numerous benefits, including value addition, preservation of seasonal surplus, reduction in post-harvest losses and creation of employment opportunities. A growing awareness of health and wellbeing, rising urbanization and shifting customer preferences have all contributed to the industry's recent steady expansion. The fruit processing sector in India has grown remarkably in the last few years due to the factors like rising disposable incomes, changing lifestyles and more urbanization (Rais *et al.*, 2013). It is quite expected that demand for processed meals that are ready to eat will rise as a result of the rise in working women and changes in their consumption patterns. This shift in the supply side is a result of the growth of organized retail food items and the expansion of global value chains for the processed food industry (Choudhary, 2023).

Because of the country's many agro-climatic zones, fruits are available year-round, supporting ongoing production and processing operations. Important companies in this sector include both big and small producers, each of which offers a range of goods designed to satisfy different customer needs. Processed fruit products also serve as convenient and nutritious options for consumers, contributing to improved dietary patterns (Vikram *et al.*, 2023). There are various methods of food processing through which fruits can be stored for longer duration, which not only provides variation in fruit products but also off-season availability of many fruits and vegetables (Rahman, 2007). Maintaining food quality is necessary to make sure that food with good nutritional values is used, which is crucial for overall public health. Therefore, the best way to maintain food quality and stop it from deteriorating is to use preservation techniques (Dwivedi *et al.*, 2017). Moreover, the export of processed fruit products enhances India's trade potential and foreign exchange earnings.

Overview of India's fruit processing industry

India has diverse climate, which ensures the availability of almost all fruits and vegetables andoffer a plentiful supply of raw materials for the fruit processing industry. India stands second in the production of fruits and vegetables in the world, after China and produced 107.24 million metric tonnes of fruits, while the area under fruits was recorded in 7.05 million hectares during 2020-21 (Anonymous, 2021). In between 2011-12 and 2020-21, the production of fruits in India increases by 34.46 per cent, processed fruits production increased by 48.63 per cent. Amongst fruits, the country ranks first in the production of banana (26.45 per cent), mango including mangosteen and guava (43.80 per cent) and papaya (39.30 per cent) (Anonymous, 2021b). The raw materials for the preservation industry in India are now growing at a compound annual growth rate of 76.22 per cent. The government's suitable fiscal policies and the evolving lifestyles of people nowaday can significantly contribute to the industry's growth (Singh *et al.*, 2021). The market for processed fruits and vegetables is estimated to be worth US$ 74,134.8 million during 2024. People's daily eating habits have altered as a result of the World Health Organization's advice to consume at least 400 grams of fruits and vegetables. Considering sustainable production, processing and packaging techniques, the processed fruit business is growing significantly. With more indigenous and local businesses joining the market, its valuation is expected to increase at a compound annual growth rate of 3.80 per cent to US$ 107,500.4 million by 2034 (Anonymous[c], 2024). The government has also taken proactive measures to reduce industrial bottlenecks in processing by prioritizing processing and storage capacities and providing funding to the industry sector (Murthy and Yogesh, 2014). By creating community canning centres, offering training, supplying the required supplies and setting up commercial processing and packaging units, the central and state governments are supporting the farmers. Although the food processing industry in India has seen significant growth over the last ten years but processing level in the country is still low when compared with massive production. It should therefore be understood that an increase in it would strengthen the Indian economy. Additionally, the government has increased its attention to this industry, given it several tax breaks and encouraged it to go into business (Matapurkar and Arnav, 2011).

This vast production base offers India's tremendous potential for export. During 2022-23, India exported fresh fruits and vegetables of worth Rs. 13185.30 crore or USD 1635.95 million, which comprised fresh fruits of worth Rs. 6,219.46

crore or USD 770.70 million. The processed fruits and vegetables including pulses exported were of Rs. 18,090.80 crore or USD 2,248.96 million, which comprised of processed fruits and juices Rs. 5,944.49 crore or USD 737.81 million in 2022-23 (Sharma *et al.*, 2016).

Majority of the fruits exported from the nation are grape, pomegranate, mango, banana and orange, etc. The United Arab Emirates, Bangladesh, Nepal, Malaysia, the Netherlands, Sri Lanka, the United Kingdom, Qatar, Oman and Iraq are among the main destinations for Indian fresh fruits. The USA, UAE, Bangladesh, China, Saudi Arabia and the Netherlands are the main destinations for Indian processed fruits and vegetables (Reddy and Patel, 2023).

India's horticultural produce is becoming more and more popular, although having a meager 1 per cent market share worldwide. This is the result of simultaneous advancements in the fields of quality control and cold chain infrastructure. In addition to significant contributions from the private sector, the government has launched APEDA to establish many centres for perishable cargoes and integrated post-harvest processing facilities across the nation. Initiatives aimed at increasing capacity among farmers, processors and exporters have also aided in this endeavor (Kumar, 2022).

When it comes to output, growth, consumption and exports, the food processing industry is one of the biggest in India. India's food processing industry includes fruit and vegetable processing, spices, plantations and other consumer product categories like mineral water, high-protein food items, confections, chocolate and cocoa products, soy-based products and mineral water industries. The government has approved plans for joint ventures, international cooperation, industrial licences and investment proposals from 100 per cent export-oriented enterprises. An estimated Rs. 7194.13 crores, or USD 895.34 million, wasinvested in the food processing industry by foreign direct investment in 2022-23 (Anonymus, 2023).

India's exports of processed food was Rs. 59,580.72 crore or USD 7,408.64 million in 2022-23, which includes the share of products like mango pulp (Rs. 1189.66 crore or USD 147.60 million), processed fruits, juices and nuts (Rs. 4754.83 crore or USD 590.21 million), cocoa products (Rs. 1242.13 crore or USD 154.54 million), otherpreparations (Rs. 8889.18 Crore or USD 1106.42 million), *etc*. The food processing sector in India is mostly focused on exports. India has the distinct benefit of being connected to Europe, the Middle East, Japan, Singapore, Thailand, Malaysia and Korea due to its unique geographic location. India's geographic advantage is demonstrated, for instance, by the importance of commerce in processed foodstuffs and agricultural products between India and the Gulf (Anonymous[a], 2023).

India is undergoing a period of transition involving one of the biggest industries in the world economy. It is projected to expand between 2022 and 2030 at a 3.0 per cent Compound Annual Growth Rate (CAGR). A primary source of the food processing industry's lack of competitiveness is the expense and caliber of its marketing channels. Superstores account for more than 72 per cent of food sales globally. India offers enormous potential and is poised for a significant shift in retail. India has the least amount of small organized retail competition and the least saturated global market overall (Shingare and Kanoi, 2020).

Product categories for local and export market

Product categories for local market:

Pickle: Pickling is the process of preserving fruit in vinegar and salt. Pickles can be made partially or completely fermented, or not at all. Spices, sugar, edible oil, jaggery and other ingredients are added to increase the product's flavour and palatability. Pickles are a great appetizer because they stimulate the flow of stomach fluids, which aid in digestion. Pickles can have different nutritional values depending on the kind of raw material and preparation technique, *e.g.*, fermentation or not. The production of pickles has grown to be a national business. The commercial items sold in India's local markets include mango pickle, lime/lemon pickle, *aonla*, *etc*. (Behera *et al*., 2020).

***Squash*:** Fruit squash is made of fruit pulp or juice that has been sweetened with cane sugar. Indian Food Laws stipulate that fruit squash must have a minimum of 25 per cent fruit juice or pulp and a maximum of 40 per cent total solublesolids. Before serving, squash is usually diluted in a ratio 1-3 with water. During summer, commercial items such as lime, lemon, mango, orange, guava, jamun, *bael* and pineapple squash are produced for local Indian markets (Punam, 2009).

Nectar: This kind of fruit beverage is heat-processed to retain at least 20 per cent fruit juice or pulp and 15 per cent total soluble solids. It is served without being diluted. It is regarded as a popular drink in the summer and is commercially accessible in the Indian local market (Nisha *et al*., 2018).

RTS: This kind of fruit beverage has a minimum of 10 per cent total soluble solids and at least 10 per cent fruit juice (5 per cent juice in the case of lime drink). It is widely accessible in all local Indian markets and is referred to as a *ready to serve drink* since it is not diluted before serving, as in the case of mango, guava and pineapple drinks, among others (Punam, 2009).

***Cordial*:** It is a sparkling, transparent, sweetened fruit juice that has been fully extracted from pulp and other suspended materials. It has a minimum of 25 per cent juice and 30 per cent total soluble solids. It may also be used with alcoholic beverages, such as lime juice cordial and is widely commercialized in local Indian markets (Malik *et al*., 2022).

***Chutney*:** One of the key items made from fruits and vegetables is chutney. In these items, preservatives and flavour enhancers such as salt, spices, sugar and acid are added. These goods are renowned for being tasty and pleasant to the palate. The raw components used to make these items include mango, apple, plum, apricot, *etc*. Chutneys made from mango, plum and mixed fruits are highly flavoured. The process for making fruit chutney is the same as that for making jam, with the addition of vinegar, spices and salt. The best option is to add the spice vinegar essence instead of the real spices. In the finished product, the chutney must have at least 40 per cent fruit (w/w), 50 per cent total soluble solids and not more than 2.1 per cent acidity. This is among the most marketable products in local market (Kumar *et al.,* 2020).

***Fruit preserve*:** These are big chunks of fruit or entire fruits covered in a thick sugar syrup that is frequently quite jellied. Almost any fruit may be turned into preserves, including apple, pear, plum, *aonla*, peach, grape, muscadine and quince. Firm-ripe fruits are preferable to soft-ripe fruit for storing it. For the fruit to cook uniformly, its bits and size should be uniform. In India, preserves are commonly found in local markets (Kamboj *et al*., 2020).

Candy: A candied fruit is a fruit that has been impregnated with glucose and cane sugar, drained and then dried. Fruit candies are mainly prepared from aonla, citrus, ber, apple, strawberries, mango, *etc*. It is quite popular in Indian markets and can be formulized in various forms such as crystallized, glazed, *etc*. (Rajanya and Singh, 2021).

***Syrup*:** It is prepared by using minimum of 25 per cent fruit juice or pulp and sweetened by using cane sugar. Syrup can be prepared from pomegranate, jamun, bael, rose, mango, pineapple, strawberry, *etc*. It is widely used to prepare juice, shake, ice-cream, cake, *etc*. This is also available in local markets of India.

***Jam*:** Jams are made of fruit pulp or juice, sugar and additional pectin. They are semi-solid fruit products. They can be made from mixed fruits, mango, apple, guava, pineapple, or any combination of fruits. Jam should have a minimum of 68 per cent sugar overall. The processed product is extensively available in both foreign and Indian markets (Fugel *et al*., 2005 and Biswas, 2021).

***Jelly*:** Jelly is crystal clear jam, produced by using clear fruit juice instead of fruit pulp. It should be mainly prepared from fruit containing higher pectin, *e.g.*, guava, papaya, mango, *etc*. Fruit jelly ismost popular product among children and is widely available in local markets of India (Souza *et al.*, 2022).

***Marmalade*:** The primary ingredient in marmalade is transparent citrus juice, containing tiny pieces of peel floating in it. Fruits such as lime, orange, grapefruit, lemon and orange are frequently utilized for making marmalade. Citrus fruit should make up at least 20 per cent of the fruit content and jam-like sugar levels are appropriate. Marmalade has both international and domestic markets in India (Fugel *et al.*, 2005).

***Dehydrated products*:** Different types of fruit are dehydrated, *viz. ber*, *lasoda/ gonda*, apple, mango, *etc*. These dried fruits can be used form making juices, soups and can be rehydrated to be used as raw fruit. It has wide market in India and can be sold where there is harsh climate (Sagar and Kumar, 2010).

***Sauce*:** Similar ingredients and preparation methods apply to sauces as well as chutneys. Except that after heating, the fruit pulp is sieved to remove skin, seeds and stalks, giving the finished product a silky consistency in sauces. Compared to fruit chutney, the sauce has a thinner consistency. There are two types of sauce. Vinegar extract of herbs, spices and other flavourings make up the majority of thin sauces with low viscosity, whereas, thick sauces with high viscosity are found elsewhere. Sauces can be prepared from wide range of vegetables such as tomato, hot pepper, pumpkin, *etc*. It is widely available in local markets of India (Garcia-Casal, 2016).

***Ketchup*:** Tomato juice is used to make ketchup, which is then concentrated with seasonings/spices, sugar and salt. A muslin cloth is used to loosely bind spices like cloves, cardamom, pepper, cinnamon and other ingredients and the spices are then added to boiling liquid in a steam jacketed kettle. As tomato ketchup is a common addition to many Indian dishes, it is a well-liked commodity in Indian local markets (Garcia-Casal, 2016).

***Tomato puree*:** To prepare tomato puree, the tomato juice is vacuum-concentrated to a total soluble solids content of 9 to 12 per cent. The mixture is poured into bottles, sealed with a crown cork, heated to a boil for thirty minutes, then let to cool. In both domestic and export markets, this product has great popularity and is frequently utilised throughout the off-season (Thakur *et al.*, 1996).

Product categories for export market

Processed fruits, juices and nuts: The fruit and vegetable processing sector in India is largely decentralized and has a large capacity. India has a wide range of agro-climatic zones that enable the cultivation of nearly all types of fresh fruits and green vegetables. In 2021, India produced 107.85 million tonnes of fruits, out of a total of 912.22 million tonnes worldwide (Anonymous[b], 2021). During 2022–2023 the country exported 388,207.95 MT of processed fruits, juices and nuts to the globe for a total value of Rs. 4,754.83 Crore or USD 590.21 million. The main destinations were the Netherlands, the United States of America, Saudi Arabia, the United Arab Emirates and Russia (Parte *et al.*, 2023).

The individual products under this sub-head are as below (APEDA, 2023).

Product name	Export destinations
Apple juice	United Arab Emirates, Bangladesh, Nepal, Malaysia, the Netherlands, Sri Lanka, the United Kingdom, Qatar, Oman and Iraq.
Dried apples	United Arab Emirates, Bangladesh, Nepal, Malaysia, the Netherlands, Sri Lanka, the United Kingdom, Qatar, Oman and Iraq.
Fruit andnuts	USA, Saudi Arab, United Arab Emirates and Russia
Grape juice	USA, Indonesia, Nepal, Brazil and United Arab Emirates.
Cherries	USA, Saudi Arab, United Arab Emirates and Russia.
Pineapple juice	USA, Saudi Arab, United Arab Emirates and Russia.
Lemon juice	USA, UK, Philippines, Germany and Thailand.
Grapefruit juice	USA, Indonesia, Nepal, Brazil and United Arab Emirates.
Dried apricots	USA, Saudi Arab, United Arab Emirates and Russia.
Jam jellies of other fruits	United States, United Arab Emirates and Netherlands.
Jam jellies of apple	United States, United Arab Emirates and Netherlands.
Mango juice	USA, Indonesia, Nepal, Brazil and United Arab Emirates.
Olives	USA, Saudi Arab, United Arab Emirates and Russia.

Mango pulp: A specific variety of fresh mango fruit is used to make mango pulp. Harvested mangoes when fully grown, are swiftly sent to the fruit processing unit where they are examined and cleaned. Whenthe mangoes are fully ripened, they are blanched, pulped, deseeded, centrifuged, homogenized, concentrated where needed, thermally treated and aseptically packed to ensure sterility. Only the best fruits are sent to the controlled ripening chambers. The steps involved in the preparation include packing, de-stoning, refining and cutting. When a product is aseptic, the pulp is packed in aseptic bags after being sterilized. Retorted and hermetically sealed, the refined pulp is likewise packaged in cans.In plate freezers, frozen pulp is pasteurized and deep-frozen.

The procedure makes sure that the finished product retains the fruit's inherent taste and perfume. Mango pulp/concentrate is ideal for processing into a wide range of beverages, including drinks, nectars, jams, fruit cheese and juices. It may also be used to create the most delectable ice creams, yoghurt and confections, as well as puddings, bakery fillings, kid-friendly fruit dinners and flavouring for the food sector.There are two primary clusters of mango pulp in the nation, each with around 65 processing units and a strong backward connectivity between the Alphonso and Totapuri varieties of mango. These clusters are situated in Tamil Nadu State of Krishnagiri and the Andhra Pradesh state of Chittoor. Some of the processing units are in the state of Maharashtra and Gujarat (Thakor, 2019; Thawab *et al.*, 2019). India is also a major exporter of mango pulp in the world. The country has exported 109501.38 MT of mango pulp to the world for the worth of Rs. 1189.66 crore or USD 147.60 million during the year 2022-23 (Anonymous, 2023).

Varieties used for processing: Alphonso, Totapuri and Kesar. Major export destinations are Saudi Arab, Netherland, USA, UK and Arab Emirates.

Cocoa products: Commercial plantation crops include cocoa. After fermentation, the cocoa fruit's seeds produce flavoured cocoa powder. Cocoa is generally cultivated as inter- or multi-crop. Its ability to coexist with tall-growing coconut and arecanut palms in alleyways and blend in with the micro-climatic conditions found in these perennial gardens facilitates its production without requiring an exclusive growing environment of its own. (Karpagalakshmi and Muthusamy, 2019). Cocoa beans, cocoa powder (coating sugar), cocoa shell husks and skins and cocoa butter fat and oil are the main processed goods made from cocoa. India's primary cocoa-growing regions include Kerala, Andhra Pradesh, Karnataka and Tamil Nadu (Anonymous, 2023). The country has exported 34,249.85 MT of cocoa products to the world for the worth Rs. 1,242.13 crore or USD 154.54 million during the year 2022-23. Major Export Destinations in 2022-23 were USA, Indonesia, Nepal, Brazil and United Arab Emirates.

Market dynamics

The demand for fruit processed products in India is influenced by several factors, including changing consumer preferences, dietary trends, health consciousness and affordability. Additionally, government initiatives promoting food processing, investments in infrastructure and technological advancements have spurred the growth of the industry. In the export market, factors such as quality standards, packaging, pricing and market access play pivotal roles in determining competitiveness (Goyal, 2019).

Trends in production, processing and wastage of fruits

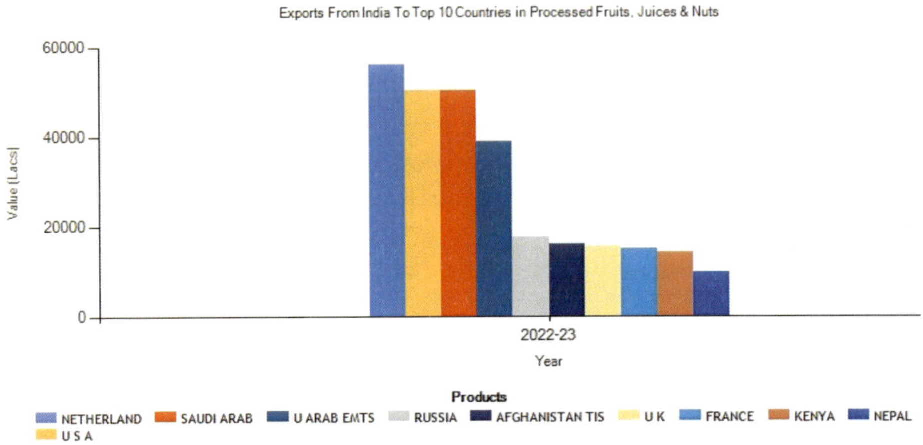

Fig. 1: Export from India to top 10 countries in processed fruits, juices and nuts (Anonymous, 2023).

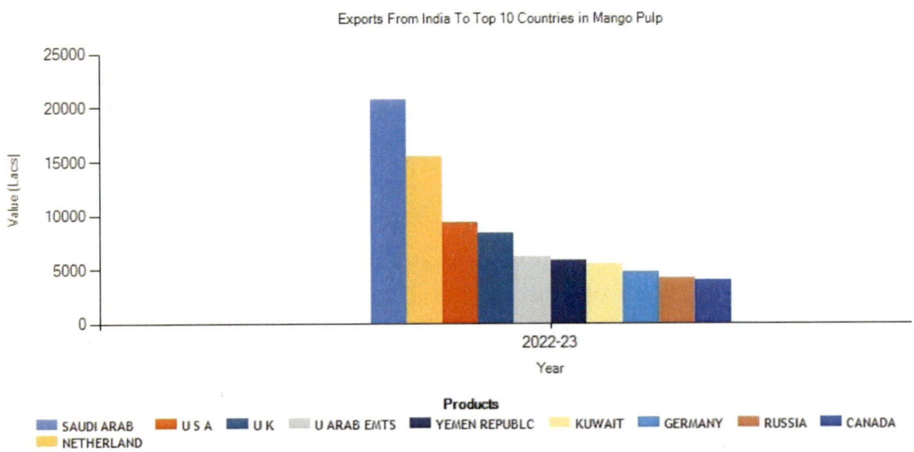

Fig. 2: Export from India to top 10 countries in mango pulp (Anonymous, 2023).

Local market dynamics: A significant portion of processed fruit products caters to the local market, where traditional preferences, cultural nuances and affordability play pivotal roles. This will examine the popular local products, regional variations and the impact of socio-economic factors on consumption patterns. Case studies of successful local brands and their market strategies will be explored (Halder and Pati, 2011).

Export market landscape: India has emerged as a key player in the global fruit processing market, with its products reaching various corners of the world. This analyzes the market demands, export trends and the regulatory landscape governing international trade. Special attention will be given to the challenges faced by Indian exporters and the strategies adopted to overcome them (Priyanka and Kerur, 2023).

Processing techniques and technologies

Understanding the intricate processes and technologies involved in fruit processing is crucial for both local and export-oriented enterprises. Fruit and vegetable processing techniques and technologies play a vital role in ensuring food safety, extending shelf-life and preserving nutritional quality. These processes involve various methods such as cleaning, sorting, cutting, blanching, pasteurization, dehydration, freezing and packaging. Cleaning and sorting remove impurities and damaged produce, ensuring product quality. Cutting and slicing facilitate uniformity and enhance presentation. Blanching helps in enzyme inactivation and colour retention, while pasteurization ensures microbial safety. Dehydration removes moisture, preserving fruits and vegetables for extended storage periods. Freezing maintains the nutritional integrity and texture of produce. Finally, packaging protects against contamination and maintains product freshness. Advancements in processing technologies, including high-pressure processing, vacuum packaging and controlled atmosphere storage, have further improved product quality and extended shelf-life, meeting consumer demands for convenient and nutritious food options. Overall, these techniques and technologies contribute to the growth and sustainability of the fruit and vegetable processing industry, meeting the diverse needs of consumers while reducing post-harvest losses (Miller *et al.*, 2023; Indriyani *et al.*, 2023).

Quality assurance and regulatory compliance

Quality assurance and regulatory compliance are paramount considerations for exporting fruit and vegetable processed products. Ensuring adherence to stringent quality standards and regulatory requirements is essential to maintain consumer confidence, meet international trade regulations and access global markets. Quality assurance measures encompass various stages of production, including sourcing raw materials, processing, packaging and transportation. These may involve implementing Good Manufacturing Practices (GMP), Hazard Analysis and Critical Control Points (HACCP) and ISO certifications to ensure food safety and quality. Additionally, compliance with international standards such as Codex Alimentarius, as well as country-specific regulations and import requirements, is crucial. Documentation, labeling and traceability systems must be in place to provide transparency and facilitate product tracking.

Regular audits and inspections by regulatory authorities and certification bodies help in verifying compliance with established standards. By prioritizing quality assurance and regulatory compliance, exporters can enhance product credibility, mitigate risks and capitalize on opportunities for market expansion and sustained growth in the global fruit and vegetable processed products trade (Sudheer and Indira, 2021; Gulati *et al.*, 2022).

Challenges and opportunities

Despite its growth prospects, the fruit processing industry in India faces challenges such as inadequate infrastructure, lack of cold chain facilities, inefficient supply chain management and quality control issues. Addressing these challenges requires concerted efforts from both the government and private sector stakeholders. Furthermore, there are immense opportunities for innovation, product diversification, value chain integration and market expansion, especially in emerging segments like organic and functional fruit products.

Future trends and recommendations

The future-prospects of the Indian fruit processing industry in both domestic and local markets are promising due to increasing demand for convenient and healthy food options, driven by urbanization and rising disposable incomes. Technological advancements in processing methods and packaging enhance product quality and shelf-life, while government initiatives support infrastructure development and value addition in the agriculture sector. Export opportunities and consumer preferences for ethnic flavours further contribute to growth potential. To capitalize on these opportunities, stakeholders must focus on quality assurance, innovation and market development strategies.

Conclusion

In conclusion, fruit processed products play a vital role in India's agriculture-based economy, offering significant economic, social and nutritional benefits. The industry's continued growth hinges on overcoming existing challenges and capitalizing on emerging opportunities. With strategic interventions, including policy support, infrastructure development and technology adoption, India can further enhance its position as a leading player in the global fruit processing market. India's fruit processing industry stands at a crucial juncture, poised for further expansion and global recognition. This chapter serves as a comprehensive guide, offering insights into the industry's intricacies, challenges and opportunities, with the aim of fostering a sustainable and thriving future for fruit processed products in India.

References

Anonymous, 2021. *Annual Report*. Agricultural and Processed Food Products Export Development Authority (APEDA), IARI, Pusa, New Delhi.

Anonymous, 2021. 3rd Advance Estimates of National Horticulture Database, India.

Anonymous, 2021b. Food and Agriculture Organization.

Anonymous, 2022. Study to Determine the Post-Harvest Losses of Agri Products in India. NABCONS. Ministry of Food Processing Industries, Government of India.

Anonymous, 2023a. Report. Ministry of Food Processing Industries, Government of India.

Anonymous, 2023. *Annual Report*. Agricultural and Processed Food Products Export Development Authority (APEDA), IARI, Pusa, New Delhi.

Anonymous, 2023b. Horticultural Statistics at a Glance. Ministry of Agriculture and Farmers Welfare, Government of India, New Delhi.

Anonymous, 2024. Processed Fruits and Vegetable Market. Processed Fruit and Vegetable Market Outlook from 2024 to 2034. Future Market Insights.

Behera S S, El-Sheikha A F, Hammami R and Kumar A, 2020. Traditionally fermented pickles: How the microbial diversity associated with their nutritional and health benefits. *Journal of Functional Foods,* 70:103971.

Birthare S, 2018. Scope, Status and importance of fruits and vegetables industries in India and role of FandV in Indian Economy. Working paper of the Department of Horticulture, Government of India, APEDA, p.14.

Biswas A, 2021. A review paper on fruit jam products. *Energy (Kcal)*, 3: 57.

Choudhury A, 2023. Role of fruits and vegetables processing in Indian economy. *Ymer*, 22(8): 222-238.

Dwivedi S, Prajapati P, Vyas N, Malviya S and Kharia A, 2017. A review on food preservation: Methods, harmful effects and better alternatives. *Asian J. Pharm. Pharmacol.*, 3(6):193-199.

Food Safety and Standards Authority of India (FSSAI), 2020. Guidelines for Registration and Licensing of Food Business.

Fugel R, Carle R. and Schieber A, 2005. Quality and authenticity control of fruit purées, fruit preparations and jams- A review. *Trends in Food Science and Technology*, 16(10): 433-441.

Future Market Insights, 2024. Processed Fruits and Vegetables Market. Retrieved from https://www.futuremarketinsights.com/reports/processed-fruits-and-vegetables-market.

Garcia-Casal M N, Peña-Rosas J P and Malavé H G, 2016. Sauces, spices and condiments: definitions, potential benefits, consumption patterns and global markets. *Annals of the New York Academy of Sciences*, 1379(1):3-16.

Goyal R K, 2017. Post-Harvest Management and Value Addition of Fruits and Vegetables. in Souvenir cum Proceedings National Symposium on Innovations in Horticulture: Production to Consumption. GB Pant University of Agriculture and Technology, Pant Nagar. 171-183.

Goyal S K, Agarwal S and Goyal M, 2019. An analysis of export growth and instability of processed foods from India.*Journal of Community Mobilization and Sustainable Development,* 14(3):579-585.

Gulati A, Ganguly K and Wardhan H, 2022. Agricultural value chains in India: Ensuring competitiveness, inclusiveness, sustainability, scalability and improved finance. *Springer Nature*, (p. 298).

Halder P and Pati S, 2011. A need for paradigm shift to improve supply chain management of fruits and vegetables in India. *Asian Journal of Agriculture and Rural Development*, 1(3): 1-20.

Indriyani I, Hasanah L N, Auliani R, Hertini E S and Suharmono S, 2023. Utilization of Fruit and Vegetable Processing Technology: Supporting Sustainability, Waste Reduction and Improved Nutrition. *West Science Interdisciplinary Studies*, 1(09):791-798.

Kamboj V, Tripathi S, Joshi U and Tewari V, 2020. Underutilized fruits crops and technologies for meeting their market needs. *Journal of Post-harvest Technology*, 8(4):64-76.

Karpagalakshmi S and Muthusamy A, 2019. A study on export performance of cocoa products in India. *International Journal for Innovative Research in Multidisciplinary Field,* ISSN, 2455-0620.

Kumar A, 2022. Enhancing Export Potential of Indian Fresh Farm Produce Through Logistics and Supply Chain Innovations. *IUP Journal of Supply Chain Management*, 19(4):132-136.

Kumar S, Gehlot R, Sindhu R and Singh R, 2020. Studies on development and evaluation of guava-papaya chutney. *Journal of Pharmacognosy and Phytochemistry*, 9(3):1282-1284.

Malik F, Nadeem M, Ainee A, Kanwal R, Sultan M, Iqbal A and Algarni E H A, 2022. Quality evaluation of lemon cordial stored at different times with microwave heating (pasteurization). *Sustainability*, 14(4):1953.

Matapurkar Aneesh and Sinha Arnav, 2011. "Cold chain: Finally warming up to India", *Infrastructure Today*, pp 22-25.

Miller F A, Brandão T R and Silva C L, 2023. New approaches for improving the quality of processed fruits and vegetables and their by-products. *Foods,* 12(7):1353.

Murthy T M S and Yogesh M S, 2014. An overview of the food processing industry in India: challenges and opportunities. *Online Interdisciplinary Research Journal*, 4(5):187-193.

Nisha C, Kurrey V K, Minz R R and Omesh T, 2018. Guava nectar as a refreshing beverage an overall review. *Plant Archives*, 18(1):1163-1169.

Parte J, Tripathi P, Thingbaijam L, Patel M and Pandey S, 2023. Export Status of Fruits and Vegetables from India. *Asian Journal of Agricultural Extension, Economics and Sociology*, 41(9):983-988.

Priyanka T and Kerur N M, 2023. Growth and instability in area, production and productivity of major fresh fruits and vegetables in India. *The Pharma Innovation Journal*, 12(9):1484-1488.

Punam P, Rakesh Gehlot R G, Singh R and Siddiqui S, 2009. Development and evaluation of bael-mango ready-to-serve drink and squash.*Haryana Journal of Horticultural Sciences*, 38 (3/4):232-2361.

Rahman M S, 2007. Food Preservation: Overview: Handbook of food preservation. 2nd Edition, pp. 3-7, London, CRC Press.

Rais M, Acharya S and Sharma N, 2013. Food processing industry in India: S and T capability, skills and employment opportunities. *Journal of Rural Development*, 451-478.

Rajanya D R and Singh G, 2021. Recent trends in osmotic dehydration of fruits: A review. *Plant Archives,* (09725210), 21(1).

Reddy R and Patel D B, 2023. Export Potential and Competitiveness of Processed Food Products from India. *Asian J. Agric. Ext. Econ. Soc.*, 41(9):722-729.

Report: Human Resource and Skill Requirements in the Food Processing Sector, 2022.

Sagar V R and Suresh Kumar P, 2010. Recent advances in drying and dehydration of fruits and vegetables: A review. *Journal of Food Science and Technology*, 47:15-26.

Sharma A, Singh B K and Anand N, 2016. Fruit processing industry in India: a short review. *Cold Chain Logistics in Horticulture and Agriculture*, 1-17.

Shingare A D and Kanoi T, 2020. India 2030: the decade ahead. *International Journal for Research in Applied Science and Engineering Technology*, 8(7):1503-1514.

Singh G, Daultani Y and Sahu R, 2021. Investigating barriers to growth in the Indian food processing sector. *Opsearch,* 59:441-459.

Souza P B A, Santos M D F, Carneiro J D D S, Pinto V R A and Carvalho E E N, 2022. The effect of different sugar substitute sweeteners on sensory aspects of sweet fruit preserves: A systematic review. *Journal of Food Processing and Preservation*, 46(3):e16291.

Sudheer K P and Indira V, 2021. Importance of Quality and Safety of Processed Products. In Entrepreneurship and Skill Development in Horticultural Processing, CRC Press, 367-394.

Thakor N J, 2019. Indian mango–production and export scenario. *Peach*, 18(107):0-12.

Thakur B R, Singh R K and Nelson P E, 1996. Quality attributes of processed tomato products: A review. *Food Reviews International*, 12(3):375-401.

Thawab M A, Thowseaf S, Millath M A and Ali K M, 2019. Reconnoitering the Impact of Economic Variables on Fruit Pulp Export from Tamil Nadu. *International Journal of Recent Technology and Engineering,* 8(4):58-65.

Vikram B, Gangwar S, Belwel A, Mathur R, Kumar P, Kishor B, Kumari D. and Sikarwar P S, 2023. A Review on Post-harvest management and value addition of horticultural crops: a source of income generation for the farmers of Bundelkhand, India. *International Journal of Environment and Climate Change*, 13(11): 4662-4672.

9

Potential of Value Added Fruit Products and their Biochemical Changes During Storage

Alok Kumar Gupta, Akath Singh and T Damodaran

ICAR-Central Institute for Subtropical Horticulture, Rehmankhera P.O. Kakori, Lucknow-226 101, Uttar Pradesh

Introduction

The horticulture sector is contributing over 30 per cent share in agricultural GDP of the country. Presently, India producing about 351.92 million tons of horticultural produce from 28.04 million hectare which has surpassed the food grain production for last five consecutive years (Anonymous, 2023). Horticulture sector has been identified by Government of India also identified horticulture as one of the vital sector for entrepreneurship development, employment generation, livelihood and nutritional security, doubling farmer's income, achieving SDGs of the United Nations.

Fruits and vegetables have become part of a modern healthy lifestyle as they possess high amounts of minerals, especially electrolytes, dietary fiber, phytochemicals, especially antioxidants, and vitamins (Yahia *et al.*, 2019; Krishnan Kesavan *et al.*, 2023). In consumption, the qualities of products play a critical role in consumers' choices, which are directly related to health problems (Tian, and Xu, 2022). Bioactive compounds are phytochemical substances present in small quantities in plant materials that can provide biological activity, being generally associated with health benefits, including hepatoprotective, anti-inflammatory, antioxidant, antithrombotic, anticarcinogenic, antidiabetic and prevention of cardiovascular and degenerative diseases (Campoli *et al.*, 2018). Therefore, maintaining high-quality fruits and vegetables is very important for the fruit and vegetable industry. This requires some methods to evaluate the quality of fresh fruits and vegetables during their post-harvest manipulation and storage and before their industrial processing as well (Tian and Xu, 2022).

Food processing is needed to guarantee preservation, availability across geography and seasons, and convenience. Moreover, processing can guarantee food products with better quality attributes, including sensorial and health aspects. Although, food processing can degrade nutrient and bioactive compounds, it can also be a valuable tool to manage digestive barriers and increase the health aspects of plant-based foods (Verkempinck *et al.*, 2020; Rojas *et al.*, 2021). In fact, processing can be seen as a series of structural, physical, chemical and biochemical transformations in food products, which impact their properties. Therefore, food processing can degrade or not the nutrients and bioactive compounds, depending on the product and processing conditions (Sicard *et al.*, 2023). Generally speaking, the quality of fresh fruits and vegetables must be evaluated according to the requirements of consumers. The comprehensive quality evaluation of fresh fruits and vegetables includes external quality evaluation and internal quality attribute evaluation (Akter *et al.*, 2024). External quality factors, such as size, shape, texture, color, surface defects, and appearance, are perceived by the consumer's senses, like sight, taste, smell, and touch. As the external quality of fresh fruits and vegetables can be directly perceived by consumers, these appearance factors directly affect the selling price of fresh fruits and vegetables and consumers' purchase behavior (Suntharalingam *et al.*, 2023). The internal qualities of fresh fruits and vegetables, such as maturity, sugar content, acidity, firmness, oil content, internal defects, and tissue breakdowns, are invisible but the most important quality attribute of agricultural products. It is often difficult to evaluate the internal quality attributes of fresh fruits and vegetables (Tian, and Xu, 2022).

Nonthermal techniques provide promising alternative to thermal processing, which is commonly used for microbial and enzyme inactivation in foods despite of its largely negative effects on bioactive compounds and undesirable changes to the sensory and nutritional characteristics of foods. The temperatures applied with nonthermal techniques do not exceed 50 °C, effectively reducing the destructive effects of high temperatures on nutrients and other quality attributes (Al-Juhaimi *et al.*, 2018). Nonthermal food processing methods are being applied with increasing frequency, and therefore, it is imperative to determine their effects on bioactive compounds in fruits and vegetables and compare their efficacy with that of conventional methods (Barbosa-Cánovas *et al.*, 2022). Therefore, the preliminary objective of this chapter was to summarize the effect of various processing conditions on the biochemical properties of fruits. Thus, both thermal and nonthermal processing conditions were taken into consideration to evaluate the effect on biochemical properties. The chapter also aims to explore and suggest some strategies for fruits and vegetables handling and processing to ensure maximum consumer health benefits. This chapter, therefore, can be helpful for the researchers and scientist to carry out their research in the fruit processing sector.

Status of fruit processing in India

The fruit processing industry in India is highly decentralized having wide capacities. The diverse agro-climatic zones make it possible to grow almost all varieties of fresh fruits in India. India is the second-largest producer of Fresh Fruits in the world (ranks next to China) and accounts for about 12 per cent of the world's production of Fresh Fruits in 2021 (Source: FAO). Since liberalization and withdrawal of excise duty on fruit products there has been a significant rise in the growth rate of the industry. Out of 912.22 million tons of fruits production (food primary, Source: FAO) in the world, India accounts for 107.85 million tons in 2021.The country has exported 388,207.95 MT of processed fruits, juices and nuts to the world for the worth of Rs. 4,754.83 Crores/ 590.21 USD Millions during the year 2022-23. The fruit processing industry in the country is remains underdeveloped, with processing accounting for only around 2 per cent of fruits and vegetables. However, India witnesses nearly 4.6-15.9 per cent wastage in fruits and vegetables annually, due to lack of modern harvesting practices and inadequate cold chain infrastructure. Further, the actual production of processed products stood at only 1.33 million tones with a less than 35 per cent capacity utilization of the installed processing units. There are 70 per cent units in home/cottage/ small scale sector and 30 per cent units in large scale sector with capacity of 250 tonnes per annum and 30 tonnes/hr, respectively. Large scale contributes 70 per cent of the production. So, there is huge opportunity to invest in this sector that will also help in reducing wastage by converting in various value added processed products. The total utilization of harvested horticultural produce during processing can be explored by linking processing, pharmaceuticals and cosmetic industries by converting waste into wealth.

Table 1: Major states contributing to fruit processing.

States	Number of processing unit
Maharashtra	182
Andhra Pradesh	143
Gujarat	85
Uttarakhand	32
West Bengal	30

Source: MOFPI

Table 2. Fruits and vegetables used for processing in different countries.

Country	% processing
Malaysia	83
Saudi Arabia	30
Philippines	78
Brazil	70
Australia	60
USA	40
UK	50
India	2.20

Fruits and vegetable processing

India is the world's second largest producer of fruits and vegetables. It has potential to grow all types of temperate, sub - tropical and tropical fruits and vegetables because of varied agro - climatic diversity. The processed products from fruits and vegetables are beverages, jams, jellies, candies, preserves, canned fruits and vegetables, dehydrated fruits and vegetables, pickles, soup mixes, sauces and ketchups. Products that have growing demand, especially in the Middle East countries include pickles, chutneys, fruit pulps, canned fruits and vegetables, concentrated pulps and juices, dehydrated vegetables and frozen fruits and vegetables. The production of pickles and chutneys has traditionally been rural level cottage industrial activity. However, in the recent years, processed foods in the form of canned fruits such as pineapple, mango slices and pulps, grapes, apple, peaches etc have increased considerably. The uses of fruits in the form of concentrated juice, dry powder, jam and jelly have also increased.

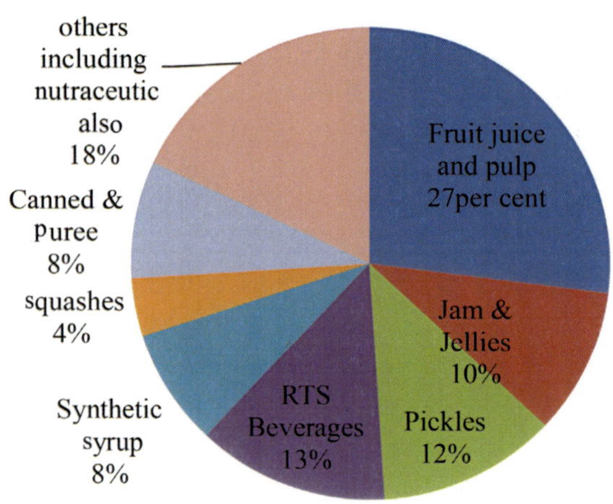

Fig. 1: Percentage production of processed fruits and vegetables.

Need for value addition in fruits:

- To improve the profitability of farmers.
- To empower the farmers and other weaker sections of society especially women through gainful employment opportunities and revitalize rural communities.
- To provide better quality, safe and branded foods to the consumers.
- To emphasize primary and secondary processing.
- To reduce post-harvest losses.
- Reduction of import and meeting export demands.
- Way of increased foreign exchange.
- Encourage growth of subsidiary industries.
- Reduce the economic risk of marketing.
- Increase opportunities for smaller farms and companies through the development of markets.
- Diversify the economic base of rural communities.
- Overall, increase farmers' financial stability.

SWOT analysis of fruit processing:

Strengths:

1. Round the year availability of raw materials.
2. Government support and Social acceptability for fruit products.
3. Vast network of manufacturing facilities within the country.
4. Vast domestic and international market.

Weaknesses:

1. High requirement of working capital.
2. Inadequately developed linkages between R and D labs and industry.
3. Lack of adequate quality control and testing methods as per international standards.
4. Inefficient supply chain due to a large number of intermediaries.
5. Less remuneration compare to other contemporary industries.
6. Seasonality of raw material.

Opportunities:

1. Products diversification due to diverse climate within the country.
2. Integration of developments in contemporary technologies such as electronics, material science, computer, bio-technology etc.
3. Export generation for additional income due to opening of global markets.
4. Huge employment generation opportunities.

Threats:

1. Competition from global players.
2. Shortage of trained manpower due to better working conditions in other industries.
3. Rapid developments in contemporary and requirements of the industry may lead to fast obsolescence.
4. Affordability and cultural preferences of fresh food.

Effects of major processing conditions on the biochemical properties of fruits

Nonthermal processing:

Pulsed electric field: Among the emerging non-thermal technologies, PEF is an innovative technology that involves the application of electrical pulses in food processing. In this approach, electric current is applied at short intervals, promoting changes in the micro-structure of the food by redistributing charges in cellular structures and macro-molecules (Cavalcanti *et al*., 2023, Brito, and Silva, 2024). These changes result in microbial and enzymatic inactivation, providing microbiological safety and kinetic stability to the beverage. Furthermore, PEF technology has various applications, such as for the recovery of bioactive compounds in food matrices and agro-food waste, molecular complexation, biopolymer modification, and serving as an assisting technology in freezing/thawing and drying processes, and others (Bernabeu *et al*., 2024, Nemli *et al*., 2024, Öztürk *et al*., 2024). The electric field can still preserve nutritional and functional products without compromising sensory characteristics. Botrytis cinerea causes post-harvest spoilage in important fruits such as strawberry. PL treatment could be an environmentally friendly post-harvest alternative to synthetic fungicides in berries. Cultivability, physiological state, the ultrastructure of *B. cinerea* suspended in peptone water and irradiated with PL (1.2–47.8 J/cm2) were investigated by Romero Bernal *et al.* (2019). Xu *et al.* (2019) the effect of a spiral continuous flow-through PL system on the inactivation of *E. coli* in red grape juice by response surface methodology. The results showed the inactivation effect increased with intensities and pulse numbers. Reductions of 2.1, 2.6 and 3.7 log were achieved in the treatment of 100 pulses, 30 ml/min flow rate, intensities of 0.13, 0.40 and 0.66 J/cm2/pulse, respectively. The results suggest that the spiral continuous flow PL could improve the inactivation of *E. coli* in red grape juices. In another study, red grapes were processed by PL at several treatment conditions before fermentation (Escott *et al.*, 2017). PL treatments decreased significantly microbial loads in grapes. In addition, PL treatments control wild micro-organisms, especially yeasts, facilitating the growth and development of the yeast starter affecting wine quality. Cao *et al.* (2017) studied the effect of PL treatments on inactivation of Salmonella on blueberries and its impact on shelf-life and quality attributes and health-benefit compounds of blueberries. For Salmonella inactivation, dry PL treatment achieved 0.9 and 0.6 log reduction of Salmonella for spot and dip inoculation, respectively; while the water-assisted PL treatment reduced Salmonella by 4.4 log and 0.8 log for spot and dip inoculation, respectively.

Irradiation: Food irradiation includes the use of ionizing or non-ionizing energy to extend shelf life of foods. It is a nonthermal food preservation method aimed at disinfecting foods by eliminating pathogenic microbes (Al-juhaimi *et al.*, 2018; Yang *et al.*, 2021). Irradiation treatment can effectively avoid the drawbacks of instability, high cost, and potential destruction of food flavor and quality caused by ultra-high temperature or pressure, which has been proven to have positive effects on improving food hygiene quality, preserving nutritional quality and flavor, and prolonging shelf-life (Gómez-López *et al.*, 2022; Shankar *et al.*, 2022; Jia *et al.*, 2023). Additionally, irradiation technology can treat foods in large quantities without chemical reagents, making it receive extensive attention in improving food quality and reducing the harmful substances in foods (Aly *et al.*, 2021).

Maraei and Elsawy (2017) found that the strawberry fruit irradiated with 600 Gy had the highest total phenolic content and antioxidant activity, and irradiation could stimulate the biosynthesis of some phenolic compounds. Khalili *et al.* (2017) found the preservation capacity of chlorophyll and vitamin C in cucumber after irradiation was increased by about 3 and 1.4 times, respectively. Meanwhile, the storage period was extended for about 1 week and the chilling injury was reduced, indicating that the proper dose of irradiation could improve the quality of food (Khalili *et al.*, 2017). Some studies also found that irradiation can cause adverse effects on bioactive substances in foods (Fruehwirth *et al.*, 2021; Yang *et al.*, 2021). Therefore, the selection of appropriate irradiation type, processing time, and doses according to the food type is very important for food preservation and quality improvement. These studies demonstrated that food irradiation can induce alterations in the structure or composition of main nutrients in the fruit, which were closely linked to the dose of irradiation, as well as the type and state of the food.

UV treatment: UV treatment is being used more frequently in food processing and similar to irradiation, its prime objective is microbial deactivation. Ultraviolet light refers to a band that exists between the X-ray (200 nm) and the visible light region (400 nm) in the electromagnetic spectrum. It is usually classified into long-wavelength UV-A (320 nm–400nm), medium-wavelength UV-B (280–320 nm) and short-wavelength UV-C (200–280 nm) (Zhang, and Jiang, 2019). The ray of sun contains all of the UV light, but only UV-A and UV-B can reach the surface of the earth, while the UV-C is completely absorbed by the ozone layer and reaches the earth without any dose. UV light of different intensities or types (A, B, and C) is applied, although UVC is the most effective for food disinfection (Urban *et al.*, 2016; Sharma *et al.*, 2023).

UV-C and UV-B treatments can both significantly increase total phenolic content in post-harvest grapes including total flavonoid, total flavanol, and total anthocyanin for 28 d, however, at the same dose (3.6 kJ m^{-2}), UV-C treatment increased more than UV-B (Sheng *et al.*, 2018). UV-B is UV light that plants can be exposed to under natural conditions and has less damage than UV-C, so it is a more promising method for post-harvest storage of fruits and vegetables (Ruiz *et al.*, 2016; Sheng *et al.*, 2018). In apple, treatment with high-dose UV-B (219 kJ m^{-2}) can significantly improve the total phenolics and total flavonoids content of peel, interestingly, the total phenolic content and total flavonoids in apple peel reduced at the initial 36 h after UV-B treatment, and significantly increased after 14th day (Assumpcao *et al.*, 2018). Similar results were reported in peaches where 36 kJ m^{-2} UV-B treatment had a negative effect on most of the phenolic substances within the first 24 h, however, the situation turned at 36 h and most of the phenolic substances initially reduced then significantly increased (Santin *et al.*, 2018).

Thermal processing:

Blanching: Blanching has been considered to be one of the important unit operations. The blanching pretreatment helps in (1) improving juice yield; (2) removing entrapped air between the cells; (3) maintaining the freshness of the color, the stabilization of the texture and nutritional quality; (4) inactivating a variety of undesirable enzymes, such as peroxidase (POD), polyphenol oxidase (PPO) and lipoxygenase (LOX), etc.; (5) facilitating the dissolution of pigment and flavor substances; and (6) destroying the micro-organisms to some extent and removing some unpleasant odors (Lan *et al.*, 2023; Tripathy and Srivastav, 2023).

Hot water blanching and steam blanching are not only the most conventional blanching techniques, but also the most widespread in the food industry. These two blanching treatments are generally performed by exposing fruits or vegetables to hot water or hot steam for several seconds or minutes (Hussain *et al.*, 2023; Tripathy and Srivastav, 2024). Recently, innovative blanching technologies including microwave blanching, ohm blanching, infrared blanching and high-pressure blanching have attracted widespread attention in the food industry (Table 3) (Lan *et al.*, 2023; Lohita and Srijaya, 2024). A number of studies proved that a reduction in the time and energy required to achieve the inactivation of enzymes (POD, PPO and pectinases) is obvious when using microwave blanching (Dorantes-Alvarez *et al.*, 2017). Furthermore, the preservation of bioactive compounds may be enhanced as a consequence of the short processing times required to achieve microwave blanching and avoid a leaching effect of water-soluble phytoc-hemicals (Langston *et al.*, 2023). It has been concluded that the applications of microwave blanching achieve better retention of biological phytochemicals as compared to other methods (Ranjan *et al.*, 2017). Ohm blanching is also an alternative to the conventional technique that uses electrical energy for food processing, which has the advantage of fast, uniform heating owing to the generation of heat within the food material (Bhat, Saini, and Sharma, 2017). Several studies suggested that ohm blanching results in a better color and yield as well as imparts the nutritional characteristics of many foods (Lan *et al.*, 2023; Kaur *et al.*, 2024). Moreover, ohm blanching requires less energy consumption in terms of the heat required compared with conventional blanching (Bhat *et al.*, 2017).

Enzymolysis: Enzymatic technology is considered an integral part of modern FVJ processing. The principal purposes of enzymatic technology in fruit processing are juice extraction and clarification (Sharma *et al.*, 2017). The pre-treatment of a variety of enzymes contributed to increasing the juice yield and the production of various juices. The primary reason for this is that the cell wall of most fruits and vegetables is mainly composed of cellulose, hemicellulose and pectin. The addition of pectinase, cellulase and hemicellulase in the processing of FVJ can soften and degrade the tissue of fruits and vegetables, and thus greatly improve the juice yield and juice extraction efficiency (Sharma *et al.*, 2017).

Table 3: Effect of blanching on the biochemical properties of fruits.

Fruit	**Blanching method**	**Blanching parameters**					**Major findings**	**References**
		Temperature (°C)	**Time (min)**	**Pressure (PSI)**	**Strength (V/cm)**	**Voltage (kV)**		
Trapiá	Thermal	-	15	-	-		Blanching increase in antioxidant activity.	Moura *et al.* (2024).
Sugarcane	Steam	-	5	15	-		Minimum changes in quality characteristics up to a storage period of 90 days.	Vats *et al.* (2024).
Mango	Ohmic	-	0.67-1.83		10, 15, 20, and 25		Blanching preserves the functionality of mango cubes.	Priyadarshini *et al.* (2023).
Kakadu plum	Thermal	90	5	-	-	-	Blanched fruit demonstrated excellent antioxidative properties by reducing MDA production.	Bobasa *et al.* (2023).
Apple	High-humidity hot air impingement.	110	1.5	-	-	-	Blanchedsample showed lower browning index and higher antioxidant capacity compared with the untreated ones.	Wang *et al.* (2022).
Banana	Cold plasma	-	0.58-2.583	-	-	4.8-6.9	CP can be a promising non-thermal process for fresh fruits blanching.	Pour *et al.* (2022).
Apricot		60-70	2	-	-	-	SHA might be as a novel green and efficient method for preserving functional compounds..	Zhang *et al.* (2021).

Furthermore, the use of enzymes such as cellulases, pectinases and a combination of these enzymes helps to promote the dissolution of nutrients and functional components from fruit and vegetable pulp, thereby increasing the TSS, TPC, total anthocyanin content (TAC), TFC, TCC and the antioxidant activity, in addition to acquiring high-quality fruit juice (FJ) (Lan *et al.*, 2023). Enzymatic treatments are widely used in FJ processing to enhance the juice yield and many potentially health-beneficial components. As a sustainable and eco-friendly technology, enzymatic pretreatment has shown highly positive effects on improving FJ yield and quality in most studies, but this is usually achieved through a careful selection of enzymes and treatment parameters for FJ processing. The enzymolysis effect depends on the type of enzyme, incubation time, incubation temperature, enzyme concentration, agitation, pH and the use of different enzyme combinations (Sharma *et al.*, 2017). Therefore, it is crucial to control and optimize the enzyme treatment parameters during FVJ production with a view to ensure the adequate sensory quality of the juice. This seems to limit its ability to obtain a common condition for processing multiple types of products. However, this is also common in other extraction techniques. In addition, simple enzymolysis has certain limitations, e.g., inadequate and uneven contact between enzyme and substrates, decreased enzyme activity caused by the release of plant protease, or protein accumulation (Łubek-Nguyen *et al.*, 2022, Umego *et al.*, 2021). Currently, there are some techniques, such as ultrasonic-assisted enzymolysis and microwave-assisted enzymolysis, that attempt to address these shortcomings (He *et al.*, 2022, Umego *et al.*, 2021). However, these solutions are still on a laboratory or pilot scale, there is still a certain distance to reach the level of commercial production.

Drying: Fresh fruits and vegetables, having moisture contents of more than 80 per cent, are classified as highly perishable commodities and their preservation through sun-drying, which dates back many centuries, have often result in poor quality and product contamination (Vidinamo *et al.*, 2021).

As such, there is a strong need for developing alternative drying technologies considering operational capacity, process control, time requirements, cost economics, product quality, safety and environmental aspects. Novel drying technologies are intelligent combinations of conventional technologies necessitated by changes in consumer and market requirement (Kumar and Karim, 2019; Fathi *et al.*, 2022). Microwave drying, infrared drying, heat pump drying, fluidized bed drying, radio frequency drying, ultrasound drying, nano spray drying and Refractance Window drying are considered as fourth generation drying technologies based on the type of raw materials they can handle and the retention of quality attributes of the intended products (Table 4) (Raghavi *et al.*, 2018).

Table 4: Effect of drying on the biochemical properties of fruits.

Fruits	Drying method	Drying conditions					Major findings	References
		Temperature (°C)	Time (h)	Pressure (Pa)	Frequency (MHz)	Power (W)		
Apricot	Convective	50	2				Functional bioactive compounds preserved during drying.	Juhnevica-Radenkova *et al.* (2024).
Wild jujube	Infrared	50, 60 and 60	1, 2, 3 and 4				Phenolic decreased, however, triterpenoid acids increased.	Xue *et al.* (2024).
Moroccan grapes	Hygrometric enclosure HCP type	60-80	-	-	-	-	Resulted in high amount of phenols.	Boudboud *et al.* (2023).
Windfall apples	Microwave	-	-	-	2.465	900	Drying maintained nutritional compounds	Çetin and Sağlam (2023).
White mulberry	Freeze	30, 50, and 70	-	60	-	-	Freeze drying preserve the phenolics and antioxidant.	Krzykowski *et al.* (2023).
Quince	Convective	50 and 70	24 and 48	-	-	-	Drying maintained the antioxidant properties.	Najman *et al.* (2023).
Red dragon fruit	Infrared	50, 60 and 60	-	-	-	300	Maintained the bioactive compounds effectively.	Bassey *et al.* (2022).
Satkara	45, 50, and 55	-	-	-	-	-	Drying reduces the loss of nutritional compounds	Roy *et al.* (2023).
Chilean papaya	Vacuum	40, 50, 60, 70, and 80	-	15000	-	-	Antioxidants were maintained in the sample.	Vega-Gálvez *et al.* (2021).
Cornelian cherry	Microwave	-	-	-	2.465	100, 300, and 500	Phytochemical compounds loss was reduced.	Zia andAlibas (2021).

According to the review by Kamiloglu *et al.* (2016) on recent findings on the effects of different drying techniques on major antioxidants of fruits and vegetables, freeze-drying is superior in comparison to the other drying methods, in terms of preserving the antioxidant compounds. Recent literature, which studied the effects of different drying techniques on phytochemical content and antioxidant activity of various fruit and vegetable products, indicated significant increase and decrease in the phytochemical contents. However, the phytochemicals and antioxidant activity tend to increase at temperatures 60 °C. This is because at higher temperatures, the cell wall breaks down and releases the cell wall bound phenolics and the flavonoid causing increase in the antioxidant properties of the dried product which results in minimizing the oxidation reactions (Khan and Karim, 2017; Khan, *et al.*, 2018).

Recently, there has been extensive research about the effect of processing on the total phenolic contents of various fruits and vegetables including the influence of drying. During the drying processes, activation of oxidative enzymes, such as polyphenol oxidase and peroxidase, may lead to the loss of phenolic compounds (Cao *et al.*, 2021). In addition, binding of phenolic compounds to proteins, changes in chemical structures or low extraction efficiencies are other factors related to the loss in total phenolic content (Fathi *et al.*, 2022). On the other hand, various drying methods resulted in up to a 10-fold increase in total phenolic content of apple. These increases in the total phenolic content can be explained by the release of phenolic compounds bound to the plant cell wall (Fathi *et al.*, 2022).

References

Akter T, Bhattacharya T, Kim J H, Kim M S, Baek I, Chan D E and Cho B K, 2024. A comprehensive review of external quality measurements of fruits and vegetables using nondestructive sensing technologies. *Journal of Agriculture and Food Research*, 101068.

Al-Juhaimi F, Ghafoor K, Özcan M M, Jahurul M H A, Babiker E E, Jinap S and Zaidul I S M, 2018. Effect of various food processing and handling methods on preservation of natural antioxidants in fruits and vegetables. *Journal of Food Science and Technology*, 55:3872-3880.

Aly A A, Maraei R W, Abd-Allah M M and Safwat G, 2021. Evaluation of physical, biochemical properties and cell viability of gamma irradiated honey. *Journal of Food Measurement and Characterization,* 15(5):4794-4804.

Assumpção C F, Hermes V S, Pagno C, Castagna A, Mannucci A, Sgherri C and de Oliveira Rios A, 2018. Phenolic enrichment in apple skin following post-harvest fruit UV-B treatment. Post-harvest *Biology and Technology*, 138:37-45.

Barbosa-Cánovas G V, Donsì F, Yildiz S, Candoğan K, Pokhrel P R and Guadarrama-Lezama A Y, 2022. Nonthermal processing technologies for stabilization and enhancement of bioactive compounds in foods. *Food Engineering Reviews*, 1-37.

Bassey E J, Cheng J H and Sun D W, 2022. Improving drying kinetics, physicochemical properties and bioactive compounds of red dragon fruit (Hylocereus species) by novel infrared drying. *Food Chemistry*, 375:131886.

Bernabeu M, Salgado-Ramos M, Barba F J, Collado M C, and Castagnini J M, 2024. Impact of sweet potato peels extracts obtained by pulsed electric fields on the growth of probiotic strains from Lactobacillus genus. *Innovative Food Science* and *Emerging Technologies*, 92:103590.

Bernal A R R, Contigiani E V, González, H. H., Alzamora, S. M., Gómez, P. L., and Raffellini, S, 2019. Botrytis cinerea response to pulsed light: Cultivability, physiological state, ultrastructure and growth ability on strawberry fruit. *International Journal of Food Microbiology*, 309:108311.

Bhat S, Saini C S and Sharma H K, 2017. Changes in total phenolic content and color of bottle gourd (Lagenaria siceraria) juice upon conventional and ohmic blanching. *Food Science and Biotechnology*, 26:29-36.

Bobasa E M, Srivarathan S, Phan A D T, Netzel M E, Cozzolino D and Sultanbawa Y, 2023. Influence of blanching on the bioactive compounds of Terminalia ferdinandiana Exell fruit during storage. *Journal of Food Measurement and Characterization*, 17(1):244-252.

Brito I P C and Silva E K, 2024. Pulsed electric field technology in vegetable and fruit juice processing: A review. *Food Research International*, 114207.

Cao H, Saroglu O, Karadag A, Diaconeasa Z, Zoccatelli G, Conte-Junior C A and Xiao J, 2021. Available technologies on improving the stability of polyphenols in food processing. *Food Frontiers*, 2(2):109-139.

Cao X, Huang R and Chen H, 2017. Evaluation of pulsed light treatments on inactivation of Salmonella on blueberries and its impact on shelf-life and quality attributes. *International Journal of Food Microbiology*, 260:17-26.

Cavalcanti R N, Balthazar C F, Margalho L P, Freitas M Q, Sant'Ana A S and Cruz A G, 2023. Pulsed electric field-based technology for microbial inactivation in milk and dairy products. *Current Opinion in Food Science*, 101087.

Çetin N and Sağlam C, 2023. Effects of ultrasound pretreatment assisted drying methods on drying characteristics, physical and bioactive properties of windfall apples. *Journal of the Science of Food and Agriculture*, 103(2): 534-547.

Chemat A, Song M, Li Y and Fabiano-Tixier A S, 2023. Shade of innovative food processing techniques: potential inducing factors of lipid oxidation. *Molecules*, 28(24):8138.

De Corato U, 2020. Improving the shelf-life and quality of fresh and minimally-processed fruits and vegetables for a modern food industry: A comprehensive critical review from the traditional technologies into the most promising advancements. *Critical Reviews in Food Science and Nutrition*, 60(6): 940-975.

Dorantes-Alvarez L, Ortiz-Moreno A, Guzmán-Gerónimo R and Parada-Dorantes L, 2017. Microwave-assisted blanching. In *The Microwave Processing of Foods* (pp. 179-199). Woodhead Publishing.

Escott C, Vaquero C, del Fresno J M, Bañuelos M A, Loira I, Han S Y and Suárez-Lepe J A, 2017. Pulsed light effect in red grape quality and fermentation. *Food and Bioprocess Technology*, 10:1540-1547.

Fathi F, N. Ebrahimi S, Matos L C, P P Oliveira M B and Alves R C, 2022. Emerging drying techniques for food safety and quality: A review. *Comprehensive Reviews in Food Science and Food Safety*, 21(2), 1125-1160.

Fruehwirth S, Call L, Maier F A, Hebenstreit V, D'Amico S, and Pignitter M, 2021. LC–MS/MS method validation for the quantitation of 1-kestose in wheat flour. *Journal of Food Composition and Analysis*, 100, 103930.

Gómez-López V M, Pataro G, Tiwari B, Gozzi M, Meireles M Á A, Wang S and Morata A, 2022. Guidelines on reporting treatment conditions for emerging technologies in food processing. *Critical Reviews in Food Science and Nutrition*, 62(21):5925-5949.

He S, Zhang Y, Gu C, Wu Y, Farooq M A, Young D J and Gao J, 2022. Impact of different processing techniques on reduction in oil content in deep-fried donuts when using kombucha cellulose hydrolysates. *Frontiers in Sustainable Food Systems*, 6: 997097.

Hussain I, Ahmad Z, Rafique N, Bashir S, Atta Khan F, Al-Asmari F and Godswill Awuchi C, 2023. Quality attributes of dehydrated pear (*Pyrus communis* L.) as affected by conventional and novel blanching techniques. *International Journal of Food Properties*, 26(1):2656-2666.

Jia W, Wang X, Zhang R, Shi Q and Shi L, 2023. Irradiation role on meat quality induced dynamic molecular transformation: From nutrition to texture. *Food Reviews International*, 39(7): 4442-4464.

Juhnevica-Radenkova K, Krasnova I, Seglina D, Kaufmane E, Gravite I, Valdovska A, and Radenkovs V, 2024. Biochemical Profile and Antioxidant Activity of Dried Fruit Produced from Apricot Cultivars Grown in Latvia. *Horticulturae*, 10(3):205.

Kamiloglu S, Toydemir G, Boyacioglu D, Beekwilder J, Hall R D and Capanoglu E, 2016. A review on the effect of drying on antioxidant potential of fruits and vegetables. *Critical Reviews in Food Science and Nutrition*, 56(Sup1), S110-S129.

Kaur M, Kumar S, Samota M K and Lalremmawii, 2024. Ohmic heating technology systems, factors governing efficiency and its application to inactivation of pathogenic microbial, enzyme inactivation, and extraction of juice, oil, and bioactive compounds in the food sector. *Food and Bioprocess Technology*, 17(2): 299-324.

Khalili R, Ayoobian N, Jafarpour M and Shirani B, 2017. The effect of gamma irradiation on the properties of cucumber. *Journal of Food Science and Technology*, 54, 4277-4283.

Khan M I H and Karim M A, 2017. Cellular water distribution, transport, and its investigation methods for plant-based food material. *Food Research International*, 99: 1-14.

Khan M I H, Joardder M U H, Kumar C and Karim M Am 2018. Multiphase porous media modelling: A novel approach to predicting food processing performance. *Critical Reviews in Food Science and Nutrition*, 58(4):528-546.

Krishnan Kesavan R, Begum, S, Das P and Nayak P K, 2023. Hurdle effect of thermosonication and non-thermal processing on the quality characteristics of fruit juices: An overview. *Journal of Food Process Engineering*, 46(6): e14310.

Krzykowski A, Dziki D, Rudy S, Polak R, Biernacka B, Gawlik-Dziki U and Janiszewska-Turak, E, 2023. Effect of air-drying and freeze-drying temperature on the process kinetics and physicochemical characteristics of white mulberry fruits (*Morus alba* L.). *Processes*, 11(3):750.

Kumar C and Karim M A, 2019. Microwave-convective drying of food materials: A critical review. *Critical Reviews in Food Science and Nutrition*, 59(3):379-394.

Lan T, Wang J, Bao S, Zhao Q, Sun X, Fang Y and Liu S, 2023. Effects and impacts of technical processing units on the nutrients and functional components of fruit and vegetable juice. *Food Research International*, 112784.

Langston F M, Nash G R and Bows J R, 2023. The retention and bioavailability of phytochemicals in the manufacturing of baked snacks. *Critical Reviews in Food Science and Nutrition*, 63(14): 2141-2177.

Lohita B and Srijaya M, 2024. Novel Technologies for Shelf-Life Extension of Food Products as a Competitive Advantage: A Review. *Food Production, Diversity, and Safety Under Climate Change*, 285-306.

Łubek-Nguyen A, Ziemichód W and Olech M, 2022. Application of enzyme-assisted extraction for the recovery of natural bioactive compounds for nutraceutical and pharmaceutical applications. *Applied Sciences*, 12(7), 3232.

Maraei R W and Elsawy K M, 2017. Chemical quality and nutrient composition of strawberry fruits treated by γ-irradiation. *Journal of Radiation Research and Applied Sciences*, 10(1): 80-87.

Moura F T D, Silva S D M, Sousa F D A R D M, Santos K M, Bastos T M D R and Araújo J, 2024. Application of pulsed electric field in reducing internal browning and maintaining the functional potential of 'Pérola'pineapple. *Revista Ciência Agronômica*, 55: e20217851.

Najman K, Adrian S, Hallmann E, Sadowska A, Buczak K, Waszkiewicz-Robak B, and Szterk A, 2023. Effect of various drying methods on physicochemical and bioactive properties of quince fruit (*Cydonia oblonga* Mill.). *Agriculture*, 13(2): 446.

Nemli E, Ozkan G, Gultekin Subasi B, Cavdar, H, Lorenzo, J M, Zhao C and Capanoglu, E. 2024. Interactions between proteins and phenolics: effects of food processing on the content and digestibility of phenolic compounds. *Journal of the Science of Food and Agriculture*.

Öztürk H İ, Buzrul S, Bilge G and Yurdakul M, 2024. Pulsed electric field for shalgam juice: effects on fermentation, shelf-life, and sensory quality. *Journal of the Science of Food and Agriculture*, 104(3):1784-1792.

Parmar N, Singh N, Kaur A, Virdi A S and Thakur S, 2016. Effect of canning on color, protein and phenolic profile of grains from kidney bean, field pea and chickpea. *Food Research International*, 89, 526-532.

Pour A K, Khorram S, Ehsani A, Ostadrahimi A and Ghasempour Z, 2022. Atmospheric cold plasma effect on quality attributes of banana slices: Its potential use in blanching process. *Innovative Food Science* and *Emerging Technologies*, *76*, 102945.

Priyadarshini A, Rayaguru K, Biswal A K, Panda P K, Lenka C and Misra P K, 2023. Impact of conventional and ohmic blanching on color, phytochemical, structural, and sensory properties of mango (*Mangifera indica* L.) cubes: A comparative analysis. *Food Chemistry Advances*, 2:100308.

Raghavi L M, Moses J A, and Anandharamakrishnan C, 2018. Refractance window drying of foods: A review. *Journal of food engineering*, 222: 267-275.

Ranjan S, Dasgupta N, Walia N, Thara Chand C and Ramalingam C, 2017. Microwave blanching: an emerging trend in food engineering and its effects on *Capsicum annuum* L. *Journal of Food Process Engineering*, 40(2): e12411.

Rojas M L, Kubo M T, Caetano-Silva M E and Augusto P E, 2021. Ultrasound processing of fruits and vegetables, structural modification and impact on nutrient and bioactive compounds: a review. *International Journal of Food Science* and *Technology*, 56(9): 4376-4395.

Roy M, Bulbul M A I, Hossain M A, Shourove J H, Ahmed S, Sarkar A and Biswas R, 2022. Study on the drying kinetics and quality parameters of osmotic pre-treated dried Satkara (*Citrus macroptera*) fruits. *Journal of Food Measurement and Characterization*, 1-15.

Santin M, Lucini L, Castagna A, Chiodelli G, Hauser M T and Ranieri A, 2018. Post-harvest UV-B radiation modulates metabolite profile in peach fruit. Post-harvest *Biology and Technology*, 139: 127-134.

Shankar S, Karboune S, Salmieri S and Lacroix M, 2022. Development of antimicrobial formulation based on essential oils and gamma irradiation to increase the shelf life of boneless chicken thighs. *Radiation Physics and Chemistry*, 192, 109893.

Sharma A, Mahmoud H, Pendyala B, Balamurugan S and Patras A, 2023. UV-C inactivation of microorganisms in droplets on food contact surfaces using UV-C light-emitting diode devices. *Frontiers in Food Science and Technology*, 3: 1182765.

Sharma H P, Patel H and Sharma S, 2014. Enzymatic extraction and clarification of juice from various fruits–a review. *Trends Post Harvest Technol*, 2(1): 1-14.

Sheng K, Zheng H, Shui S, Yan L, Liu C and Zheng L, 2018. Comparison of post-harvest UV-B and UV-C treatments on table grape: Changes in phenolic compounds and their transcription of biosynthetic genes during storage. Post-harvest *Biology and Technology*, 138: 74-81.

Sicard J, Barbe S, Boutrou R, Bouvier L, Delaplace G, Lashermes G and Tonda A, 2023. A primer on predictive techniques for food and bioresources transformation processes. *Journal of Food Process Engineering*, 46(5): e14325.

Suntharalingam C, Rathakrishnan T and Safari S, 2023. Determinants of fruit purchasing decision among Singaporean consumers: An empirical study. *Journal of International Food* and *Agribusiness Marketing*, 35(3): 336-366.

Tian S and Xu H, 2022. Nondestructive methods for the quality assessment of fruits and vegetables considering their physical and biological variability. *Food Engineering Reviews*, 14(3): 380-407.

Tripathy S and Srivastav P P, 2023. Effect of dielectric barrier discharge (DBD) cold plasma-activated water pre-treatment on the drying properties, kinetic parameters, and physicochemical and functional properties of Centella asiatica leaves. *Chemosphere*, 332: 138901.

Tripathy S and Srivastav P P, 2024. Synergistic Effects of Dielectric Barrier Discharge (DBD) Cold Plasma Pretreatment Combined with Microwave Drying on the Physicochemical and Functional Properties of Centella asiatica Leaves. *Food and Bioprocess Technology*, 1-20.

Umego E C, He, R, Ren W, Xu H and Ma H, 2021. Ultrasonic-assisted enzymolysis: Principle and applications. *Process Biochemistry*, 100, 59-68.

Urban L, Charles F, de Miranda M R A and Aarrouf J, 2016. Understanding the physiological effects of UV-C light and exploiting its agronomic potential before and after harvest. *Plant Physiology and Biochemistry*, 105: 1-11.

Vats S, Kaushal M, Vaidya D and Gupta A, 2024. Stability of Sugarcane Juice with Blanching and Acidification: Biochemical Analysis, Sensory Evaluation and Storage Studies. *Sugar Tech*, 1-11.

Vega-Gálvez A, Poblete J, Rojas-Carmona R, Uribe E, Pastén A and Goñi M G, 2021. Vacuum drying of Chilean papaya (*Vasconcelleapubescens*) fruit pulp: Effect of drying temperature on kinetics and quality parameters. *Journal of Food Science and Technology*, 58: 3482-3492.

Vidinamo F, Fawzia S and Karim M A, 2021. Effect of drying methods and storage with agro-ecological conditions on phytochemicals and antioxidant activity of fruits: a review. *Critical Reviews in Food Science and Nutrition*, 62(2): 353-361.

Wang J, Chen Y, Wang H, Wang S, Lin Z, Zhao L and Xu H, 2022. Ethanol and blanching pretreatments change the moisture transfer and physicochemical properties of apple slices via microstructure and cell-wall polysaccharides nanostructure modification. *Food Chemistry*, 381: 132274.

Xu F, Wang B, Hong C, Telebielaigen S, Nsor-Atindana J, Duan Y and Zhong F, 2019. Optimization of spiral continuous flow-through pulse light sterilization for Escherichia coli in red grape juice by response surface methodology. *Food Control*, 105: 8-12.

Xue T T, Ruan K H, Xu H B, Liu H B, Tang Z S, Yang Y G and Song Z X, 2024. Effect of different drying methods on the drying characteristics, chemical properties and antioxidant capacity of *Ziziphus jujuba* var. Spinosa fruit. *LWT*, 115873.

Yahia E M, García-Solís P and Celis M E M, 2019. Chapter 2-Contribution of Fruits and Vegetables to Human Nutrition and Health. *Post-harvest Physiology and Biochemistry of Fruits and Vegetables*.

Yang J, Pan M, Han R, Yang X, Liu X, Yuan S and Wang S, 2023. Food irradiation: An emerging processing technology to improve the quality and safety of foods. *Food Reviews International,* 1-23.

Yang J, Rainville P, Liu K and Pointer B, 2021. Determination of lactose in low-lactose and lactose-free dairy products using LC-MS. *Journal of Food Composition and Analysis,* 100, 103824.

Zhang H S, Guo P H, Zhang Q A, Wu D D and Zheng H R, 2021. Effects of saturated hot air pretreatment compared to traditional blanching on the physicochemical properties of Apricot (*Prunus armeniaca* L.) kernels and its skin during removing skin. *Lwt*, 150, 111947.

Zhang W and Jiang W, 2019. UV treatment improved the quality of post-harvest fruits and vegetables by inducing resistance. *Trends in Food Science* and *Technology*, 92: 71-80,

Zia M P and Alibas I, 2021. The effect of different drying techniques on color parameters, ascorbic acid content, anthocyanin and antioxidant capacities of cornelian cherry. *Food Chemistry*, 364:130358.

10

Waste to Wealth for Nutritional Security

Dinesh Kumar and Anamika Thakur

Division of Food Science and Post-harvest Technology, ICAR-Indian Agricultural Research Institute, PUSA, New Delhi-110 012

Waste is defined as anything that doesn't add value to their operations. To the average person, waste is simply stuff that's unwanted or useless. But from a scientific perspective, there's really no such thing as waste. Nearly all the things we throw away can be transformed or processed in ways that make them useful again.

Waste comes in various forms, ranging from household trash to industrial byproducts. These include domestic waste, factory waste, e-waste, construction waste, agricultural waste, and more. We can categorize waste into different types:

Solid waste: This includes organic materials like vegetable and kitchen waste, as well as general household waste.

E-waste: Discarded electronic devices such as computers, TVs, and music systems fall into this category.

Liquid waste: This comprises water used in various industries, as well as waste from tanneries, distilleries, and thermal power plants.

Plastic waste: Items like plastic bags, bottles, and containers make up this category.

Metal waste: Unused metal sheets, scraps, and other metal items fall into this group.

Nuclear waste: Materials left over from nuclear power plants are considered nuclear waste.

Furthermore, we can group these types of waste into two main categories: wet waste, which is biodegradable, and dry waste, which is non-biodegradable.

In 2016, it was estimated that India generated a staggering 277 million tonnes of municipal solid waste annually. This figure constitutes over 80% of the total 334 million tonnes of waste produced in South Asia and approximately 13 per cent of the worldwide waste output.

Agro-food processing wastes

The traditional perception of waste as having no utility is evolving towards viewing waste as a resource that can be transformed into secondary materials through modification (Gillian, 1992). This shift allows for wastes to be converted into valuable resources that can be utilized domestically or even traded for economic gain. Waste recycling involves collecting discarded materials such as husks, peels, poultry droppings, cow dung, biomass, etc., and processing them to create new products. Recycling essentially transforms waste into usable forms, thereby contributing to resource recovery and minimizing environmental exposure and associated health risks.

Waste minimization sits at the top of waste management hierarchies globally, as it is considered a fundamental technique for waste management (Anonymaus, 2001). The objective of waste recycling is to repurpose discarded materials into useful products, leading to numerous benefits:

- Reduction in the amount of waste needing disposal.
- Conservation of natural resources, including non-renewable ones like petroleum.
- Decrease in the energy required to manufacture new products.
- Mitigation of pollution and destruction associated with obtaining new raw materials.
- Creation of employment opportunities.
- Contribution to the national economy by reducing the need for importing raw materials.

Modern agriculture, characterized by the extensive use of fertilizers, pesticides, and other chemicals, has increased agricultural productivity but also led to a rise in agricultural waste. These wastes include crop residues, pruning residues, weeds, animal droppings, and food processing wastes (Akaranta, 2006). Such wastes constitute a significant portion of agricultural production, ranging from 30 to 60 per cent of consumable products and animal feed. They are generated through various agricultural processes, including food manufacturing, animal rearing, and household activities.

Industrial wastes, on the other hand, originate from diverse industrial activities and are also categorized into solid and liquid wastes. The increase in waste generation over the years is attributed to factors like population growth, urbanization, industrialization, and intensified agricultural production.

Agricultural wastes, particularly rich in lignocellulosic materials or animal residues, have the potential to serve as raw materials for the production of various useful products within a waste-to-wealth framework (Odocha, 1994). Understanding the characteristics of these wastes is crucial for leveraging their potential in such schemes.

Food waste

Food is a basic necessity for humans, yet food waste poses a significant global challenge. UNICEF's 2011 report highlighted that around 21,000 individuals die every day due to issues related to hunger. Shockingly, one out of every nine people worldwide goes to bed hungry. At the same time, an alarming amount of food, nearly one-third of what is produced, ends up in landfills. This wastage exacerbates the problem of hunger and underscores the urgency of addressing food waste on a global scale (Singh *et al*., 2022). Food waste stems from multiple factors such as food production, distribution, lifestyle choices, industrial processes, and consumer behavior. Although "food waste" and "food losses" are often used interchangeably, they refer to slightly different concepts. "Food loss" occurs within the supply chain after harvesting, encompassing stages like handling, transportation, storage, and distribution. It specifically refers to the decrease in the quantity of food before it reaches the consumer (Kour *et al*., 2023). Food waste pertains to the discard of food intended for human consumption, which subsequently deteriorates, becomes lost, and ends up in landfills. This issue is largely driven by consumer habits and retail practices. Often, the food is still suitable for other uses but is disposed of before being consumed. Developed countries are particularly affected by food waste, with a significant portion occurring at the retail or consumer level (Rather *et al.,* 2022b). Reducing food waste not only conserves resources but also lessens the demand for crop production, consequently reducing water usage in agriculture. Additionally, water resources are often compromised due to industrial effluents or inadequate sanitation facilities (Rather *et al.,* 2022b). Numerous studies have explored various uses for waste materials derived from fruits and different carbon nanomaterials in treating wastewater. Developing nations experience significant food loss because of inadequate post-harvest infrastructure. The majority of food waste occurs within the supply chain during manufacturing or various processing stages (Solangi *et al.,* 2021).

Waste generation

Food waste is increasing globally and is not limited to just industrialized nations. The Food and Agriculture Organization has supplied data indicating that food waste generation is happening at similar levels in both industrialized and developing countries across various regions of the world (FAO, 2011). Despite similarities in food waste generation between developed and developing nations, significant differences persist. In developing countries, over 40 per cent of food losses happen during post-harvest and processing stages, while in industrialized nations, 40 per cent of losses occur at the retail and consumer levels. Consequently, on a per-capita basis, industrialized countries waste considerably more food than their developing counterparts (Khan *et al.,* 2021). The issue of food waste generation involves two main aspects: prevention at the production level and segregation of waste at its source downstream. Preventing the generation of waste is the primary step in effectively controlling food waste. However, some amount of waste is inevitable, and thus proper source separation becomes crucial. The food supply chain comprises a complex network involving farmers, customers, and various food processing and marketing companies. With globalization and industrialization, these supply chains have become longer and more intricate, increasing the risk of food spoilage. There are numerous causes of food waste, often specific to particular stages of the food chain, but certain reasons are universal across all levels of the chain. Understanding these sources of waste is essential for devising strategies to prevent it. Food waste is typically categorized based on where it occurs within the food supply chain.

Food loss and waste

According to the United Nations Development Programme, India wastes up to 40 per cent of its food production. This includes approximately 21 million tonnes of wasted wheat alone. Shockingly, 50 per cent of all food produced worldwide suffers the same fate, never reaching those who need it most. The agriculture ministry reports that INR 50,000 crores worth of food produced is wasted annually in India.

Wastage of fresh water: Around 25 per cent of fresh water used in food production ends up being wasted. This is concerning because many people still lack access to clean drinking water.

Increase in hunger: The number of hungry people in India has risen by 65 million, exceeding the population of France. A survey by Bhook in 2013 found that 20 crore Indians go to bed hungry on any given night.

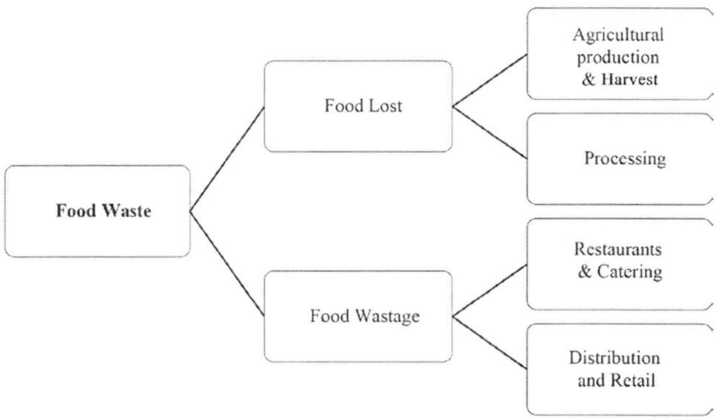

Fig. 1: Broad classification of food waste.

Food wastage is a significant problem for several reasons:

Land degradation: Large areas of land are deforested to cultivate food crops. Approximately 45 per cent of India's land is degraded, primarily due to deforestation, unsustainable farming practices, and excessive groundwater extraction to meet food demands.

Energy consumption: A significant amount of energy, approximately 300 million barrels of oil, is used in food production processes, only for the food to end up being wasted.

These factors highlight the environmental, social, and economic impacts of food wastage, underscoring the urgency to address this issue.

Production of food wastes

In the food supply chain (FSC), primary production encompasses both pre-harvest activities (such as animal husbandry, fisheries farming, and agriculture) and postharvest operations (including handling, processing, and storage). Factors like adverse weather conditions, spoilage, and pest infestation are major contributors to food waste at the farm level. Developing countries experience significant agricultural losses, often due to outdated harvesting methods, transportation challenges, inadequate storage facilities, poor infrastructure, and limited preservation techniques. These issues exacerbate postharvest losses in third-world nations (Bos-Brouwers *et al.,* 2013).

Availability of agro-waste

Agro-waste comes in various forms depending on its origin and availability, primarily composed of cellulose, hemicellulose, and lignin (Donner *et al.*, 2020). Lignocelluloses, due to their complex structure, are challenging to break down. This waste includes crop residue, aquaculture waste, and other agricultural byproducts. Crop residue and agro-industrial waste are the largest categories, generated in significant quantities daily by agricultural and food processing industries. However, the lack of proper management techniques for these wastes, often stemming from limited access to information, presents a substantial obstacle.

Cattle manure, a major source of harmful gases, micro-organisms, and odors, raises concerns for public health and environmental pollution. Aquaculture solid waste primarily consists of unconsumed feed and undigested substances expelled by fish as feces, posing risks of water contamination if not managed properly. Utilizing farm waste effectively can improve irrigation and erosion control. It's important to distinguish agricultural residue from other solid fuels like charcoal based on availability and distinctive characteristics (Sorathiya *et al.,* 2014).

Handling and storage of wastes

These types of waste occur due to mishandling of food commodities during transportation from farms to stores or from stores to markets/processing units. This issue is particularly prevalent in many third-world countries due to insufficient infrastructure and knowledge about handling various commodities (Mishra *et al.,*1021). The lack of proper handling and storage facilities for fresh produce, poultry, and meat significantly contributes to food loss, especially in underdeveloped nations. These countries are responsible for 19 per cent of fruit and vegetable loss, often exacerbated by the quick spoilage of these foods in warmer climates, necessitating adequate cold storage post-harvesting or slaughtering. Additionally, failures in grain storage, or pulse crops stored in air-tight containers where pests and moisture can enter, can contaminate the crop with mold, toxins, or pests. This often forces farmers to sell their entire crop immediately after harvest, resulting in lower prices due to oversupply.

Food waste processing

These wastes primarily stem from inefficient processing operations aimed at separating edible and inedible parts of food, alongside inefficient handling practices. Most of these processing wastes consist of parts like peel, eyes, and skin, which are deemed inedible. Interestingly, these inedible parts contain valuable nutrients like proteins, carbohydrates, antioxidants, and fats, along with substances that demand chemical or biological oxygen. The food goes through various processing stages such as trimming, peeling, cutting, slicing, milling, cooking, boiling, and packaging. However, inefficiencies and the absence of robust quality standards lead to significant wastage. Moreover, faults in equipment involved in applying or removing heat, like distillation, sterilization, or cooling, heighten the risk of bacterial contamination, thus spoiling the food. Food manufacturers, under pressure to meet retailer specifications regarding appearance, size, weight, and quality, often discard items that fail to meet these standards and aren't visually appealing to consumers.

Furthermore, the causes of waste in manufacturing and processing are multifaceted, including issues such as unhealthy animals, contamination, quality rejections, regulatory factors, and management-related challenges. In meat production, for instance, major contributors to waste are inadequate forecasting, inventory management, and planning. Certain by-products like fruit pomace from wine or juice production and whey from cheese production are inherent to the processes. While these could potentially be repurposed as animal feed, their utilization is limited due to their low nutrient content. Firstly, it's crucial to consider the environmental impacts of food waste when devising prevention strategies, focusing not just on volume but also on the environmental footprint. Certain products like meat, fish, rice, and cheese have more pronounced negative environmental effects than vegetables, thus targeting the reduction of waste from high-impact products is prudent. Secondly, since opportunities for waste reduction exist throughout the entire food supply chain, policies addressing food waste should encompass the entire system. The study also reveals some data quality concerns. For example, the total estimated food waste in the industrial sector is approximately 9000 MT, with the beverage industry contributing the highest waste estimate (about 6000 MT), and the meat sector the lowest (about 1000 MT). In terms of quantity wasted, the beverage industry again tops the list (around 1300 MT), while the industrial sector in meat production has the lowest waste (about 15 MT). In terms of percentage, the industrial sector accounts for 2.6 per cent of total waste, with the highest percentage waste coming from fruit and vegetable production (4.5%), and the lowest from grain and starch products (1.7%).

Food wastes distribution

The detailed breakdown of global food waste from production to consumption, both in developing and developed nations, is depicted in Figure 2. It outlines the percentages of waste occurring during consumption, production, storage, packaging, and retail and distribution stages. Inefficient stock planning often leads to market overstocking, driven by factors like consumer preferences, larger portion sizes, and a general reluctance to consume leftovers or discarded food. The distribution stage involves distribution centers, wholesalers, retailers, and transportation, where perishable goods pose a significant risk of waste during storage and transit. Products requiring refrigeration or freezing for extended shelf life are particularly challenging to manage due to inadequate cold chain practices, which contribute to significant food loss. Additionally, damage or defects during handling render some goods unsuitable for further distribution.

At the retail level, food waste is even more complex and varies depending on the product. Factors such as overstocking, inaccurate projections, poor shelf-rotation, expiration of "best before" dates, interruptions in the cold chain, limited storage, and inadequate packaging contribute to waste. Inadequately trained staff may also contribute to waste through improper stock management practices. Retailers can mitigate these issues and the associated economic impact by offering special promotions and discounts. The issue of food waste has garnered unprecedented global attention, particularly with growing populations and more advanced economies. The World Bank and FAO estimate that over 1.33 billion tons of food are lost or wasted annually worldwide, with projections suggesting this figure could rise to 2.2 billion tons if current trends persist. Fig. 2 illustrates the percentage of food waste at each stage of the value chain, revealing variations between countries with higher and lower incomes, such as the proportion of waste attributed to dairy products versus roots and tubers.

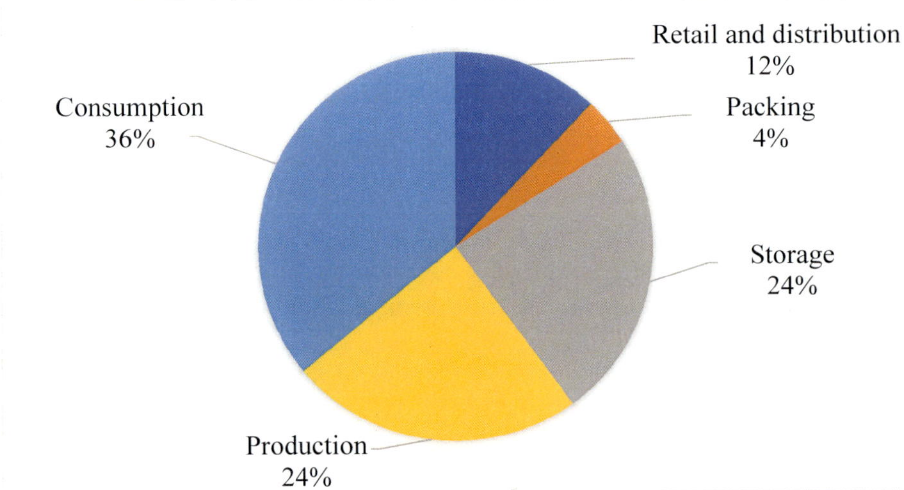

Fig. 2: Distribution of food waste from production to consumption in developing and developed countries.

Consumer/household food wastes

Household food waste percentages tend to be significantly higher in industrialized nations compared to third-world countries, mirroring trends seen in agricultural waste. In industrialized countries, the ability to afford such wastage is identified as a primary cause, according to FAO (2011), while developing countries often lack the financial means to engage in such practices. FAO data indicates relatively little difference between developed and developing nations in terms of waste and losses per capita at the final stages of consumption, except in south-east Asian nations. However, the significant disparity lies in the amount of food wasted at the retail level. For instance, in the European Union and North America, this amounts to between 95 and 115 grams per person per year, while in South-East Asia and Sub-Saharan Africa, it's only about 6–11 kilograms per person annually. Throughout the food production, processing, distribution, retail, and consumption phases, wastage occurs at alarming rates. In developing nations, the majority of food waste occurs during production, processing, and distribution phases, whereas in developed countries, it primarily happens at the retail and consumption stages due to consumption patterns and expectations rather than infrastructure deficiencies. Studies show that developed countries generate more household waste due to their higher income per capita compared to developing nations. This waste typically arises from unplanned and excessive purchasing, discarding items based on labels or expiry dates, improper storage practices, especially for perishables, and disposal habits.

Impact of food waste

Food waste poses significant threats to both human health and the environment, as valuable resources are squandered in the production of unconsumed food. Inadequate management of decomposing food waste results in the release of greenhouse gases (GHGs) and contamination of water bodies by mineral runoff and leachate. These pollutants can act as vectors for diseases and pose various health hazards to both humans and ecosystems.

Global reasons for concern about food security

It's often asserted that there's enough food to adequately feed the global population, measured in calories per person, if distribution were more equitable. According to this viewpoint, achieving distributive justice in terms of food security is a matter of crucial importance. The challenge lies in establishing more efficient markets and mechanisms to overcome barriers to effective distribution and enhance productivity for the future. Sustainable food systems are essential for ensuring food security, with sustainability serving as the foundation for both food production and promoting healthy consumption habits. Addressing climate change necessitates both adaptation and mitigation strategies. Challenges such as water scarcity, competing land use pressures, finite fossil fuel resources, the need to enhance soil fertility, and the projected rise in the global population to nine billion by 2050 are key concerns.

Waste management and sustainability concepts

The waste hierarchy principles were initially established in European policy during the 1970s, notably with the 1975 waste directive and the EU's 1977 second environment action program. The concept gained global recognition as the primary waste management framework after being clearly defined by European legislation in the community strategy for waste management in 1989. This waste hierarchy serves as a guide for identifying the most environmentally friendly End-of-Life (EoL) treatment options. Asian countries, such as Japan, have also adopted similar frameworks for waste management, emphasizing the three Rs: reduction, reuse, and recycle.

The hierarchy aims to categorize solutions that achieve the best overall environmental outcomes. "Prevention" sits at the top of the pyramid, signifying the most advantageous option, while "disposition" at the bottom is the least desirable. Prevention involves efforts to keep edible food edible, which can be achieved through various means such as improved processes and tools at manufacturing, processing, and retail levels, or by educating consumers and encouraging better utilization habits.

Below prevention is the level of reuse, where surplus food is directly reused for consumption or incorporated into prevention strategies. However, due to safety and hygiene standards governing the reuse of perishable food, the amount of food reused is often limited, potentially leading to increased food waste.

Recycling constitutes the third level of the hierarchy and is sometimes confused with recovery, although they are distinct categories. For instance, anaerobic digestion serves both as energy recovery and recycling. Food redistribution falls under recovery, while food waste is often utilized as fodder.

Despite the pressing need to consider social, economic, and environmental impacts, the European waste system directive emphasizes waste hierarchy as a framework to prioritize the most environmentally preferable options. However, some economists argue that the waste hierarchy should serve as a flexible protocol for developing waste management strategies rather than a rigid directive.

Valorization technologies for agro-waste

Agro-waste requires proper pretreatment before it can be effectively utilized. Modified agro-waste tends to be more efficient and valuable in its applications compared to untreated waste (Dey *et al.,* 2021). Converting the complex molecular structures of agro-waste into simpler monomers is commonly seen as an essential pre-treatment step. Table 1 provides a breakdown of the advantages and disadvantages of different pretreatment methods.

Table 1: Application based pre-treatment methods and related pros as well as cons

Application	Pre-treatment methods	Pros	Cons
Agriculture sector Animal manure and biofertilizer. Biorefinery and animal feed.	Enzyme Bacteria Fungi	Moderate circumstances are present, and minimal effort is necessary. Economical and requiring only mild reaction conditions. Inexpensive, destroys lignin and hemicellulose, minimal energy needs.	Low hydrolysis rate and large sterile space requirement.
Food and pharma Sector Antibiotic production Antioxidant properties Antibacterial and anticancer properties.	Ionic liquids Deep eutectic Solvents Natural deep eutectic solvents	Effective at dissolving copious amounts of cellulose and recovering usable cellulose from lignin. Conditions re modest but environmentally friendly and safe. Low-cost, ready able, highly modifiable, and less hazardous.	Possible toxicity, prohibitively expensive method, and a lack of practicality for mass production. Creates undesirable contaminants and higher viscosity on occasion.
Agriculture sector Biofuels and manure Enzymatic digestibility Ethanol production	Grinding Ultrasonic Steaming explosion	From biomass, a fine powder with a crystallinity of up to 0.2 mm is produced. Easing the process of g down a variety of lignocellulosic materials. Minimal need for energy.	Lack of long-term viability in technique calls for a lot of energy. Collisions between particles during prolonged sonication could result in an antagonistic effect. Incomplete lignin-carbohydrate matrix cleavage, xylan fraction destruction, creation of hydrolysis, and fermentation inhibitors.
Food Sector Bioactive compounds. Nutraceuticals Ethanol and enzyme production.	Microwave Pyrolysis Irradiation	Easily functional, and with efficiency in handling large agro-waste with fewer inhibitors being formed. The highest possible rate of cellulose sugar conversion. The surface area was increased, crystallinity was reduced, hemicelluloses were hydrolysed, and the structure of lignin was altered.	This causes both a rise in temperature and an increase in the amount of electricity used. High-cost technique. Expensive method.

Pre-treating wheat straw for biodegradation can be done using mechanical, thermal, and sonic methods, which increase the kinetics of degradation. However, these methods require high energy consumption. They reduce the size of the biomass and increase its surface area, improving hydrolysis by enhancing heat and mass transmission. Chemical pretreatments, such as using acids, alkalis, ozonolysis, organosolv, and wet oxidation, break the internal linkages of lignin and hemicellulose. Pre-treatment has been shown to decrease the weight of cassava bagasse and significantly increase cellulose yield compared to untreated bagasse, with the maximum cellulose yield achieved with H_2SO_4 pre-treatment. Biological processes involving microbes facilitate hydrolysis and saccharification mediated by exoenzymes. Fermentation using whole-cell systems or enzymatic digestion of solid or liquid substrates is necessary, which is more energy-efficient, environmentally friendly, and produces no inhibitors. Green pretreatment methods include using ionic liquids (ILs), deep eutectic solvents, and natural deep eutectic solvents. Pre-treatment with ILs enhances glucose production from enzymatic hydrolysis by increasing cellulose crystallinity and reducing particle size and lignin content. Improved breakdown of lignin and cellulose occurs by breaking intramolecular and intermolecular hydrogen bonds. These pre-treatment approaches make enzymes more accessible for hydrolysis, increasing surface area while reducing operational costs. Fig. 3 illustrates possible approaches for converting agro-waste into useful bioproducts and biochemicals.

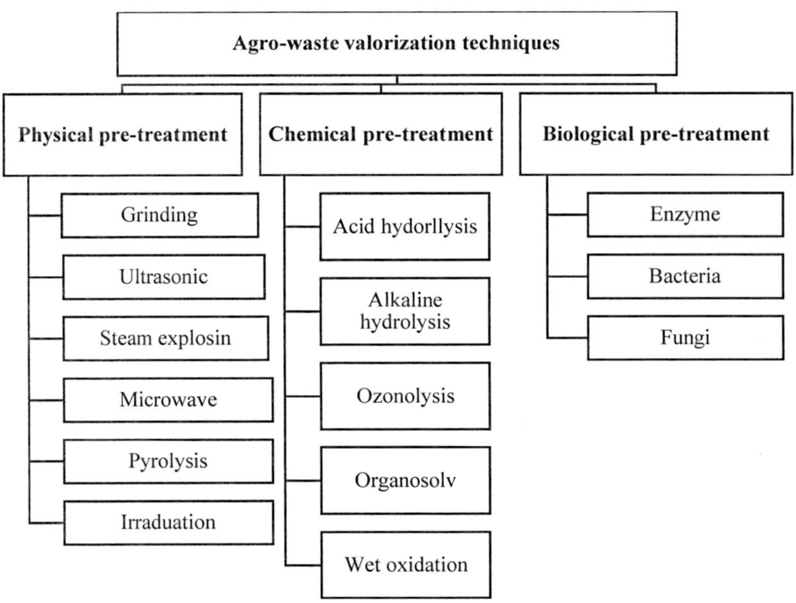

Fig. 3: Agro-waste valorization methods for the development of value-added bioproducts.

Bioactive compounds from agro-waste

Bioactive compounds (BCs) are naturally occurring substances capable of interacting with living tissues and producing various effects. Sometimes, there's confusion regarding the definitions of dietary supplements, nutraceuticals, and functional foods. Nutraceuticals and functional foods, unlike dietary supplements and food additives, play essential roles. Essentially, functional foods contain bioactives that may offer health benefits at higher levels compared to regular foods. These substances are believed to provide superior health benefits compared to conventional foods. Food additives are substances incorporated during food processing to enhance food quality and prolong shelf life. With consumers increasingly recognizing the health-promoting effects of nutraceuticals and dietary supplements, there's been a significant surge in interest in adding functional and natural food additives in recent years. Nutraceuticals and dietary supplements encompass a wide range of products known for their health benefits. However, nutraceuticals stand out as a distinct subset of dietary supplements. They contain pharmaceutical-grade substances but aren't subject to the same rigorous testing requirements as medications. Interest in bioactive compounds (BCs) for various culinary uses continues to grow, driven by ongoing research efforts to explore their health properties and potential applications. These compounds, primarily sourced from natural origins like plants, vegetables, and microorganisms, have garnered attention from both researchers and consumers due to their pharmacological properties. Initially developed as medications, they are now increasingly utilized in commercial products (Daliri and Lee, 2015).

Application of agro-waste in agri-food-pharma

Current research is heavily focused on the biobased concept due to its significant potential to enhance efficiency, reduce costs, and increase yields, all while promoting environmental sustainability. Various valuable bioproducts can be derived from agricultural waste biomass through potential bioconversion pre-treatment methods. These bioproducts span across different sectors, including agriculture (such as vermicomposting, biofertilizers, biochar, and wood vinegar), bioenergy (biofuels), and pharmaceuticals (antimicrobials, antioxidants, and antibiotics). Agro-waste holds promise for the development of a wide range of bioactives and bioproducts, as depicted in Fig. 4. Through biotransformation, agricultural wastes can be converted into high-value-added products.

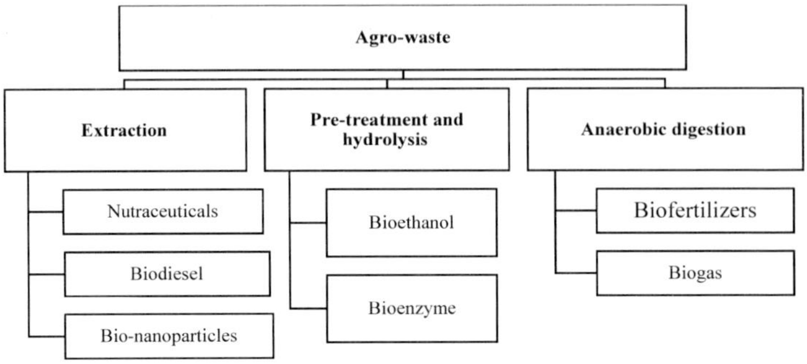

Fig.4: Agro derived bioproducts and bioactive compounds.

Food sector

Agro-waste serves as a rich source of bioactive compounds suitable for various food applications (Fig. 5). These compounds, extracted from agricultural waste or byproducts, include polyphenolic compounds, vitamins, minerals, fatty acids, volatiles, anthocyanins, and pigments, all of which offer valuable health benefits. Extracted bioactive compounds can be utilized in food fortification, leading to the development of innovative functional and health foods. The diverse array of bioactive compounds from agro-waste serves as a source of natural antioxidants and antimicrobials, finding broad applications in the food and beverage industries. In meat processing, bioactive compounds are employed to inhibit lipid oxidation in meat products. Furthermore, micro- and nanoencapsulation technologies are utilized to encapsulate these bioactive compounds, expanding their usability in the food sector. Various technologies are being employed to enhance the applicability of agro-waste in food production.

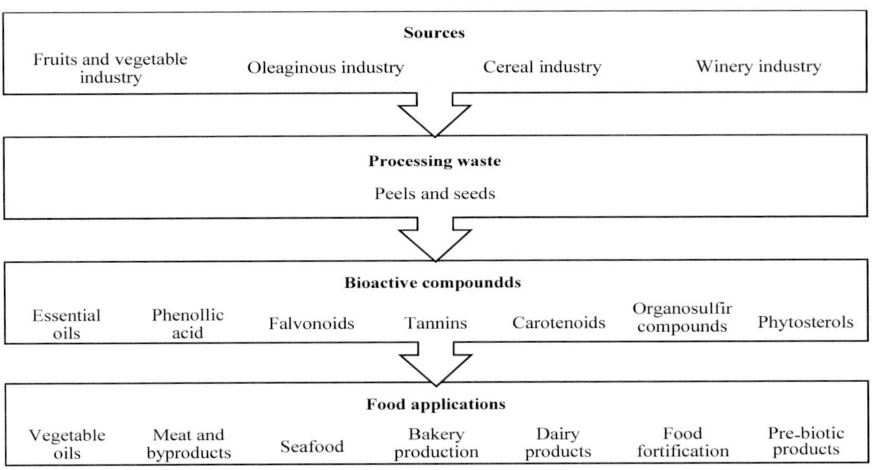

Fig. 5: Agro-waste as natural sources of bioactive compounds for food applications.

Bioactives from agro-waste: Micro/nano formulation and food application

The primary distinction between microencapsulation and nanoencapsulation lies in the particle size, with microencapsulation typically involving particles ranging from 1 micrometer (μm) to 1 millimeter (mm), and nanoencapsulation involving much smaller particles. Microencapsulation has been widely applied across various industrial sectors, including agriculture, environmental, household, and personal care products. Both nanoencapsulation and microencapsulation technologies rely on particle size and distribution uniformity for effectiveness.The use of supercritical carbon dioxide (CO_2) has shown promising advantages in controlling particle size design and drug-loading procedures for both microencapsulated and nanoencapsulated pharmaceuticals. Nanoencapsulation offers unique advantages such as preventing the degradation of active pharmaceutical ingredients and improving medication delivery accuracy by coating or conjugating the surface for proper cell entrance. Nanoencapsulated pharmaceuticals can also be labeled with fluorescent probes for imaging, aiding in assessing therapeutic efficacy during research. In agriculture, nanofertilizers are advantageous for nutrient management due to their potential to enhance nutrient utilization efficiency. These nanofertilizers release nutrients more slowly than conventional fertilizers, improving nutrition management.Various techniques are available for microencapsulating food components, each with its own benefits and limitations. The choice of technique depends on factors such as target particle size, physicochemical properties of the materials, release mechanisms, and cost considerations. Nanoencapsulation is particularly promising for protecting food elements from physiological conditions or deterioration, improving stability, bioavailability, solubility, and delivery of bioactive compounds.Advancements in nanotechnology have led to significant progress in the food industry, with nanoencapsulation offering promising options for enhancing food formulations and public health (Fig. 6). These methods can create protective shields to withstand harsh environments, mask undesirable flavors and odors, and improve nutrient absorption, release, and delivery. When selecting a nanocarrier for bioactive molecules, factors such as release patterns, safety, toxicity concerns, and economic considerations are carefully considered.

Various initiatives around the world are addressing the issue of food wastage and converting it into valuable resources:

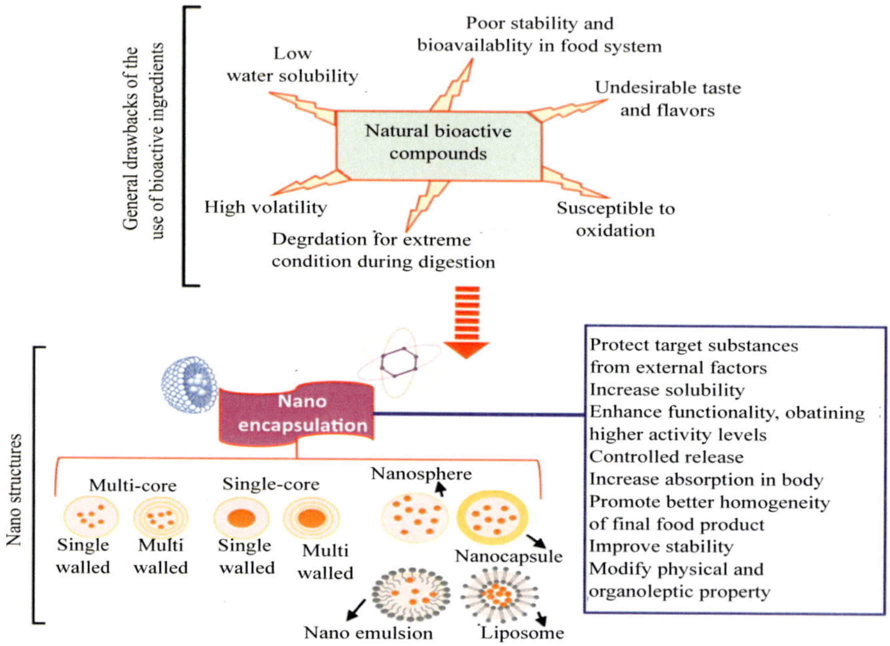

Fig. 6: Limitations of bioactive compounds and application of nanoencapsulation to increase bioactive compounds applicability in food industry.

Food waste into plant food (Kowloon Hospital, Hong Kong): The hospital efficiently handles around 400-500 kg of food waste per day, mostly rice, vegetables, and fruit peels. Through fermentation aided by enzymes, this waste is converted into a clear, transparent liquid fertilizer. This approach has reduced fertilizer requirements by 20 per cent and effectively manages over 90 per cent of the waste, providing proven health benefits for plants.

Food waste as fuel for biogas plant (NIT-T, Trichy, India): The National Institute of Technology in Trichy has established a biogas plant, NISARGRUNA, in collaboration with Bhabha Atomic Research Centre, Mumbai. This plant utilizes food waste as feed to produce biogas, which is then supplied to hostels for cooking. This pilot project aims to demonstrate the benefits of biogas plants to villages and towns, promoting sustainable waste management practices.

Additionally, various countries have implemented measures to combat food wastage:

France: Introduced a national pact against food waste with a goal of reducing waste by at least half by 2025. Legislation requires supermarkets to donate unsold food to charity or send it to farmers for feed or fertilizer.

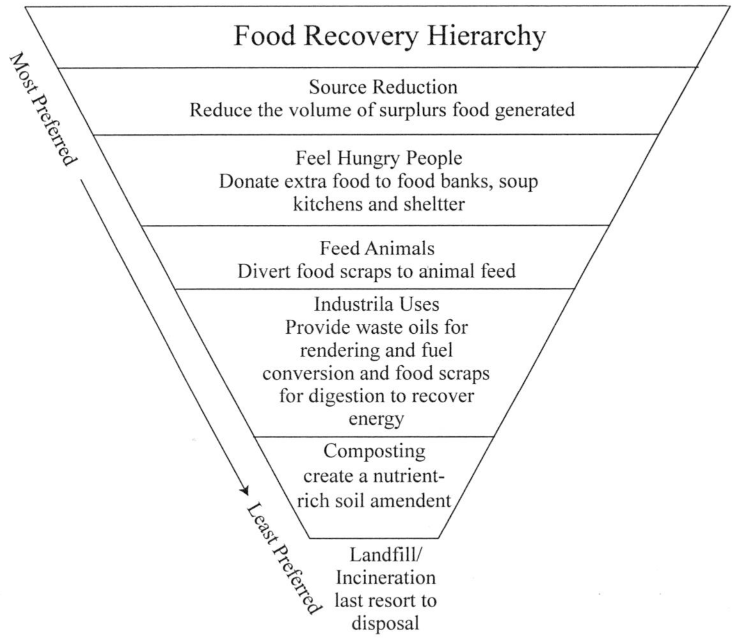

Fig. 7: Food Recover Hierarchy.

Canada: Utilizes food rescue organizations to redistribute unspoiled food from retailers, restaurants, and caterers to charities. Nanotechnology is employed to prevent fruit spoilage, and initiatives like Flashfood facilitate the resale of surplus food.

Switzerland: Implements innovative measures such as charging food-wasting customers extra at restaurants to discourage wastage.

Italy: Enacted legislation offering incentives to businesses donating food to charities and implementing programs to reduce waste in schools, hospitals, and public canteens.

India: Initiatives like PM Modi's Mega Food Parks aim to reduce food wastage by providing adequate storage and processing facilities. State governments like Karnataka have also mandated measures to store and distribute excess food from marriage halls to needy individuals.

These efforts underscore the global commitment to reducing food wastage and transforming it into valuable resources for sustainability and societal benefits.

The government has launched the comprehensive 'SAMPADA' program to enhance the economic viability of the food processing sector. Several schemes have been implemented under this program to reduce food waste and benefit farmers:

Mega food parks: This scheme aims to establish modern food processing units in clusters, linking agricultural production to markets. Financial assistance, in the form of a capital grant of 50-75 per cent, is provided for project implementation. As of March 2019, 42 parks were in various stages of implementation.

Cold chain, value addition and preservation infra: Integrated cold chain and preservation infrastructure facilities are provided across the food processing supply chain. Financial support is offered for storage, transport infrastructure, and value addition facilities. By March 2019, 299 cold chain projects were approved.

Creation of food processing and preservation capacities: This scheme focuses on modernizing processing and preservation capacities to reduce wastage. Capital grants of 35-50% are provided for approved projects. Till December 2018, 134 projects were approved.

Creation of backward and forward linkages: Financial assistance is extended for establishing primary processing centers, collection centers, and modern retail outlets. Capital grants of 35-50% are provided for eligible projects. By December 2018, 70 projects were approved.

Food safety and quality assurance infra: Infrastructure for safety and quality assurance services is developed to enhance India's competitive edge in the market. Financial assistance is provided for laboratory equipment, civil work, and compliance with standards. By November 2018, 76 food testing labs were established.

Agro processing cluster: Cluster-based infrastructure development encourages entrepreneurs to establish food processing units. Grants-in-aid of 35-50 per cent of the project cost are provided, up to a maximum limit. As of December 2018, 33 projects were approved under this scheme.

These schemes aim to modernize the food processing sector, enhance value addition, and reduce wastage, thereby benefiting farmers and promoting sustainable development.

In terms of future research directions for food waste management, biodigesters show significant promise. It's crucial to implement comprehensive life cycle planning for strategic food waste management, starting from responsible food purchasing and extending to minimizing waste throughout the entire life cycle, including responsible disposal.

Efforts to reduce food waste require collaboration across the food industry and beyond, targeting not only business practices but also consumer behavior. Addressing this issue effectively will necessitate collective action and a shift in societal norms surrounding food consumption and waste.

Innovative technologies like artificial intelligence (AI) offer new possibilities for food waste prevention. Startups such as Phood are leveraging AI and computer vision (CV) to develop solutions for tracking and managing food waste in public settings like restaurants. Their products, like PHOOD X and PHOOD XL, utilize AI to detect and monitor waste, enabling businesses to identify the root causes of waste and implement preventative measures. Additionally, these technologies facilitate food waste management in alignment with the EPA's food recovery hierarchy, promoting more sustainable practices across the industry.

Effective utilization of agro-food wastes in alternative energy generation, such as producing biogas, biofuels, and other forms of renewable energy

This area of utilizing agro-food wastes for alternative energy generation is becoming increasingly crucial, especially considering the escalating costs of energy supply and waste disposal, along with growing concerns about environmental degradation (Brennan *et al.,* 2010). From the standpoint of sustainability, which emphasizes operational efficiency, minimizing environmental impact, and socio-economic factors, it's evident that continued reliance on fossil fuels is unsustainable due to dwindling reserves and their associated greenhouse gas emissions. Hence, there is a significant push towards developing alternative renewable energy sources, including solid, liquid, and gaseous biofuels, which have the potential to be carbon neutral. One notable example of utilizing agro-food wastes for energy production is seen in rice processing plants. Rice husks, which pose environmental challenges in disposal, can be collected and used as a fuel source for boilers during the processing of paddy, saving the plant considerable costs (Zhang *et al.,* 2007). Similarly, groundnut shells from processing plants can be used as an energy source during groundnut oil production, and various residues from cereals, pulses, spices, and condiments processing plants can also be used as fuel for heat generation (Heo *et al.,* 2004). Additionally, domestic food wastes can serve as sources of biogas through processes like anaerobic digestion, which is particularly suitable for organic wastes due to their high moisture content. Several studies have demonstrated the feasibility and effectiveness of anaerobic digestion in converting food wastes into biogas, providing both energy and organic fertilizers for plant growth. However, there are challenges that need to be addressed for effective waste reuse, including variations in the types and quantities of wastes generated across different industries and regions, as well as technological limitations (Wang *et al.,* 2005). Moreover, while utilizing agricultural waste for energy production offers promise, concerns exist regarding potential conflicts with food production and land use. In addressing these concerns, third-generation biofuels derived from microalgae are emerging as a promising alternative. Microalgae offer several advantages, including rapid growth rates, minimal land requirements, and the ability to produce lipids, proteins, and carbohydrates that can be processed into biofuels and other valuable products in biorefineries (Lusk, 1988). This approach aligns with the concept of biorefinery, which aims to sustainably process biomass into a range of marketable products, including fuels, chemicals, and materials, thereby maximizing resource utilization and minimizing waste (Fredga and Maller, 2010).

Utilization of fruit waste

Banana fiber, a byproduct of banana cultivation, is derived from the plant's pseudo stem. This fiber boasts excellent physical, mechanical, and tensile properties. Various methods are employed to extract banana fibers, including water retting, chemical treatments, mechanical decortication, mechanical extraction, and combinations of these techniques. Among these, researchers have found that mechanical decortication stands out as a quick, straightforward, and cost-effective process that yields a large amount of high-quality fiber compared to water and chemical retting methods (Ameil *et al.*, 2013). The center core of a banana plant, a byproduct of banana plantations, is rich in fiber and offers significant health benefits. A technique has been developed to add value to this waste by drying and powdering it for use in biscuits.

Dinesh *et al.* (2023) worked on citrus waste disposal. The study investigated the presence of flavonoids, phenols, and the antioxidant capacity in fallen fruits of the underutilized citrus species known as pomelo (Citrus grandis (L.) Osbeck). Hesperidin content varied between 1.22 and 2.83 per cent in fruits measuring 10 mm to 14 mm, while naringin was more prevalent in fruits sized 10 mm and 12 mm, with percentages of 60.61, 60.77, 47.76, and 45.87 per cent in freeze-dried (FD) and hot air oven-dried (HAOD) samples respectively. Antioxidant assays showed that freeze-dried samples had the highest levels of ABTS and DPPH, ranging from 9.679 to 10.416 mmol L-1 Trolox and 14.825 to 16.432 mmol L-1 Trolox. The Ferric Reducing Antioxidant Power (FRAP) assay indicated higher antioxidant content in 10 mm samples for both FD and HAOD methods, with values of 4.578 mmol L-1 Trolox and 3.730 mmol L-1 Trolox. Total phenol content was highest in fruits with diameters between 10 mm and 18 mm, ranging from 48.479 to 54.498 mg GAE L-1 in FD samples and 45.757 to 51.159 mg GAE L-1 in HAOD samples. The smallest, immature fruits had the highest content, which decreased when dried by HAOD. There was a significant positive correlation between flavonoids, antioxidants, and total phenols at $p<0.01$ and $p<0.05$. The study concluded that immature, dropped fruits of the underutilized citrus species *Citrus grandis* could be a valuable source of flavonoids, total phenol concentration, and antioxidant potential. Freeze drying is recommended to preserve the most bioactive substances from these fruits for use in the pharmaceutical and nutraceutical industries, helping to manage the environmental impact of dropped citrus fruits.

Dinesh *et al.* (2018) studied quantification of flavonoids, phenols and antioxidant potential from dropped *Citrus reticulata* Blanco fruits influenced by drying techniques. Immature Citrus reticulata Blanco fruits, often considered waste and discarded in citrus orchards, actually contain valuable bioactive compounds such as flavonoids, antioxidants, and total phenols. A study was conducted to identify and quantify these compounds and to assess how different drying methods, specifically freeze drying and hot air oven drying, affect the flavonoids (notably flavanone glycosides), antioxidant potential, and total phenol content in these immature fruits. High-performance liquid chromatography (HPLC) was used to quantify flavonoids, while antioxidant activity was measured using three assays: ABTS, DPPH, and FRAP. The total phenol content was also determined.

Dinesh *et al.* (2018) conducted a study on underutilized citrus sp. Pomelo (*Citrus grandis*) and Kachai lemon (*Citrus jambhiri*) exhale in phytochemicals and antioxidant potential. The phytochemical content and antioxidant potential of sixteen citrus varieties, including mandarins, limes, sweet oranges, and underutilized fruits, were evaluated. The levels of limonoids, ascorbic acid, and carotenoids varied significantly among these varieties. The antioxidant potential was measured using four assays: ABTS, DPPH, FRAP, and TPC. Among these varieties, the Kachai lemon exhibited high antioxidant capacity, particularly in the DPPH assay (9.38 mM L-1 Trolox), and recorded the highest total phenolic content (TPC) at 13.57 mM L-1 Trolox. The pomelo demonstrated significant antioxidant potential, with the highest values in the ABTS (4.49 mM L-1 Trolox) and FRAP (1.92 mM L-1 Trolox) assays, comparable to that of grapefruit. There was a notable correlation between the DPPH and TPC assays, as well as between FRAP and TPC. These findings suggest that Kachai lemon and pomelo, as underutilized citrus fruits, show enhanced antioxidant capacity and could be valuable for energy, nutrient, and health supplement purposes.

The study found that freeze-dried samples of 12 mm and 14 mm fruits retained the highest hesperidin flavonoid content (27.03% and 27.20%) compared to hot air dried samples (17.99%). These freeze-dried samples also had higher phenolic content, ranging from 50.54 to 54.19 mg GAE L-1. The antioxidant activity in freeze-dried fruits ranged from 12.21 to 13.55 mM L-1 Trolox (ABTS assay) and 15.27 to 16.72 mM L-1 Trolox (DPPH assay), with FRAP values between 7.31 and 9.07 mM L-1 Trolox. There was a significant positive correlation between the flavonoid hesperidin, the antioxidant assays, and total phenolic content.

The results indicated that these discarded citrus fruits could be a valuable source of bioflavonoids, especially hesperidin, and antioxidants, making them useful for the pharmaceutical and nutraceutical industries.

Essential oil, a byproduct of citrus fruits, has garnered significant interest. These volatile, fragrant oils are found in various parts of plants such as buds, flowers, leaves, fruits, seeds, and roots. They capture the complete scent profile of a specific plant species. Recently, citrus essential oil has been discovered in different parts of the fruit, particularly in the flavedo, and is typically extracted through methods like expression, solvent extraction, or distillation. These oils are highly volatile and consist of a diverse range of organic compounds with various functional groups and molecular structures, including terpenes and their oxygenated derivatives, aromatic benzenoid compounds, aliphatic hydrocarbons, and compounds containing nitrogen or sulfur.

Every year, over 50 million metric tons of apples are produced, and apple pomace is often discarded as waste in the food industry. This byproduct of apple juice and cider production is a sustainable raw material that can be transformed into valuable products like nutraceuticals and pectin. Rich in antioxidant compounds, apple pomace offers numerous health benefits, making its valuable components useful in the food and pharmaceutical industries. Various traditional and innovative technologies are employed to extract these valuable products from apple pomace, with a particular emphasis on new and environmentally friendly techniques.

Tropical fruits like mango, jackfruit, papaya, passion fruit, and dragon fruit are favored for their delicious taste and rich nutritional content. However, a significant amount of these fruits, including their peels and seeds, often goes unused, resulting in considerable waste. The valorization process aims to extract bioactive compounds from these fruit wastes using methods such as solvent extraction, supercritical fluid extraction, or enzymatic extraction. These extracted compounds can be added to various products to boost their nutritional value or functionality. For instance, extracts from fruit peels that are rich in antioxidants can be used as natural preservatives to extend the shelf life of food products. Seed extracts with antimicrobial properties can be utilized in creating antimicrobial films for food packaging, which helps reduce food spoilage and enhance food safety. Additionally, bioactive compounds from tropical fruit wastes can be turned into dietary supplements or skincare products to promote health and wellness.

Conclusion

Food waste occurs at various stages of the food supply chain, but the highest proportion happens in households. Governments and professionals are prioritizing efforts to reduce food waste at home due to its detrimental effects on the economy, environment, and society. This article explores different types of food waste and their unique treatment methods. Both researchers and business professionals are increasingly interested in recycling materials, especially food scraps, into usable fuel. Various waste management technologies like incineration, composting, anaerobic digestion, landfill, and animal feeding are discussed, but sustainable methods are crucial for environmental protection. Achieving sustainable development goals (SDGs) 2 (zero hunger) and 12 (climate action) requires significant progress in reducing food loss and waste (FLW) to ensure sustainable consumption and production patterns. Food waste significantly impacts the financial, environmental, and social aspects of sustainability. More scientific research is needed to provide accurate estimates of food loss. Anaerobic digestion is a widely used method for treating food waste as it generates energy and biogas, diverting waste from landfills and reducing methane emissions. Microbial fuel cells (MFCs) are also being studied as an energy source due to the challenging nature of anaerobic digestion. Despite the significant amount of food waste generated annually, it indicates that there is enough food produced to support the world's population. Bioenergy production from food waste can diversify agricultural systems, reduce greenhouse gas emissions, achieve fossil-fuel independence, and mitigate climate change concerns. Changes in consumer lifestyles due to the COVID-19 pandemic have the potential to improve the efficiency of domestic food production processes and reduce food waste.

References

Akaranta O, 2006. Raw Materials for Surface Coatings from Agricultural Wastes. *Surface Coatings International*. 79(4):152-154.

Amel B A, Paridah M T, Sudin R, Anwar U M K and Hussein A S, 2013. Effect of fiber extraction methods onsome properties of kenaf bastfiber. *Industrial Crops and Products,* 46:117-123.

Bos-Brouwers H E, Scheer F P, Nijenhuis M A, Kleijn F and Westerhof M, 2013. FoodBattle: reductie milieudruk voedselverspilling op het snijvlak van supermarkt and consument. Food Biobased Res Wageningen UR

Brennan L, Owende P and Brennan A F, 2010. Biofuels from microalgae-A review of technologies for production, processing, and extractions of biofuels and co-products, Renew. *Sust. Energ. Rev*. 14(2):557-577.

Daliri E B, M and Lee B H, 2015. Current Trends and Future Perspectives on Functional Foods and Nutraceuticals. In Beneficial Microorganisms in Food and Nutraceuticals; Springer International Publishing: Cham, Switzerland, pp. 221–244.

Dey T, Bhattacharjee T, Nag P, Ghati A and Kuila A, 2021. Valorization of agro-waste into value added products for sustainable development. *Bioresour. Technol. Rep*. 16:100834.

Dinesh Kumar, Ladaniya M S, Gurjar M, Kumar S and Mendke S, 2021. Quantification of Flavonoids, Phenols and Antioxidant Potential from Dropped Citrus reticulata Blanco Fruits Influenced by *Drying Techniques Molecules* 26(14):4159

Dinesh Kumar, Ladaniya M S, Manju Gurjar, Sachin Mendke, Sunil Kumar and Dilip Ghosh, 2023. Elucidation of flavanones, phenols and antioxidant capacity influenced by drying methods from physiologically dropped underutilized Citrus grandis fruits. Front. Plant Sci., 10 July 2023 Sec. Plant Nutrition. Volume 14 - 2023 | https://doi.org/10.3389/fpls.2023.1193635

Dinesh Kumar M S, Ladaniya M and Gurjar, 2018. Underutilized Citrus sp. Pomelo (Citrus grandis) and Kachai lemon (Citrus jambhiri) exhale in phytochemicals and antioxidant potential. *Journal of Food Science and Technology*, 56 (1):217-223.

Donner M, Gohier R and de Vries H, 2020. A new circular business model typology for creating value from agro-waste. *Sci. Total. Environ.* 716, 137065.

FAO G, 2011. Global food losses and food waste–extent, causes and prevention. Save FOOD: Initiative Food Loss Waste Reduction 9:2011

Fredga O A and Maller R, 2010. Plant residues as soil amendments and feedstock for bioethanol production. *Waste Manag.* 28(6):737–748.

Gillian D, 1992. Wasting our Natural Resources, *The Outreach Magazine* 1, (2):13-17.

Heo N H, Park S C and Kang H, 2004. Effects of mixture ratio and hydraulic retention time on single-stage anaerobic co-di*gestion of food was*te and waste activated sludge. *J. Environ. Sci. Health.* A39(7):1739–1756.

Khan A H, López-Maldonado E A, Khan N A, Villarreal-Gómez L J, Munshi F M, Alsabhan A H and Perveen K, 2021. Current solid waste management strategies and energy recovery in developing countries-State of art review. Chemosphere 291:133088. https:// doi.org/10.1016/j.chemosphere.2021.133088

Kour R, Singh S, Sharma H B, Naik T S, Shehata N, Pavithra N, Ali W, Kapoor D, Dhanjal DS, Singh J and Khan A H, 2023. Persistence and remote sensing of agri-food wastes in the environment: current state and perspectives. Chemosphere 14:137822. https://doi.org/10.1016/j.chemosphere.*2023.137822*

Lusk P, Methane 1998. Recovery from animal manures the current opportunities casebook. National Renewable Energy Laboratory, NREL/ 1998; SR-580-25145.

Mishra G, Panda B K, Ramirez W A, Jung H, Singh C B, Lee S H and Lee I, 2021. Research advancements in optical imaging and spectroscopic techniques for nondestructive detection of mold infection and mycotoxins in cereal grains and nuts. *Compr Rev Food Sci Food Saf* 20(5):4612–4651

Odocha J N K, 1994. Waste Generation and Management in Depressed Economy. A lecture delivered to environmental faculties, Abia State University Auditorim, Uturu.

Rather R A, Ara S, Sharma S, Padder S A, Lone F A, Mir S A, Ahmad Baba Z, Ayoub I, Bhat T A and Baba T R, 2022a. Seasonal changes and determination of heavy metal concentrations in Veshaw River of Western Himalaya. *Front Environ Chem* 3:1018576. https://doi.org/10.3389/fenvc.2022.1018576

Rather R A, Bano H, Firoz A, Mohammed Ali H, Bhat MA, Padder S A, Nafees H and Hakeem K R, 2022b. The assessment of morphological diver*sity* of Colchicum luteum L., an economically important threatened medicinal plant of Kashmir Himalaya. Sustainability 14(3):1327. https://doi.org/10.3390/su14031327

Safeguard International Training Institute, 2001. Environmental management Training Modules, Port Harcourt, Nigeria.

Singh A, Singhania R R, Soam S, Chen C W, Haldar D, Varjani S, Chang J S, Dong C D and Patel A K, 2022. Production of bioethanol from food waste: status and perspectives. Bioresour Technol 20:127651. https://doi.org/10.1016/j.biortech.2022.127651

Solangi N H, Kumar J, Mazari S A, Ahmed S, Fatima N and Mubarak N M, 2021. Development of fruit waste derived bio-adsorbents for wastewater treatment: a review. J Hazard Mater 416:125848. https://doi.org/10.1016/j.jhazmat.2021.125848

Sorathiya L M, Fulsoundar A B, Tyagi K K, Patel M D and Singh R, 2014. Eco-friendly and modern methods of livestock waste recycling for enhancing farm profitability. *Int. J. Recycl. Org. Waste Agric.* 3:1–7.

Wang J Y, Zhang H, Stabnikova O and Tay J H, 20*05. Comparison of* labscale and pilot-scale hybrid anaerobic solid–liquid systems operated in batch and semi-continuous modes. *Process Biochem*. 40(11):3580–3586.

Zhang R, El-Mashad H M, Hartman K, Wang F, Liu G, Choate C and Gamble P, 2007. Characterization of food waste as feedstock for anaerobic digestion. 2007;98:929–935.

11

Food Safety Management of Fruits Processed Products: Perspectives of Pesticide Residues

Partha P Choudhury and Kaushik Banerjee*

ICAR-Indian Institute of Horticultural Research, Bangalore-560 089, Karnataka
**ICAR-National Research Centre for Grapes, Pune-412 307, Maharashtra*

Introduction

A wide range of agricultural pesticides is currently being used in India. In the financial year of 2015, the size of the crop protection market in India was of about $4.5 billion (FICCI, 2016). A large number of these pesticides are also being used in considerable quantity on horticultural crops, including fruit crops. In general, judicious application of a pesticide following Good Agricultural Practices (GAP) does not leave its residues in crop produces above unacceptable limit. But the GAP recommendations are not available for all crop-pesticide combinations. The usage pattern of pesticides also sometimes gets modified at farm level due to sudden climate changes, and under such situations, the indiscriminate use of pesticides to combat disease and insect complex might result in the contamination of pesticides in fruits.

It is very difficult to protect horticultural crops from diseases and insect pests without applying pesticides. Farmers do not have many safer alternatives of hazardous pesticides especially under emergency situations, when saving the crop from economic loss appears utmost important. For the obvious reason, the use of pesticides on the commercial cultivation of horticultural crops is in increasing trend.

The load of pesticides either in the environment or in edible crop produce should be mitigated by different means. A substantial quantity of pesticides inside the fruits is managed by their own plant-metabolic activities, which are mostly enzyme-depended and enzyme mediated. Despite primary and secondary metabolism processes, still the edible portion of a fruit might carry some quantity of pesticides, the extent of which depends upon numerous factors that affect the pesticide absorption, penetration and degradation, and differ from one category of fruit to another. The rate at which pesticides are moved and dissipated is closely related to the physico-chemical parameters of pesticide itself and surrounding environmental conditions. In general, the residues of pesticides adhered on to the surface of fruits can easily be dislodged and removed by washing and peeling of the commodity. This has been documented extensively worldwide, however the chemistry of the pesticides inside the fruits and their processed items are not much investigated. This is especially important since fruits are mostly consumed as fresh food items.

The retention of pesticides depends on the physiochemical properties of the pesticide molecules as well as food. In fruits, most of the pesticide residues are retained on peel surface (Awasthi 1993). This is the reason that majority of the residues are removed by washing, peeling or treatments with chemical solutions like vinegar, turmeric, sodium bicarbonate, common salt or alcohol. Washing of commodities with water was found very effective to reduce pesticide residues, as found in the cases of malathion in okra (Nath *et al.,* 1975), malathion on bell peppers (Dash *et al.,* 2001), and chlorpyrifos and its degradation products on eggplant, tomato, okra, spinach, cauliflower and potato (Randhawa *et al.,* 2007). Many reports are available on the decontamination by washing of fruits and vegetables contaminated with different pesticides like HCB, *p,p*-DDT, lindane, pirimiphos-methyl, dimethoate, profenofos, bifenthrin, chlorothalonil, chlorpyrifos, cypermethrin, malathion, mancozeb, maneb, methyl-parathion, oxadixyl, procimidone, propineb, thiophanate-methyl and λ-cyhalothrin on tomato (Abou Arab 1999; Chavarri *et al.,* 2004, 2005; Cengiz *et al.,* 2007; Rani *et al.,* 2013; Chauhan *et al.*, 2012, 2012a; Kwon *et al.,* 2015; Reiler *et al.*, 2015), dithionan and chlorantraniliprole on eggplant (Kumar and Agarwal 1991, Vijayasree *et al.,* 2015); and trichlorfon, dimethoate, dichlorvos, fenithrotion and chlorpyrifos on cucumber (Liang *et al.,* 2012); and bupirimate, pirimicarb, tetraconzole, deltamethrin, pyraclostrobin, λ-cyhalothrin, folpet, boscalid, iprodione, trifloxystrobin, fludioxonil, cyprodinil, α-cypermethrin, fenhexamid, acetamiprid and chlorpyrifos from strawberries (Lozowicka *et al.*, 2016).

When commodities are dried in the sun, oven or a food dryer, the procedure has been found to reduce pesticide residues considerably. On exposure to sun drying, photodegradation plays the most significant role in degrading pesticide residues. Photodegradation is an abiotic process in which reactive oxygen species (ROS) like OH*, O_3, and 102 either specifically or nonspecifically oxidize the functional groups in a pesticide molecule through molecular excitation brought on by light energy absorption. For example, a sun or hot air drying eliminated 20-30 per cent of chlorpyrifos and fenitrothion residues from red pepper (Lee 2001). Industrial dehydration process reduces phosalone levels in apples by over 80 per cent (Mergnat *et al.,* 1995). Sun drying causes greater reductions in pesticide residues than the oven process, as observed in bitertanol in apricots, dimethoate in raisins making (when fresh grapes are dried), iprodione in prunes (Cabras *et al.,* 1998, 1998a, 1998b), and methamidophos in grapes (Athanasopoulos *et al.,* 2005). In addition, sun drying has been reported to cause greater reductions in pesticide residues than the oven process, as observed in dimethomorph, famoxadone, cymoxanile, pyraclostrobin and metiram in grapes (Shabeer *et al.,* 2015, 2015a).

Pasteurization and sterilization as commercial thermal processing techniques used for food preservation can cause some reduction in residues of various pesticides. This thermal process reduces the level of pesticides like chlorpyrifos reduction in peach puree (Marudov *et al.,* 1999) and malathion and parathion in tomato (Severini *et al.*, 2003). Processes like blanching, steaming, boiling, cooking, frying, baking or roasting have roles in pesticide residues reduction.

During the processing of fruits to their products, majority of the commonly used pesticides dissipate below their safety limits. However, in some cases, particularly in drying processes, pesticides may concentrate in unit weight of the final product due to the moisture loss.

Use of pesticides in fruit crops

Pesticides are the substances used to protect crops against various pests, including insects, weeds, and fungi. In contemporary agriculture, pesticides are now one of the most important inputs for guaranteeing nutrition security, especially in emerging nations where population expansion outpaces agricultural productivity.

Since fruit crops are mostly perennial in nature, with a very few exceptions like strawberry, they are vulnerable to a variety of insect pests and diseases. Insecticides and fungicides are frequently used round the year, particularly during fruit bearing season to improve production and quality. Of all the South Asian nations, India consumes the most pesticides for all kinds of crops, ranking third in the world. India ranks second in the fruit production just after People's Republic of China. In India, the Central Insecticide Board and Registration Committee (CIBRC), under the Ministry of Agriculture and Farmers Welfare is authorised to register and recommend pesticides. Around 295 different pesticides are registered in India for crop protection. For the pest management in fruit crops, the number of registered pesticides is 95, out of which 30 are insecticides, 53 fungicides, 3 herbicides and 9 growth regulators (CIB and RC 2024). In many occasions, the application of non-recommended pesticides is observed in different fruit crops, which is reflected as their residues in fruits.

Pesticide residues in fruits

The residue of any pesticide is defined by the World Health Organization (WHO) as, "Any substance or mixture of substances in food for man or animals resulting from the use of a pesticide and includes any specified derivatives, such as degradation and conversion products, metabolites, reaction products, and impurities that are considered to be of toxicological significance." Pesticide residues in fruits are a significant concern for both human health and the environment. In the case of vegetables, pesticide residues degrade during cooking in a large extent. We consume majority of the fruits as fresh commodity. So, almost entire quantity of pesticides present in the fruits gets entry into our body and may cause adverse effect to human being. It is essential to follow GAP for the management of pesticide residues in fruits.

Government of India has taken several measures for ensuring proper use of pesticides by the farmers in the country. The Department of Agriculture, Cooperation and Farmers Welfare, Ministry of Agriculture is regularly monitoring the pesticide residues in food commodities and environmental samples under the central sector scheme, "Monitoring of Pesticide Residues at National Level" which was initiated during 2005-06 and has 25 participating laboratories representing Ministry of Agriculture, Indian Council of Agricultural Research, Ministry of Health and Family Welfare, Ministry of Environment and Forest, Council of Scientific and Industrial Research, Ministry of Chemical and Fertilizer, Ministry of Commerce and State Agricultural Universities. The participating laboratories collect food commodity samples, which include fruits also from various Agriculture Produce Marketing Committee markets, public distribution systems, farm gates. The samples are analyzed for the presence of possible pesticide residues in various fruits along with other food commodities. The pesticide residue data generated under the above scheme are shared with State Governments and concerned Ministries/Organizations to initiate corrective action for the judicious and proper use of pesticides on crops with an integrated pest management approach and to generate awareness amongst farmers.

During 2017-18, only 25 (1.1%) of the 2,274 samples of different fruits, including apple, banana, pear, grapes, orange, pomegranate, guava, sapota, and mango, that were collected and examined had residues found in them, out of which 494 (21.7%) samples had FSSAI MRLs exceeded. No pesticide traces were found in fruit samples from 1780 (78.3%). In 277 fruit samples (12.8%), the non-approved insecticides were found (MPRNL 2019).

Since farmers are becoming conscious about the selection of pesticides for pest management, presence of residues can be avoided in fruits by following recommended application rates and right time of harvesting of crops. During the time of harvesting, harvested fruits should not carry pesticide residues above permissible limit. The safe time for the harvest of a fruit is defined as waiting period or post-harvest interval (PHI), which is available in the label claim for a formulation or in the recommendation given by CIBRC (CIB and RC 2024).

Fate of pesticides during processing of fruits and vegetables

The fate of pesticides during the processing of fruits and vegetables is a significant concern for both food safety and public health. Processing methods such as washing, peeling, and cooking can significantly reduce the levels of pesticide residues in fruits and vegetables. The extent to which pesticides are removed or degraded during processing depends on several factors, including the type of pesticide, its physicochemical properties (like solubility and vapor pressure), and the processing technique used. On the contrary, the per cent concentration of pesticides increases during the process like drying of fruits.

Washing: The most popular type of processing, and an essential first step in both home and commercial preparation, is washing. There are various washing techniques, which can dislodge pesticides that are retained loosely; nevertheless, the effectiveness of these techniques varies greatly.

There are different ways that pesticides can get into fruits and vegetables: direct deposition, absorption through leaves, and uptake through roots. Typically, plants absorb systemic insecticides through their leaves, roots, or both. A significant amount of the systemic insecticides that are sprayed penetrate the plant and are translocated either through phloem or through xylem (Fig. 1). Some of it might settle within the fruits. Fruits absorb and translocate very little of the contact pesticides. The majority of the contact pesticides that are sprayed deposit on the leaves and fruits' surfaces.

Pesticide absorption is expected to increase with a plant's ability to retain more of the spray. The characteristics of the target plant and the spray solution affect spray retention. A few traits that influence spray retention are leaf waxiness, pubescence (hairiness), and orientation. Compared to less waxy leaves, waxy leaf surfaces reject water-based spray solutions, making it easier for spray droplets to flow off. While thick pubescence can maintain spray droplets above the leaf surface and decrease spray contact with the leaf, sparse pubescence or hairs on the leaf may aid in retaining spray droplets (Fig. 2).

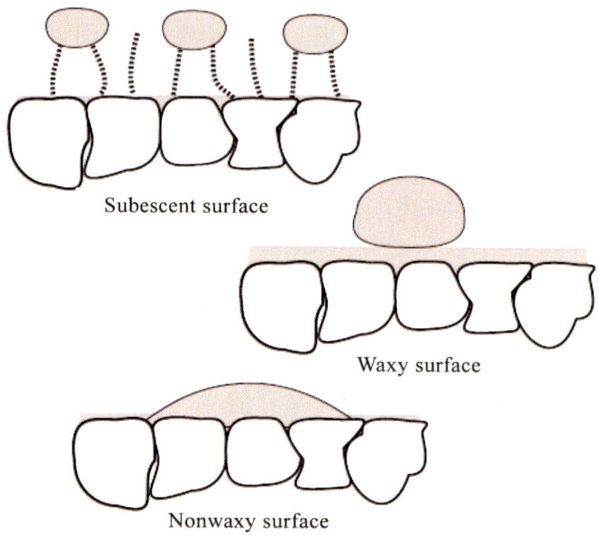

Fig. 1: Spray droplet spread on a leaf surface as influenced by leaf pubescence and waxiness.

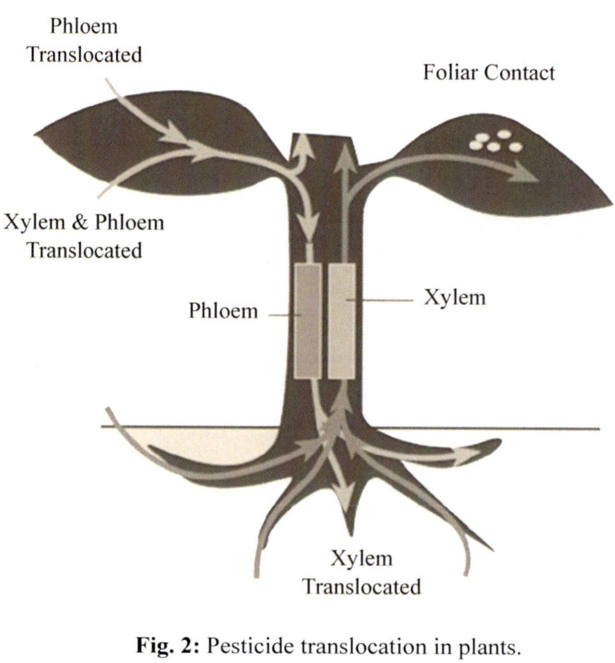

Fig. 2: Pesticide translocation in plants.

According to Heshmati *et al.* (2020), Gonzalez Rodriguez *et al.* (2011), Lozowicka *et al.* (2016), and Polat (2021) contact pesticides do not seep into the commodities. As a result, the washing process may quickly eliminate these residues. However, systemic pesticides have the potential to seep into other plant parts, making it very challenging to eradicate them from various plant sections (Acoğlu *et al.*, 2018; Lozowicka *et al.*, 2013). When it comes to lowering pesticide residues on fruits, water solubility is crucial. Higher soluble pesticides can be easily removed, with a few exceptions (Krol *et al.,* 2000; Lozowicka *et al.,* 2016; Randhawa *et al.,* 2014b).

A simple washing under running water or dipping in water for 15 to 20 seconds is a basic method to wash away some pesticides. The washing of mango fruits by dipping in water for 10 minutes reduced the level of dimethoate and fenthion by 66-68 per cent and fenvalerate by 21-27 per cent (Awasthi 1993). Similarly, washing of golden delicious apples reduced phosalone residues by 30-50 per cent (Mergnat *et al.,* 1995). The use of water with common salt or sodium bicarbonate, non-toxic acidic solution, ozonated water, natural or synthetic surfactants and ultrasonic cleaning may enhance the efficiency of water washing, which are given in Table 1.

Table 1: Effects of different washing treatments on pesticide residues.

Commodity	Washing treatments	Pesticides	Range of reduction	Reference
Apple	Water, sodium hypochlorite, peroxyacetic acid, tween 20.	Thiabendazole Diphenylamine Pyrimethanil Thiabendazole Diphenylamine.	50.97 88.8 40.36 49.04 46.9	Al-Taher *et al.,* 2013
Lemon	Tap water, sodium hypochlorite, peroxyacetic acid, tween 20	Imazalil.	41.68	Al-Taher *et al.,* 2013
Orange	Tap water, sodium hypochlorite, peroxyacetic acid, tween 20.	Imazalil Thiabendazole Abamectin.	64.71 78.05 2.38	Al-Taher *et al.,* 2013 Acoglu and Omeroglu 2021
	Tap water, sodium carbonate, sodium chloride, acetic acid, apple vinegar-water, grape vinegar-water.	Buprofezin Etoxazole Imazalil Thiophanate-methyl.	24-59 5-46 5-61 39-82	Acoglu and Omeroglu 2021
Peach	Water, sodium hypochlorite, peroxyacetic acid, tween 20.	Fludioxonil.	71.63	Al-Taher *et al.,* 2013

Commodity	Washing treatments	Pesticides	Range of reduction	Reference
Strawberries	Tap water, ozone eater, ultrasonic cleaning.	Bupirimate Cyprodinil Fenhexamid Fludioxonil Folpet Iprodione Pyraclostrobin Tetraconazole Trifloxystrobin Acetamiprid Alpha-cypermethrin Chlorpyrifos Deltamethrin Lambda-cyhalothrin Pirimica.	6-57 15-54 16-57 12-60 10-66 19-65 20-89 2-85 11-52 24-63 35-81 42-79 14-72 6-58 14-65	Lozowicka *et al.*, 2016

Peeling and trimming: Peeling and trimming are important steps in food processing, especially for fresh-cut fruits. this process is very effective since many pesticides are concentrated in the peel (Bajwa and Sandhu, 2014). Peeling is the removal of the outer skin or peel of fruits and vegetables. It is required to improve the appearance, texture, and taste of the final product. Peeling can be achieved through various methods, including (Toker and Bayiindirli, 2003):

Steam peeling: Uses high-pressure steam to loosen the skin.

Knife peeling: Involves manual or mechanical peeling with a knife.

Abrasion peeling: The produce is rubbed against an abrasive surface to remove the skin.

Caustic peeling: Involves the use of chemicals to soften and remove the skin.

Flame peeling: Direct exposure to flames to char and remove the skin.

Trimming: Trimming refers to the removal of unwanted or inedible parts from raw fruits such as stems, cores, and damaged areas. This step is crucial for ensuring the quality and safety of the final product. Trimming can also help in reducing microbial load and improving the overall hygiene of the food.

The deposition of pesticides in fruits occurs in three different routes: 1. Pesticides are absorbed by the root and translocated to fruits; 2. Pesticides are absorbed by leaf and translocated to fruits; 3. Pesticides directly deposit on fruits. The first two routes lead the deposition of systemic pesticides inside the fruits; thus, it is highly difficult to remove systemic pesticides from fruits

(Acoğlu *et al.,* 2018, Lozowicka *et al.,* 2013). Contact pesticides do not penetrate into the commodities (Heshmati *et al.,* 2020, Gonzalez Rodriguez *et al.,* 2011, Lozowicka *et al.,* 2016, Polat, 2021). Therefore, these residues could easily be reduced through peeling process. In an experiment on guava, three pesticides, viz. imidacloprid, deltamethrin and carbendazim, of different group were applied at recommended rate on guava tree. During the processing study, it was observed that peeling of guava reduced almost the entire load of all three pesticides and the products prepared from it were found pesticide free (Choudhury, 2024).

Comminution: Comminution is a fundamental process in food processing that involves reducing the size of solid materials into smaller pieces. This is achieved through various operations such as crushing, grinding, milling, mincing, and dicing. Each of these operations is suited to specific applications and food types, for example:

- Milling is often used for cereals.
- Mincing is typically associated with meat.
- Dicing is used for cutting tubers into cube-shaped pieces.
- Grinding is commonly applied to spices.

The purpose of comminution is to produce smaller particles that are desired either because of their large surface area or because of their shape, size, and number. This can aid other processes such as extraction and can shorten heat treatments like blanching and cooking. The reduction in size can also enhance the rate of physical or chemical processes by increasing the area of contact between solid particles and other phases involved in reactions. The milling process is not required during fruit processing, though it removes a substantial quantity of pesticide residues from cereals (Sgarbiero *et al.,* 2002, Uyga *et al.,* 2005). However, During Mincing and grinding steps, the scope for pesticide dissipation is less, but the increased surface may expose pesticide molecules towards the physico-chemical factors, which may dissipate pesticides in subsequent steps of processing.

Juicing: The process of juicing fruits can significantly reduce the amount of pesticide residues. During juicing, the combination of supplementary procedures such as washing, pressing, sterilization, and enzymatic treatment greatly helps in reducing pesticide residues in the fruit matrix (Nguyen *et al.,* 2020). For example, a study has shown that the amount of pesticide residue in oranges was reduced by 63–100 per cent during fruit juice processing (Omeroglu *et al.,* 2022). Similarly, various food processing techniques, including juicing, have been found to be effective in removing pesticide residues from apples, with juicing showing a reduction range between 63 and 100 per cent (Stowik-Borowiec and Szpyrka 2020).

Cooking: Thermal Processing and Pesticide Dissipation: Thermal processing techniques, such as cooking, baking, and canning, can significantly affect pesticide residues in fruits. The effectiveness depends on the type of pesticide and the duration of treatment. Exposure to heat during cooking can lead to the degradation of pesticides. High temperatures break down chemical bonds, rendering pesticides less toxic. Some pesticides undergo hydrolysis (chemical breakdown due to water) during cooking, leading to their dissipation. Volatile pesticides may evaporate during cooking.

Apple fortified with daminozide at the tolerance level of 30 ppm and boiled for 30 min contained 1.53 ppm of dimethylhydrazine. Daminozide was found to decompose to dimethylhydrazine when boiled in the presence of apple homogenate (Newsome, 1980). The boiling process of strawberry in tap water for five minutes reduces boscalid, bupirimate, chlorpyrifos, cyprodinil, fludioxonil, and iprodione by 41, 49, 42, 43, 34, 42 per cent, respectively; however, the process reduces pyraclostrobin by 91 per cent (Lozowicka and Jankowska, 2016).

In one experiment on the estimation of pesticide residues in papaya, and its processed products, the dynamics of four pesticides, viz. lambda cyhalothrin, tebuconazole, difenconazxole and trifloxistrobin were studied (Choudhury, 2024). In osmotically dehydrated (OD) product, pesticides dissipated through physical exclusion due to osmosis. Lambda cyhalothrin, difenconazole and trifloxystrobin were not detected in the OD bar of papaya. The dissipation of tebuconazole in OD product was also found around 95 per cent. In oven-dried bar, pesticide dissipation was found above 85 per cent for all four pesticides. Similarly, the process of preparing OD and oven-dried bar from pesticide-contaminated mango pulp could reduce the load of acephate, imidacloprid and carbendazim by over 90 per cent.

Plasma activated water (PAW) is generated using atmospheric air plasma discharge. It contains reactive species like nitrates, nitrites, and hydrogen peroxide. PAW effectively reduces pesticide residues while maintaining fruit quality and nutritional parameters (Sarangapani *et al.,* 2020). Studies have shown reductions of up to 79 per cent on grapes and 69 per cent on strawberries when treated with plasma-activated water (PAW), a novel approach for pesticide reduction. Another pesticide, carbaryl, showed reductions of 86 per cent on grapes and 73 per cent on strawberries when treated with PAW.

Drying: Drying fruits is a preservation method that involves removing water from the fruit, which inhibits the growth of microorganisms and reduces decay. Here are some common drying processes used for fruits:

Sun drying: This is the most traditional method, where fruits are laid out in the sun to dry naturally. It's cost-effective but can be influenced by weather conditions.

Hot air drying: Fruits are dried in ovens or dehydrators using artificially generated hot air. This method provides more control over the drying process compared to sun drying.

***Freeze drying*:** Also known as lyophilization, this process involves freezing the fruit and then reducing the surrounding pressure to allow the frozen water in the fruit to sublimate directly from the solid phase to the gas phase. Freeze drying preserves the most nutrients but is also the most expensive.

***Vacuum drying*:** This method dries fruits in a vacuum, which lowers the boiling point of water and allows drying at lower temperatures, thus preserving the fruit's flavor and nutritional value.

***Microwave drying*:** Utilizes microwave energy to heat the water inside the fruit, causing it to evaporate quickly. This method is fast but may not be suitable for all types of fruits.

***Osmotic drying*:** Involves soaking fruits in a concentrated sugar or salt solution before drying. This process not only removes water but also infuses the fruit with additional flavors.

The dissipation of pesticides during the drying process of fruits is an important factor in ensuring the safety and quality of the final product. Pesticides are degraded at higher rates during certain drying processes. Compared to the fresh fruit, the dried apricot fruit had fewer pesticide residues. Bitertanol residues decreased by approximately 50 per cent after drying in the sun, from 0.50 parts per million at harvest. Nevertheless, oven dying resulted in a 26 per cent increase in residues; this could be explained by photodegradation under sunlight drying (Cabras *et al.,* 1998a). Raisins sun-dried to reduce dimethoate residues by 81 per cent, while raisins oven-dried, which required washing first, to reduce residues by 72 per cent. Heat was blamed for the decrease in dimethoate since it can lead to evaporation and degradation (Cabras *et al.,* 1998b). 64.2–71.9 per cent of the methamidophos in grapes was lost during the drying process, probably as a result of the pesticide evaporating (Athanasopoulos *et al.,* 2005). Sun drying was found more effective than the oven process in the dissipation of pesticide residues, as observed in dimethomorph, famoxadone, cymoxanile, pyraclostrobin and metiram during the processing of grapes to resins (Shabeer *et al.,* 2015, 2015a).

Pesticides are degraded at higher rates during certain drying processes. For instance, microwave drying (700 W, 4 min) of jujube fruit showed a degradation rate of pesticides from 67 to 93 per cent, compared to hot air drying (Nguyen *et al.,* 2020). In oranges, pesticide residues were mostly found on the peels, and washing with tap water decreased the residue levels by 26 to 84 per cent. The amount of residue in oranges was reduced by 63 to100 per cent during fruit juice processing, while residues were removed by 90-100 per cent after jam processing (Omeroglu *et al.,* 2022).

These findings indicate that the type of drying process and the physical and chemical properties of the pesticides significantly affect the dissipation rate. It is essential to consider these factors to minimize pesticide residues in dried fruits effectively.

Fermentation: Fermentation is a traditional practice for extending the shelf-life of food and enhancing its nutritional value. The process involves microbial cultures that metabolize food components, leading to desirable biochemical changes (FEMSmicroBlog, 2021). These changes can result in the production of inhibitory metabolites like organic acids, which, along with a decrease in water activity, help preserve the food. Additionally, fermentation can improve food safety by inhibiting pathogens and removing toxic compounds, as well as enriching the food's nutritional profile (Siddique *et al.,* 2023).

Fermentation is a process where microorganisms like bacteria, yeasts, and molds change the sensory and functional properties of food to produce an end product that is desirable to consumers. It's one of the oldest methods developed to extend the storage time of foods and includes techniques such as salting, cooking, smoking, and sun drying. A variety of foods are commonly fermented, which include dairy products, vegetables, fruits, meats, and grains. The key micro-organisms associated with these foods are also identified, ensuring the right culture is used for each type of food. During the fermentation processes, microbes release extracellular enzymes, which are mostly responsible for the degradation of biomolecules. Some of these enzymes may incidentally degrade pesticides present in the matrix, which is called incidental metabolism. Sometimes microbes develop enzymes through mutation to detoxify pesticides for their survival.

Pesticide-contaminated grape pulp was incubated with *Saccharomyces cerevisiae* var ellipsoids and the fermented mixture was further clarified to get wine (Choudhury, 2024). During this fermentation process of grape pulp containing three fungicides, viz. fluopyram, tebuconazole and trifloxystrobin, the range of degradation was found within 63 to 84 per cent in wine. In another experiment with 30 pesticides, most of the pesticides were found to be degraded by the microbial process below safety limits. After ageing of 100 days, the concentrations of almost all pesticides went below safety limits.

Experiments on the pesticide dynamics during the preparation of probiotics from guava and pineapple were conducted by using *Lactobacillus helveticus* (Choudhury 2024). There is not much change in the concentration of pesticides during the preparation of guava-based probiotic. Only chlorpyrifos was degraded by 20 per cent during the pasteurization process. A similar result was found in the case of pineapple-based probiotic formulation.

Processing factor

The majority of the time, pesticides dissipate during fruit processing. However, in many instances, the concentration of pesticides increases, as observed in drying process of fruits. To gain a complete understanding of the hazards of pesticide exposure to human health from the consumption of fruit-based products, it is essential to evaluate changes in the concentration and degradation of regularly used pesticides in fruits during different processing steps. A conventional index called processing factor (PF) is used to quantify changes in pesticide residue levels that occur during food processing. The processing factor is defined as the ratio of the residue found in the processed commodity to the residue in the raw commodity before processing (Scholz *et al.,* 2018). Pesticide residue concentration factors in processed goods are often represented by PF values more than 1, while reduction factors are represented by PF values less than 1. Accurate pesticide residue data for food exposure risk assessment and a theoretical basis for processing technology optimization are made possible by an understanding of PF variations (Li *et al.,* 2023). Pesticide physicochemical characteristics, including solubility in water, melting point, degradation point, molecular weight, vapor pressure and other physicochemical parameters, are said to be the main determinants of pesticide PF values (Hrynko *et al.,* 2023, Jankowska and Lozowicka, 2022).

In general, the processing of fruits, particularly thermal process or fermentation process decreases the pesticide load sufficiently resulting in a PF < 1 (Kaushik, Satya and Naik, 2009). However, the drying process may result in a processing factor >1 for many pesticides (George *et al.,* 2013, Pathan *et al.,* 2009). During the preparation of resins from grapes, the PF values were found above 1.00 for some pesticides indicating concentration of the residues, which is due to weight loss on account of moisture loss during the drying process (Table 2).

Table 2: PF values for pesticides during the processing of grapes to resins.

Pesticide	PF value	Reference
Cyhalothrin	0.94	Choudhury, 2024
Cymoxanil	1.99	Ahmmed Shabeer *et al.*, 2015
Deltamethrin	< 0.4*	Choudhury, 2024
Dimethomorph	1.03	Ahmmed Shabeer *et al.*, 2015
Famoxadone	1.95	Ahmmed Shabeer *et al.*, 2015
Flupyram	0.36	Choudhury, 2024
Metalaxyl	< 0.32**	Choudhury, 2024
Tebuconazole	< 0.51**	Choudhury, 2024

*Based on LOQ at 10 ppb; ** Based on LOQ at 20 ppb.

Conclusion

Managing pesticide residues in fruit processing is crucial for ensuring food safety and quality. Properly washing and peeling fruits can significantly reduce pesticide residues. Scrubbing fruits under running water helps remove surface residues. Peeling can further reduce residues, but it also removes valuable nutrients. Blanching fruits (briefly immersing them in boiling water) before processing can help inactivate enzymes and remove some pesticide residues. However, blanching may also lead to nutrient loss. Heat treatments such as pasteurization, canning, or cooking can reduce pesticide residues. These processes break down pesticides and enhance food safety. However, excessive heat can also affect the nutritional content of fruits. Enzymes can be used to break down pesticide molecules during processing. For example, enzymes like cellulase can help degrade certain pesticides.

Implementing GAP during fruit cultivation helps minimize pesticide use. Proper application techniques, adherence to recommended doses, and following pre-harvest intervals are critical. IPM focuses on sustainable pest control by combining biological, cultural, and chemical methods. It aims to reduce pesticide use while maintaining crop health.

Regular monitoring and testing of raw fruits, intermediate products, and finished goods are essential. Analytical methods such as gas chromatography-mass spectrometry (GC-MS) and liquid chromatography-mass spectrometry (LC-MS) can detect and quantify pesticide residues. Comply with local and international regulations regarding pesticide residues. Maximum residue limits (MRLs) are set for various pesticides in different fruits. Ensure that processed fruits meet these standards.

A holistic approach, combining various strategies, is essential for effective pesticide residue management in fruit processing. The processing methods must be optimized in order to maximize nutrient content and the dissipation of pesticide residues. An optimization of processing methods that results in significant pesticide residue dissipation while preserving the majority of vital food nutrients requires significant attention. This will address the delicate balance between the two crucial parameters of food quality and safety. Furthermore, for the majority of pesticides, comprehensive research on the fate of metabolites during food processing is absent. Therefore, future study must give this research gap enough attention.

References

Abou-Arab A A K, 1999. Behavior of pesticides in tomatoes during commercial and home preparation. *Food Chemistry*, 65: 509–514.

Acoglu B and Omeroglu Y P, 2021. Effectiveness of Different Type of Washing Agents on Reduction of Pesticide Residues in Orange (*Citrus sinensis*). *LWT-Food Science and Technology*, 147:111690.

Al-Taher F, Chen Y, Wylie P and Cappozzo J, 2013. Reduction of Pesticide Residues in Tomatoes and Other Produce. *Journal of Food Protection*, 76:510-515. doi: 10.4315/0362-028X.JFP-12-240.

Athanasopoulos PE, Pappas C, Kyriakidis NV and Thanos A, 2005. Degradation of methamidophos on soultanina grapes on the vines and during refrigerated storage. *Food Chemistry*, 91: 235-240. doi:10.1016/j. foodchem.2003.10.018

Awasthi M D, 1993. Determination of insecticide residues on mango by washing and peeling. *Journal of Food Science Technology*, 30:132-133.

Bajwa U and Sandhu K S, 2014. Effect of handling and processing on pesticide residues in food- a review. *Journal of Food Science Technology*, 51:201–220. https://doi.org/10.1007/s13197-011-0499-5

Cabras P, Angioni A, Garau V L, Melis M, Pirisi M F, Cabitza F and Cubeddu M, 1998b. Pesticides residues on field-sprayed apricots and in apricot drying processes. *Journal of Agricultural and Food Chemistry*, 46: 23062308. doi:10.1021/jf980059d

Cabras P, Angioni A, Garau V L, Menelli E V, Cabitza F and Pala M, 1998. Pesticide residue in raisin processing. *Journal of Agricultural and Food Chemistry*, 46: 2309-2311. doi:10.1021/jf980058l

Cabras P, Angioni A, Garau V L, Pirisi M F, Brandolini V, Cabitza F and Cubeddu M, 1998a. Pesticide residue in prune processing. *Journal of Agricultural and Food Chemistry*, 46: 3772-3774. doi:10.1021/jf980098p

Cengiz M F, Certel M, Karakas B and Gocmen H, 2007. Residue contents of captan and procymidone applied on tomatoes grown in greenhouses and their reduction by duration of a preharvest interval and post-harvest culinary applications. *Food Chemistry*, 100: 1611-1619.

Chauhan R, Monga S and Kumari B, 2012a. Dissipation and decontamination of bifenthrin residues in tomato (*Lycopersicon esculentum* Mill). *Bulletin of Environmental Contamination and Toxicology*, 89: 181-186. doi:10.1007/ s00128-012-0629-4

Chauhan R, Monga S and Kumari B, 2012. Effect of processing on reduction of λ-cyhalothrin residues in tomato fruits. *Bulletin of Environmental Contamination and Toxicology*, 88: 352-357. doi:10.1007/s00128-011-0483-9

Chavarri Ma and Herrera, Antonio and Arino Agustin, 2005. The decrease in pesticides in fruit and vegetables during commercial processing. *International Journal of Food Science and Technology*, 40: 205 - 211. 10.1111/j.1365-2621.2004.00932.x.

Chavarri MJ, Herrera A and Arino A, 2004. Pesticide residues in field-sprayed and processed fruits and vegetables. *Journal of the Science of Food and Agriculture*, 84:1253-1259. doi:10.1002/jsfa.1791

Choudhury P P, 2024. Personal communication.

CIB and R C, 2024. Major Uses of Pesticides. A document compiled by Central Insecticide Board and Registration Committee, under the Ministry of Agriculture and Farmers welfare, Government of India.

Dash B, Kapoor S and Singh B, 2001. Persistence of malathion residues on/in bell pepper (*Capsicum annuum* Linn.). *Pesticide Research Journal*. 13:99.

FEMSmicroBlog, 2021. Food preservation through microbial fermentation on the FEMSmicroBlog (fems-microbiology.org)

FICCI, 2016. Next generation Indian agriculture-role of crop protection solutions: a report on Indian agrochemical industry. Tata Strategic Management Group, New Delhi, India, 2016, 10-16.

George T, Beevi S N, Xavier G, Kumar N P and George J, 2013. Dissipation kinetics and assessment of processing factor for chlorpyrifos and lambda-cyhalothrin in cardamom. *Environmental Monitoring and Assessment*, 185: (6) 5277-84. doi: 10.1007/s10661-012-2943-z.

Gonzalez-Rodriguez R M, Cancho-Grande B, Torrado Agrasar A, Simal-Gandara J and Mazaira-Perez J, 2009. Evolution of tebuconazole residues through the winemaking process of Mencia grapes. *Food Chemistry*, 117: 529-537. doi:10.1016/j.foodchem.2009.04.030

Gonzalez-Rodriguez R M, Rial-Otero R, Cancho-Grande B, Gonzalez-Barreiro C and Simal-Gandara J, 2011, A review on the fate of pesticides during the process within the food production chain. *Crit Rev Food Sci Nutr*, 51: 99-114.

Heshmati A, Nili-Ahmadabadi A, Rahimi A, Vahidinia A and Taheri M, 2020 Dissipation behavior and risk assessment of fungicide and insecticide residues in grape under open-field, storage and washing conditions. *Journal of Cleaner Production*, 270: 122287. 10.1016/j.jclepro.2020.122287.

Hrynko I, Kaczyński P, Łuniewski S and Łozowicka B, 2023. Removal of triazole and pyrethroid pesticides from wheat grain by water treatment and ultrasound-supported processes. *Chemosphere*, 138890.

Jankowska M and Łozowicka B, 2022. The processing factors of canning and pasteurization for the most frequently occurring fungicides and insecticides in apples and their application into dietary risk assessment. *Food Chemistry*. 371. doi: 10.1016/j.foodchem.2021.131179.

Krol W J, Arsenault T L, Pylypiw H M and Mattina M J I, 2000. Reduction of pesticide residues on produce by rinsing. *Journal of Agricultural and Food Chemistry*, 48: 4666–4670.

Kumar U and Agarwal H C, 1991. Degradation of dithane M-45 residues in brinjals during cooking. *Bulletin of Environmental Contamination and Toxicology*, 47(5):725-731. doi:10.1007/BF01701141

Kuna Aparna. Food Fermentation. In Principles of the Food Processing and Preservation. Edt Narendra Raju Panjagari. http://ebooks.inflibnet.ac.in/ftp1/

Kwon H, Kim T K, Hong S M, Se EK, Cho N J and Kyung K S, 2015. Effect of household processing on pesticide residues in field-sprayed tomatoes. *Food Science and Biotechnology*, 24(1):1-6. https://doi.org/10.1007/s10068-015-0001-7

Lee M G, 2001. Reduction of chlorpyriphos and fenitrothion residues in red pepper peel by washing and drying. *Food Science and Biotechnology*, 10:429-437.

Li K, Chen T, Shi X, Chen W, Luo X, Xiong H, Tan X, Liu Y and Zhang D, 2023. Residue behavior and processing factors of thirteen field-applied pesticides during the production of Chinese traditional fermented chopped pepper and chili powder. *Food Chem.* doi: 10.1016/j.fochx.2023.100854.

Liang Y, Wang W, Shen Y, Liu Y, and Liu X J, 2012. Effects of home preparation on pesticide organophosphorus residues in raw cucumber. *Food Chemistry*, 133: 636-640.

Lozowicka B and Jankowska M, 2016. Comparison of the effects of water and thermal processing on pesticide removal in selected fruits and vegetables. *Journal of Elementology*, 21:99-111. doi:10.5601/ jelem.2015.20.2.917

Lozowicka B, Jankowska M, Hrynko I and Kaczynski P, 2016. Removal of 16 pesticide residues from strawberries by washing with tap and ozone water, ultrasonic cleaning and boiling. *Environmental Monitoring and Assessment*, 188:51. doi:10.1007/ s10661-015-4850-6

Lozowicka B, Kaczynski P, Putkowska E, Jankowska M and Hrynko I, 2013. Evaluation of pesticide residues of fruits from Poland and health risk assessment. *Agricultural Science*, 4:106-111.

Marudov G, Radeva S, Pietrowska E and Nikolov I, 1999. Influence of processing peaches into puree and nectar on contents of pesticides residues. *Fluessiges-Obst*, 66: 171-172.

Mergnat T, Fritsch P, Saint Joly C, Truchot E, and Saint Blanquat G, 1995. Reduction of phosalone residue levels during industrial dehydration of apples. *Food Additives and Contaminants*, 12:759-767. doi:10.1080/02652039509374368

MPRNL, 2019. Retrieved from https://fssai.gov.in/upload/advisories/2019/10/5 da705b31ca 78Letter_ Report_Pesticides_MRL_16_10_2019.pdf

Nath G, Jat NR and Srivastava BP, 1975. Effect of washing, cooking -and dehydration on removal of some insecticide from Okra. *Journal of Food Science & Technology*, 12:127.

Nguyen TT, Rosello C, Belanger R and Ratti C, 2020. Fate of residual pesticides in fruit and vegetable waste (FVW) processing. *Foods*, *9:*1468. https://doi.org/10.3390/ foods9101468

Omeroglu PY, Acoglu Celik B and Koc Alibasoglu E, 2022. The effect of household food processing on pesticide residues in oranges (*Citrus sinensis*). *Foods*, 11:(23)3918. doi. org/10.3390/foods11233918

Pathan, Abdul Rashid Khan, Parihar NS and Sharma BN, 2009. Effect of drying on the residues of dicofol, ethion and cypermethrin in chilli (*Capsicum annum* L.). *Pest Management in Horticultural Ecosystems,* 15:167-169.

Polat B and Tiryaki O, 2019. Assessing washing methods for reduction of pesticide residues in Capia Pepper with LC-MS/MS. *J. Environ. Sci. Health Part B Pestic. Food Contam. Agric. Wastes,* 55(1): 1–10.

Randhawa M, Anjuma F, Ahmedb A and Randhawa M, 2007. Field incurred chlorpyrifos and 3,5,6-trichloro-2-pyridinol residues in fresh and processed vegetables. *Food Chemistry*, 103: 1016-1023. 10.1016/j.foodchem.2006.10.001.

Rani M, Saini S and Kumari B, 2013. Persistence and effect of processing on chlorpyriphos residues in tomato (*Lycopersicon esculantum* Mill.). *Ecotoxicology and Environmental Safety*, 95: 247-252. doi:10.1016/j. ecoenv.2013.04.028

Reiler E, Jors E, Baelum J, Huici O, Alvarez Caero MM and Cedergreen N, 2015. The influence of tomato processing on residues of organochlorine and organophosphate insecticides and their associated dietary risk. *Science of the Total Environment*, 15:527-528:262-269. doi:10.1016/j.scitotenv.2015. 04.081

Sarangapani C, Scally L, Gulan M and Cullen PJ, 2020. Dissipation of pesticide residues on grapes and Strawberries Using Plasma-Activated Water. *Food and Bioprocess Technology*, 13:1728–1741.

Sayah I, Gervasi C, Achour S and Gervasi T, 2024. Fermentation Techniques and Biotechnological Applications of Modified Bacterial Cellulose: An Up-to-Date Overview. *Fermentation,* 10:(2) 100; https://doi.org/10.3390/fermentation10020100

Scholz R, 2018. European database of processing factors for pesticides. EFSA supporting publication 2018:EN-1510. 50 pp. doi:10.2903/sp.efsa.2018.EN-1510

Severini C, de Pilli T, Petruccelli A, Baiano A and Scapicchio P, 2003. Effects of technological processes on pesticide residues in canned tomatoes. *Industrie Alimentari*, 42:487-495.

Sgarbiero E, de Baptista GC and Trevizan LRP, 2002. Evaluation of degradation and persistence of pirimiphos-methyl residues in wheat grain and derived products. *Revista Brasileira de Toxicologia*, 15(1): 5-8.

Shabeer TPA, Banerjee K, Jadhav M, Girame R, Utture S, Hingmire S and Oulkar D, 2015. Residue dissipation and processing factor for dimethomorph, famoxadone and cymoxanil during raisin preparation. *Food Chemistry*, 170: 180-185. doi:10.1016/j.foodchem.2014.08.008

Shabeer TPA, Girame R, Hingmire S, Banerjee K, Sharma AK, Oulkar D and Jadhav M, 2015a. Dissipation pattern, safety evaluation, and generation of processing factor (PF) for pyraclostrobin and metiram residues in grapes during raisin preparation. *Environmental Monitoring and Assessment*, 187:31. doi:10.1007/s10661-015-4268-1

Siddiqui S, Erol Z, Rugji J, Taşçı F, Kahraman H, Toppi V, Musa L, Di G, Giacomo, Bahmid N, Mehdizadeh M and Castro-Muñoz R, 2023. An overview of fermentation in the food industry - looking back from a new perspective. *Bioresources and Bioprocessing*, doi:10. 10.1186/s40643-023-00702-y.

Slowik-Borowiec M and Szpyrka E, 2020. Selected food processing techniques as a factor for pesticide residue removal in apple fruit. *Environmental Science and Pollution Research*, 27: 2361-2373.

Toker I and Bayindirli A, 2003. Enzymatic peeling of apricots, nectarines and peaches. *Lebensmittel-Wissenschaftund-Technologie*, 36(2): 215–221.

Uygun U, Koksel H and Atli A, 2005. Residue levels of malathion and its metabolites and fenitrothion in post-harvest treated wheat during storage, milling and baking. *Food Chemistry*, 92: 643-647.

Vijayasree V, Bai H, Beevi SN, Mathew TB, George T and Xavier G, 2015. Persistence and effect of processing on reduction of chlorantraniliprole residues on brinjal and okra fruits. *Environmental Monitoring and Assessment,* 187:299. doi:10.1007/s10661-015-4530-6.

Index

Anthocyanins 43, 46, 47, 48, 55, 56, 81, 92, 97, 103, 109, 111, 115, 152, 164, 168, 231, 240, 296
Ascorbic Acid 9, 10, 16, 19, 24, 57, 68, 69, 77, 79, 80, 81, 82, 114, 123, 141, 164, 166, 239, 240, 241, 303

B

Blueberries 43, 47, 55, 230, 231, 232, 271

C

D

N

O

P

T

U